Slipper Orchids of Vietnam

With an Introduction to the Flora of Vietnam

Paphiopedilum vietnamense CAROL WOODIN

Slipper Orchids of Vietnam

With an Introduction to the Flora of Vietnam

Leonid Averyanov

Phillip Cribb

Phan Ke Loc

Nguyen Tien Hiep

WATERCOLOURS:

Carol Woodin

ROYAL BOTANIC GARDENS KEW

First published 2003

ISBN 1 84246 047 1

Production Editor Suzy Dickerson

Cover design and page make up by Media Resources,
Information Services Department
Royal Botanic Gardens, Kew

Printed in the European Union by
Compass Press Limited

CONTENTS

FOREWORD

Vietnam's remarkable and diverse flora is still being explored and described. This is particularly apparent in the orchid family. It is probably the largest family in the flora and may exceed 1000 species in Vietnam and more are being described every year as remoter areas are explored. It seems likely that 8% of all the flowering plants in Vietnam are orchids and they are found in every habitat from the coastal mangrove forests to the highest mountain peaks. The complex life cycle of orchids which involves specific fungi to germinate the seeds and fertilisation of the flowers by specific pollinating insects, means that they are particularly vulnerable to disturbance. Thus, they are sensitive indicators of the health of the natural environment. The slipper orchids epitomise the exuberance of Vietnam's orchid flora and, at the same time, emphasise its vulnerability. Although almost half have been discovered within the past ten years, many of them are already threatened with extinction by the combined problems of forest destruction and commercial collecting.

This book, written by an international team of experts, details the diversity of this remarkable group of orchids within the country and provides the first detailed account of their habitats and ecology. On a more urgent note it also highlights the imminent threat of extinction faced by many of them. Further urgent measures are necessary both within Vietnam and abroad to protect what remains.

Prof. Phan Ke Loc from Hanoi National University (HNU), and Dr. Nguyen Tien Hiep from the Institute of Ecology and Biological Resources (IEBR), Vietnam National Centre for Natural Science and Technology (NCST) have conducted extensive government-sponsored field surveys over many years. At the invitation of the Institute of Ecology and Biological Resources Professor Leonid Averyanov and Dr. Phillip Cribb joined some of their expeditions and have contributed their expertise to these surveys. The government of Vietnam was very happy to support and approve all the work necessary to ensure that this important project could be completed. The results of their field studies, detailed here for the first time in a popular book, provide a unique insight into the rich flora of Vietnam where so many of the orchids are found. This book is a significant contribution to our understanding of plant diversity in Vietnam and to the threats to it.

Vietnam is proud of its rich biological diversity. This book will enthral readers with its descriptions of Vietnam, its unique environment, its rich flora and its endemic plants. It will also serve to alert all of us to the dangers that threaten our unique heritage.

Dr. Nguyen Ba Thu
Director of Forest Protection Department (FPD) of Vietnam
Chief — CITES Vietnam Management Authority

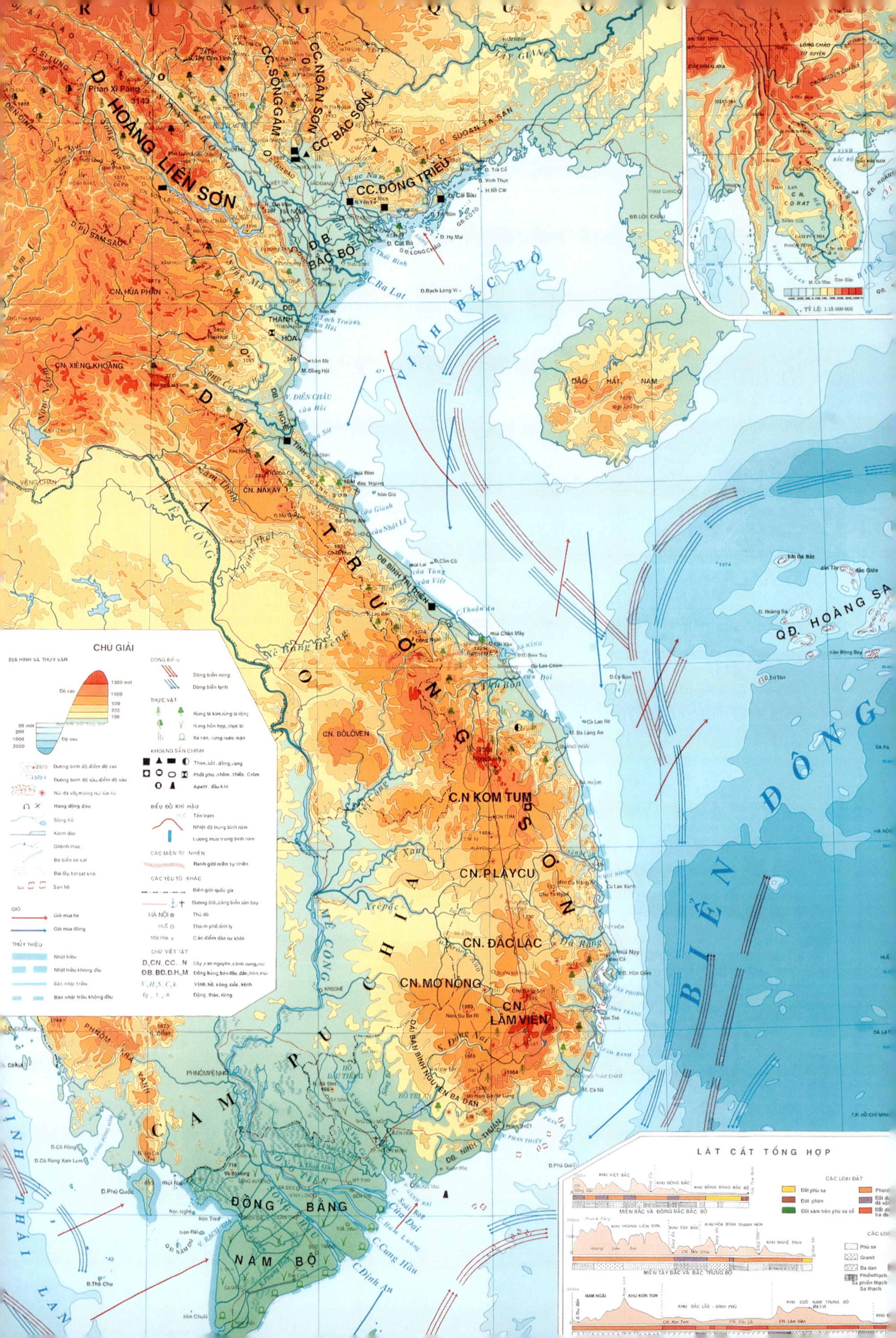
D. HOÀNG LIÊN SƠN
Phan Xi Păng
3143
CC. SÔNG GÂM
CC. NGÂN SƠN
CC. BẮC SƠN
CC. ĐÔNG TRIỀU
D. PU SAM SAO
CN. HỦA PHĂN
ĐB BẮC BỘ
VỊNH BẮC BỘ
ĐB THANH HÓA
CN. XIÊNG KHOẢNG
ĐẢO HẢI NAM
CN. NAKAY
D. TRƯỜNG SƠN
ĐB. BÌNH TRỊ THIÊN
CN. BÔLÔVEN
C.N KON TUM
CN. PLÂYCU
CN. ĐẮC LẮC
CN. MƠ NÔNG
CN. LÂM VIÊN
ĐB. NINH THUẬN
QĐ. HOÀNG SA
BIỂN ĐÔNG
CAMPUCHIA
ĐỒNG BẰNG NAM BỘ
VỊNH THÁI LAN
CHÚ GIẢI
Độ cao
1500 mét
1000
500
200
100
Độ sâu
Dòng biển nóng
Dòng biển lạnh
Gió mùa hạ
Gió mùa đông
LÁT CẮT TỔNG HỢP
Tỷ lệ: 1:15 000 000

INTRODUCTION

Until very recently, the orchid flora of Vietnam was considered to be rather depauperate in comparison with that of Thailand, China and SE Asia. However, that perception has been well and truly debunked by the publication of accounts of the orchid floras by Averyanov (1988f, 1990d, 1991e, 1994) and Seidenfaden (1992). Since then, many Vietnamese novelties have appeared, often on the international orchid market and in collections and shows worldwide. These include a new genus of vandaceous orchid, *Christensonia*, and many new species. The most amazing of the new species are undoubtedly the slipper orchids that include some of the most spectacular and attractive species in the genus. It is truly astounding that such showy orchids could have remain undiscovered for so long. However, Vietnam is large and still botanicially rather poorly explored, particularly in the karst country that occupies much of the north of the country.

Vietnam is currently surging forward in industrial development and utilisation of its natural resources. Tourism is also booming. In contrast, botanical exploration is progressing rather slowly because of restraints in manpower and funding. Deforestation is proceeding at an alarming rate, especially in the fragile forests of the north and many plants and animals are being pushed into the small fragments of forest that remain. Undoubtedly, some are becoming rare and may indeed be threatened with extinction.

We have included a great deal of new ecological information on the orchids gathered during field-work over a period of more than 25 years. The habitats of slipper orchids are often remote and difficult to reach. Not only are the orchids little known but the same applies to the associated flora. Indeed, on a recent field trip the authors discovered a conifer that has proven to belong to a new genus, *Xanthocyparis* (Farjon *et al.*, 2002). Most of this work has been undertaken under the auspices of the Institute of Ecology and Biological Resources in Hanoi and we thank them for their co-operation.

A major reason for writing this book has been our concern that some orchids are on the verge of extinction. In particular, some of the slipper orchids that are highly prized by horticulture and other orchids that are collected for traditional medicine are now very difficult to find in the wild and some may already be extinct. We highlight here Vietnam's amazing diversity of slipper orchids that is unmatched in any other country within the range of the genus. We also hope that by raising issues concerning their conservation we will alert the relevant agencies to the plight of these handsome plants and, thereby, provide just cause for their remaining habitats to be protected. We also hope that, knowing the plight of these orchids in the wild, growers will exercise restraint in buying wild-collected Vietnamese orchids. Those that have already succumbed we hope will learn enough of their ecology to be able to grow better the rare material they have.

Vietnam is truly a slipper orchid paradise. We hope that it will remain so and that the rich diversity of the country will be enjoyed by generations to come.

Leonid Averyanov
Phillip Cribb
Nguyen Tien Hiep
Phan Ke Loc
May 2002

Fig. 1 (opposite). Topographical map of Vietnam.

PART I. VIETNAM

GEOGRAPHY

The Socialist Republic of Vietnam is situated along the south-eastern margin of the Indo-Chinese Peninsula, extending from latitudes 8°30'N to 23°30'N for more than 1600 km from its southernmost point of Cape Camau to its northern extremity on the Chinese border . The country is relatively narrow from west to east, about 600 km wide in the northern part of the country and only little more than 40 km at its narrowest in Quang Binh Province. The westernmost point of Vietnam lies at the junction of the borders of Laos, Vietnam and China (102°10'E), and the easternmost point is the extremity of Cape Nai (109°27'E). The country covers 325,360 sq. km. Ha Noi city is the capital of the country (Fig. 1). Currently, the country includes 57 provinces and 4 city territories (Fig. 2).

The population of Vietnam is estimated at about 75 million residents in the year 2000. Some 80% of the population is rural, the greatest concentrations of population being in the Song Hong (Red River) and Me Cong deltas, particularly in the coastal plains. The largest national group is the Vietnamese (Kinh) with 54 million people. The largest minority groups, extensively spread over the Truong Son Range mountain region, are the Tay, Khmer, Thai, Muong, Nung, Sedan and Dao. The largest minorities in the montane regions in northern Vietnam are Homon (Meo) and Zao. Until the present time the hill tribes practise shifting cultivation, clearing the forest to plant hill rice and maize (MacKinnon & Cox, 1991).

Along its northeastern and southwestern frontiers Vietnam is washed by waters of the Gulfs of Bac Bo (Tonkin) and Thailand (Siam) of the South China Sea. The coastline of mainland Vietnam is about 3440 km long. A large number of islands mainly of continental origin adjoin the territory of mainland Vietnam and there are also numerous small coral atolls dispersed in the sea. The Annamite mountain chains, mainly Den Dinh, Sam Sao, Hua Phan and Truong Son Ridges, form the natural western Vietnamese boundary with Laos and Cambodia. The northern border with China lies mostly through the mountain systems of southeastern Yunnan and Guangxi (Fig. 1). The inland border of Vietnam with these three countries extends for about 4639 km (Averyanov, 1990c; Rundel, 1999).

Three-quarters of the country is hilly or mountainous (Averyanov, 1990c). The highest mountains of Indochina, Fan Si Pan (3143 m), Ta Giang Pinh (3096 m), Si Lung (3076 m), Luong (2985 m), Phoung Chang (2825 m), and Lang Kung (2817 m), are situated in the northwest of the country (Fig. 1). The largest mountain systems are made up mainly of granite, gneiss, shale, schist and sandstones. Ancient highly metamorphosed hard limestone, from which karst formations have typically developed, is very widely distributed in the north of the country. Toward the east the mountains gradually give way to vast alluvial plains, often not much above sea level. Southwards the mountain regions have relatively level dissected plateaus, where average elevations are mostly from 500 to 1500 m above sea level. Finally, the southern part of the country contains the vast, partially swamped, lowland plain of the Me Cong River delta (Fig. 1).

The hydrographic network in the country is well developed. Here, besides a great number of small and medium-sized rivers run some of the largest rivers of Southeast Asia such as the Song Hong (Hong Ha, or Red River) and the Me Cong (Fig. 1).

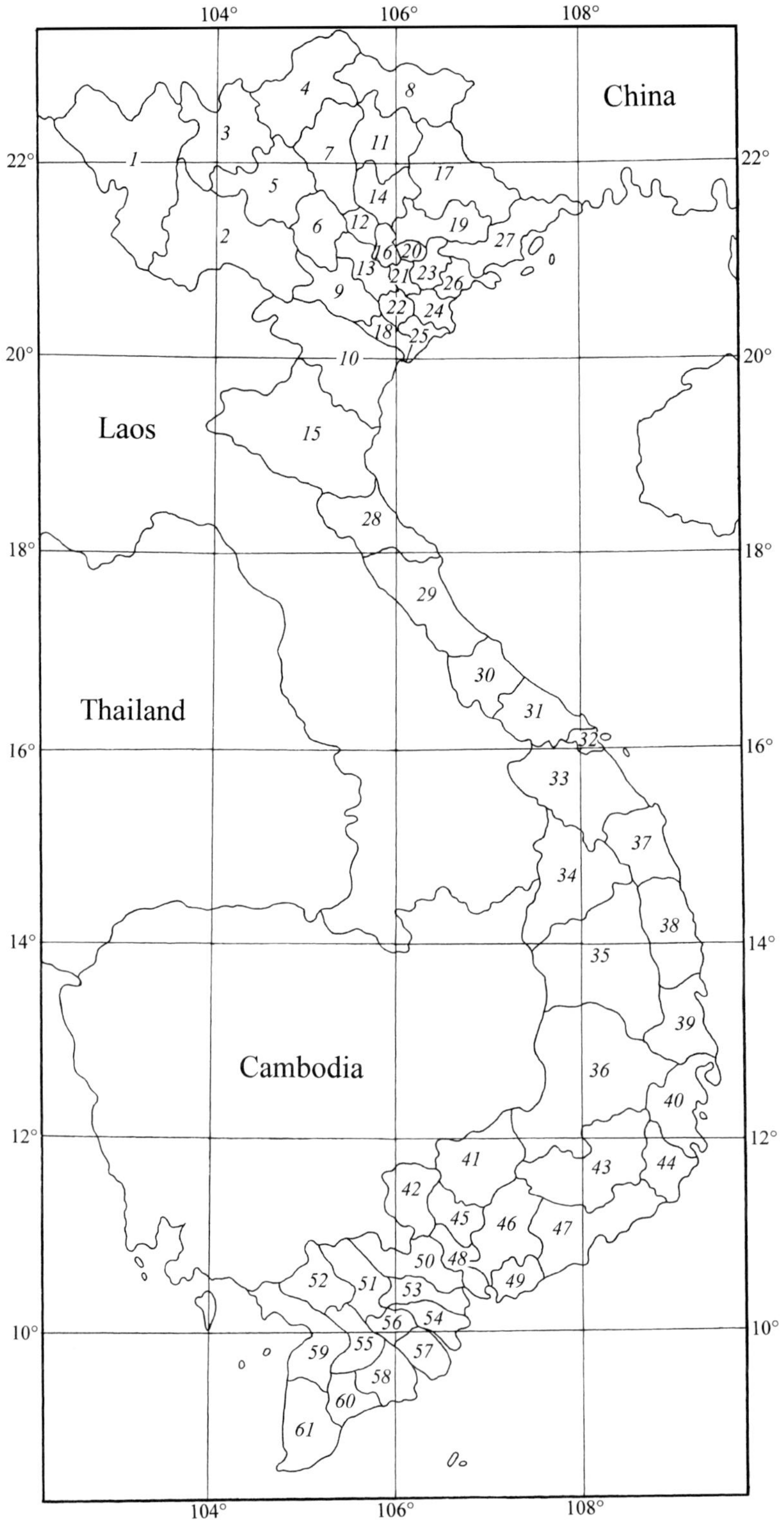
104°
106°
108°
China
Laos
Thailand
Cambodia
22°
20°
18°
16°
14°
12°
10°
1
2
3
4
5
6
7
8
9
10
11
12
13
14
15
16
17
18
19
20
21
22
23
24
25
26
27
28
29
30
31
32
33
34
35
36
37
38
39
40
41
42
43
44
45
46
47
48
49
50
51
52
53
54
55
56
57
58
59
60
61

GEOLOGICAL HISTORY

Geologically, mainland Southeast Asia forms a primary geomorphological unit that is composed of Pre-Cambrian crystalline rock that consolidated to form a stable core area in Late Triassic times. This unit known as the Southeast Asian or Sundaland prong of the Eurasian Plate extends in a SSE direction and includes the modern Indochinese and Malay Peninsulas, Sumatra and the western portions of Java, Borneo and Sulawesi. The Sunda Shelf, a shallow oceanic area of the Gulf of Thailand and the South China Sea, is also a core area of Sundaland. It was exposed many times during the Mesozoic and Tertiary as well as during the Pleistocene glacial periods when sea levels were lower (Rundel, 1999).

One of the central portions of this Pre-Cambrian plate represents a crystalline nucleus known as Indosinia, around which Indochina landforms were historically formed (Fromaget, 1952). Three main periods of tectonic folding have been identified around this ancient central block during the Paleozoic and early Mesozoic eras.

The earliest period of tectonic activity of Devonian age, known as Caledonian Folding, occurred about 400–350 million years ago and produced a large stable block covering central and southern Vietnam, most of central and southern Laos, northeastern Cambodia and possibly most of northeastern Thailand. This core region, which has been termed Annamia, appears to have remained relatively stable up to the present (Fontaine & Workman, 1978; Rundel, 1999). In peripheral zones of this block especially along its northern margin numerous mountain systems, oriented in a northwest to southeast direction, were established. Remnants of these ancient formations known as Tonkinides are observed in northern Vietnam, mainly in the north of Hoang Lien Son and in the northern part of the Truong Son Range. The stable ancient Indosinian block was dissected at this period into a number of flat solid platforms that appear now in central and south Vietnam as a number of separate plateaus. The Kon Tum, Playcu, Dac Lac, Mo Nong, Lam Vien (Dalat) plateaus of southern Vietnam represent, for example, such crystalline platforms uplifted to different elevations (Fig. 1). Along the faults bordering these plateaus later magmatic (mostly granite) intrusions and basalt flows occurred. During the Caledonian Folding the sea generally regressed. However, after the Caledonian orogenesis seas again covered large areas of Indosinia (Schzeglova, 1957; Fridland, 1961; Dovzikov *et al.*, 1965a).

The oldest rocks of Indochina are the metamorphic crystalline complexes of Pre-Cambrian and early Paleozoic age that have remained intact at least since the Devonian. Large exposed areas of these basement rocks exist today as a number of plateaus in the central part of southern Vietnam. Their impressive outcrops may be seen as high cliffs and bluffs of deep river canyons on east-facing slopes of these plateaus along the continuous geological break that outlines the seashore. At the same time Mesozoic sediments covered plateaus on their western flatter slopes (Schzeglova, 1957; Rundel, 1999) facing Laos.

Fig. 2 (opposite). Administrative divisions of Vietnam.

1 — Lai Chau; **2** — Son La; **3** — Lao Cai; **4** — Ha Giang; **5** — Yen Bai; **6** — Phu Tho; **7** — Tuyen Quang; **8** — Cao Bang; **9** — Hoa Binh; **10** — Thanh Hoa; **11** — Bac Kan; **12** — Vinh Phuc; **13** — Ha Tay; **14** — Thai Nguen; **15** — Nghe An; **16** — Ha Noi; **17** — Lang Son; **18** — Ninh Binh; **19** — Bac Giang; **20** — Bac Ninh; **21** — Hung Yen; **22** — Ha Nam; **23** — Hai Duong; **24** — Thai Binh; **25** — Nam Dinh; **26** — Hai Phong; **27** — Quang Ninh; **28** — Ha Tinh; **29** — Quang Binh; **30** — Quang Tri; **31** — Thua Thien Hue; **32** — Da Nang; **33** — Quang Nam; **34** — Kon Tum; **35** — Gia Lai; **36** — Dac Lac; **37** — Quang Ngai; **38** — Binh Dinh; **39** — Phu Yen; **40** — Khan Hoa; **41** — Binh Phuoc; **42** — Tay Ninh; **43** — Lam Dong; **44** — Ninh Thuan; **45** — Binh Duong; **46** — Dong Nai; **47** — Binh Thuan; **48** — Ho Chi Minh; **49** — Ba Ria Vung Tau; **50** — Long An; **51** — Dong Thap; **52** — An Giang; **53** — Tien Giang; **54** — Ben Tre; **55** — Can Tho; **56** — Vinh Long; **57** — Tra Vinh; **58** — Soc Trang; **59** — Kien Giang; **60** — Bac Lieu; **61** — Ca Mau.

General uplifting of the Indosinian area and extensive regression of the sea during later Devonian and Carboniferous times, about 350–300 million years ago (Dovzikov *et al.*, 1965a), were connected with Hercynian orogenesis. Extensive mountain formations were formed in broad areas of the central part of Vietnam from the coastline to the boundary with Laos as a result of Hercynian folding. Formation of considerable portions of the Truong Son (Annamite) Range in the south of Vietnam was also connected with Hercynian orogenesis. Devonian formations of Hercynian folding were consolidated by extensive granite intrusions during the middle Carboniferous Period. These granites are now exposed in many areas along the coastline of central and southern Vietnam (Schzeglova, 1957; Fridland, 1961). The Permian was characterized by numerous wide depressions and transgressions of the sea that once again covered large areas of Indosina.

The most intensive orogenic activity in the area occurred during the early Mesozoic. Marine basins around the northern, western and southwestern margins of the Indosinia block were the focus of strong folding in a series of tectonic activities from the late Permian to the Triassic. This folding, termed the Indosinian, is most evident today in large mountain areas of northern Thailand and Laos, and along the boundary of Vietnam with Cambodia and Laos. Extensive areas of granite exposed in northern Vietnam are intrusions formed primarily in the late Permian and Triassic during this folding. Formation of a number of mountain systems during Indosinian Folding, so-called indosinides (Schzeglova, 1957), was connected with tectonic activity that resulted in the wide distribution of magmatic rocks like rhyolites, dacites and other related silicate rocks. In the late Triassic, the Indochina segment of continental crust was consolidated and uplifted leading to final regression of the sea and the formation of the modern landform configuration of mainland Southeast Asia. A remarkable event at this time was the separation of Borneo from Indosinia (Schzeglova, 1957). Beginning at this time, about 180 millions years ago, up to the present, highland areas of Indosinia have experienced steady erosion (Schzeglova, 1957; Rundel, 1999) and sedimentation of considerable alluvial deposits, mainly sands. These Mesozoic sands cover large areas of southern Laos and reach 2000 m in depth, covering the ancient crystalline basement rocks. Mesozoic sands also occur along the coastline in the southern Vietnam. Some local intrusive and volcanic activity was also present over this period and later, but it had rather limited effect (Schzeglova, 1957).

Shallow seas covered much of central and northern Indochina during the late Paleozoic and early Mesozoic, leading to the formation of extensive deposits of limestone. The spectacular karst topography that has developed in northern and central Vietnam is formed mainly by limestone of these ages. They are mainly of Devonian, Carboniferous, Permian and Triassic origin. However, limestones of Precambrian, Cambrian and Ordovician ages are also locally present (Dovzikov *et al.*, 1965a, b). Numerous marine transgressions also continued during the Triassic, with seas covering eastern Cambodia and parts of southern Vietnam. Continental sedimentary formations characterized as red beds (terrain rouge) were also deposited over wide areas at this time in Cambodia and southern Vietnam (Fontaine & Workman, 1978).

Throughout the later Mesozoic and Cenozoic, the Southeast Asian subcontinent and adjacent areas of southeastern China reacted as a single unit as surrounding areas exhibited considerable tectonic activity (Fontaine & Workman, 1978). In the west, Indosinia was bordered by the Cretaceous-Tertiary orogenies of western and central Myanmar, and to the south by the Mesozoic-Cenozoic orogenies of Sumatra, Java, Borneo and the Philippines. At the same time Indosinia itself represents a distinctive geomorphologic unit from an older structural block of south China that extended into northern Vietnam. The dividing line between these blocks, which roughly follows the present course of the Song Hong River, experienced differential movements from late Cretaceous to Eocene times. Tertiary tectonic activity was associated in Indosinia with the steady uplift of montane regions up to present-day elevations of 2000–3000 m to the west and north, and a subsidence of the Sunda shelf area to below sea level in the southeast (Rundel, 1999). At the same time formation of deep depressions and breaks also occurred.

Enormous lava fields and basalt flows of Tertiary age cover thousands of square kilometres of land area in southern Vietnam along geological breaks due to Cenozoic volcanic activity. The largest of these areas are associated with the Pleicu, Dac Lac, Mo Nong, Haute Cochinchine and Djiring-Blao Plateaus of southern Vietnam (Rundel, 1999). Smaller outliers of basalt also occur in central Vietnam. A number of eroded volcanic cones of Tertiary age are still to be seen on the Kon Tum plateau. Volcanic eruptions have occurred as recently as 1923 in the South China Sea off the southeastern coast of Vietnam (Schzeglova, 1957).

Obvious uplift of the north-western montane part of Vietnam and some portions of the Truong Son Range, and clear depression of the northeast and southern areas of the country are observed from the later Tertiary up to the present time. Formation of the highest mountain systems of Vietnam, such as Fan Si Pan or Ngoc Linh, is the result of this general uplift. On the other hand large depressions have led to the accumulation of large alluvial deposits in the Song Hong River delta and the delta of the Me Cong (Schzeglova, 1957). Moderate uplift has also been experienced in some central parts of the country along the coastline. Thus, prominent Pleistocene marine terraces extend discontinuously along nearly 100 km of the coastline near Phan Rang town in southeastern Vietnam (Rundel, 1999).

Mountain erosion and extensive sedimentation of alluvial deposits has been typical in the later Tertiary and Quaternary history of Indochina. Today, Quaternary alluvial deposits cover broad areas in Cambodia, northern and southern Vietnam. Older alluvium of the lower Pleistocene reaches a thickness of more than 500 m in the eastern Me Cong delta. The Song Hong River basin around Hanoi is entirely covered by Quaternary alluvial deposits. Most recent Holocene deposits of pure quartz sand often extend inland from modern beach sands along the coast of central Vietnam and alluvial terraces of this age are common in the country along much of the coast over limestone substrates (Rundel, 1999).

GEOMORPHOLOGY

Over three-quarters of Vietnam is mountainous (Fig. 1). The mountain areas fall into three distinct regions: **1.** The mountains of Vietnam north-east of the Song Hong River; **2.** the Hoang Lien Son Ridge of north-western Vietnam north-west of the Song Hong River that represents a southern extension of the Himalayas; and **3.** the Truong Son or Annamite Range that forms the backbone of Southeast Asia along the western border of Vietnam (Schzeglova, 1957; Rundel, 1999). Other distinct regions of the country are the coastal plains and the broad flat alluvial valleys of the Me Cong and Song Hong Rivers.

Northeastern montane region

Northeastern Vietnam to the northeast of the Song Hong River represents a southern extension of the South China Platform, otherwise comprising western Guangxi, southern Guizhou, and eastern Yunnan, and differs structurally from the southern parts of Southeast Asia (Dovzikov *et al.*, 1965b; Rundel, 1999). This area has experienced repeated uplift and subsidence but relatively little deformation through its history. The landscape is formed by a combination of exposed ancient metamorphic basement rocks and eroded marine sediments deposited in the late Paleozoic and early Triassic. Most of these sediments are limestone that have been uplifted and highly eroded to form an extensive karst topography that can be seen as a system of narrow remnant ridges and column-like mesas with very steep slopes and vertical bluffs. In Vietnam, these are usually 300–700 m high, however toward the Chinese border they often reach 1400–1600 m. Numerous caves, sinkholes, underground rivers and small lakes that disappear during dry season are typical for this area.

Tectonic activity in the late Triassic during Indosinian folding and later resulted in local intrusions (Dovzikov *et al.*, 1965b; Rundel, 1999), forming a number of isolated granite mountain systems, such as Tay Con Linh (2419 m), Pu Tha Ca (2274 m), Pia Da (1980 m) and Pia Oac (1930 m). These are the highest peaks in the area, but rise well above the surrounding landscape. There is a prominent depression extending from Cao Bang town to the coast (Fig. 1).

The valley of the Song Hong River separates the northeastern region from the northwestern montane region of Vietnam. Wide limestone terraces, extensive alluvial plains, and low hills bound the delta area of the Song Hong River and other major rivers of the region. The northeast coast is dotted with hundreds of small shale, sandstone and limestone islands. Highly eroded column-like limestone hills are also found in some inland places on the alluvial plains and low terraces of the Song Hong River delta.

Northwestern montane region

The eastern Himalayas (Hengduan Mountains) extend from Yunnan into northern Vietnam as a number of high ranges — Hoang Lien Son, Si Lung and Den Dinh (Fig. 1). Mostly above 2000 m, they form a large montane region in the northwest of the country. A number of peaks are above 2500 m, including Fan Si Pan (3143 m), Ta Giang Pinh (3096 m), Si Lung (3076 m), Luong (2985 m), Phoung Chang (2825 m), and Lang Kung (2817 m). Most of them were formed by granitic or quartzite intrusions in the late Mesozoic. To the south of the main ridges lies an area of complex folded rocks, the Song Da and Dien Bien Phu synclines and the Song Ma anticline.

A series of Devonian and Triassic limestone plateaus, such as Son La and Moc Chau with elevations up to 1500–1700 m, separated in this area by narrow canyons or very deep river valleys, spread along the Song Da River (Black River) from the Chinese border to the coast. On the periphery of plateaus, the limestone appears as a spectacular area of highly eroded karst topography (Schzeglova, 1957; Dovzikov *et al.*, 1965b; Rundel, 1999). It can be seen in many places in Lai Chau, Son La, Hoa Binh, Thanh Hoa, Nghe An and Quang Binh Provinces (Dovzikov *et al.*, 1965b).

Truong Son Range region

The Truong Son or Annamite Range, in its broad sense, extends for nearly 1200 km from the north to the south along the Vietnamese border with Laos and Cambodia and includes the southern plateau areas in the south of Vietnam. It represents a complicated structure with a long history of multiple folding of marine sediments and metamorphic rock, uplifted basement rock, and later volcanic activity with extensive basalt flows.

The main uplift of the Truong Son Range occurred with the Indosinian orogenies of the late Paleozoic and early Mesozoic. The major portion of the range is composed of deformed Paleozoic marine sedimentary rocks and late Paleozoic — early Mesozoic granite intrusives. The highest mountain, situated on the Vietnam-Laotian border in the northern part, is Xai Lai Leng which reaches 2711 m. It and other high peaks in the area, such as Phu Hoat (2452 m), Rao Co (2235 m) and Muong Tip (2040 m), are composed of Triassic granite (Dovzikov *et al.*, 1965b).

The northern part of Truong Son Range, the narrowest part of Vietnam, lies to the south of Vinh town. It is relatively low with the highest peaks reaching only rarely about 1300 m while some passes are lower than 500 m (Fig. 1). These ancient eroded mountains, in a belt which extends from Laos across central Vietnam around Hoi An to the south of Danang, were formed by Hercynian foldings covered by thick continental sandstones of late Triassic age. They are rich in plant fossils (Schzeglova, 1957; Rundel, 1999). Large areas here are also covered by highly metamorphosed marine limestone sediments, mainly of Devonian, Carboniferous and Permian age in which marine fossils are abundant. The main belt of these deposits crosses the Truong Son Range and extends into Laos in a north-westerly direction from the coast. To the northwest of Vinh and west of Dong Hoi towns, erosion of the eastern edges of these limestone plateaus has resulted in spectacular and numerous crowded narrow ridges and mesas towering over flat alluvial valleys (Dovzikiv *et al.*, 1965a,b).

The southern portion of Truong Son Range is dominated by an ancient crystalline platform composed of Precambrian gneiss. During the Paleozoic and early Mesozoic, this platform was dissected into a number of distinct plateaus each uplifted to a different elevation. The Kon Tum and Lam Vien (Dalat) plateaus at 1000–1500 m are the highest of these. The Playcu, Dac Lac, Mo Nong (Djiring) and other plateaus are lower, between 100–770 m. The southern plateaus form a complex mosaic composed of late Paleozoic and early Mesozoic granite intrusives and broad areas of Tertiary and Quaternary volcanic deposits that usually cover ancient basement rocks. Scattered areas of granite outcrops and chains of highly eroded volcanic cones are also present on some plateaus (Schzeglova, 1957; Rundel, 1999).

The Truong Son Range rises into the large Kon Tum Massif to the southwest of the town of Da Nang. This, the largest plateau of basement rock, extends for approximately 250 km to south and inland for 200 km. The granitic peak of Ngoc Linh (2598 m) at the northern margin of the Kon Tum Massif is the highest point of the Truong Son Range in central Vietnam. Chu Dang Sin (2405 m) and Bi Dup (2287 m) peaks, the highest points of the southern highland portion of Truong Son Range, are also composed of granite. They tower above the Lam Vien (Dalat) plateau. They are composed of schist and quartzite, as well as extensive areas of granites and volcanic rhyolites, andesites and dacites. The central portions of the Lam Vien (Dalat) highlands are dominated by dacites, including Lang Bian peak which reaches 2163 m (Rundel, 1999).

Subsequent to the formation and uplift of the southern Truong Son Range, extensive lava flows in the late Tertiary and Quaternary formed large plateaus of resistant basalt to form the Playcu, Dac Lac, Lam Vien (Dalat), Mo Nong (Djiring) and other plateaus in the south of the country. Predominant elevations here are 200–700 m. The first three are sometimes collectively referred to as the Tayngyuyen Plateau.

The Playcu Plateau is bounded to the north by the Kon Tum Massif, and to the east by the Da Rang River and Cheo Reo Depression. The latter has extensive alluvial plains at about 150 m. The basaltic soils of the Dac Lac Plateau are most weathered and deepest to the south and shallower to the north, while the soils of the western margin of the Playcu Plateau are also deep and rich (Rundel, 1999).

The relatively low plateaus, 300–600 m high, at the extreme southern part of the Truong Son Range are derived from the peneplain of leveled Hercynian Folding of old schist and sandstone largely overlain by basalts of varying ages. The older basalts lie to the north and younger basalts to the south. Scattered areas of granite outcrops are also present. Extensive rolling hills to the southeast, with occasional elevations of 1000–1600 m, are largely formed of weathered granites. A terrace region that gives way to the Me Cong delta marks the southern margin of Truong Son Range.

It is remarkable that in the south of Vietnam, the hydrologic drainage patterns run largely to the west toward Cambodia, isolating river valleys from the wetter eastern slopes in Vietnam. The very steep east slopes of Truong Son Range are the result of an extensive geological fault, which cut the solid crystalline blocks of ancient plateaus during the late Triassic age. In some places this fault is seen as a vertical cliff more than 1000 m high. This area is rich in narrow canyons formed by water that eroded ancient crystalline rocks (mainly granite and gneiss) and the area is abundant in showy waterfalls. Rivers flowing to the east from the Truong Son Range slopes are short but torrential and very rapid.

Coastal plains

Irregular flat coastal plains, almost at sea level, connect the fertile river valleys of northern and southern Vietnam in a narrow belt along the coastline. The coastal plain in some places reaches 50 km wide, but usually it is much narrower (Fig. 1). In some places, for example, near Ngang, Chan May and Nai Capes, branches of the Truong Son Range interrupt the coastal plain and, where they approach the coast, appear as high steep rocky cliffs.

Hilly parts of the coastal plain adjacent to the mountains are formed of Quaternary alluvium deposited from the Truong Son Range. Old, highly eroded, coastal terraces can usually be observed here. The youngest terraces, composed mainly of stabilized sand deposits and dunes of Quaternary age, are also present along the coastline near the sea.

Alluvial valleys of the Me Cong and Song Hong Rivers

The two largest rivers of Southeast Asia — the Song Hong (Red River) and the Me Cong form broad deltas covered by alluvial river deposits of Quaternary age.

The delta of the Song Hong River, or Bac Bo lowland, covers more than 15,000 square km, predominantly at an elevation of only 2–6 m and, in some places, lying below sea level. The Song Hong River carries large amounts of loess and silt (about 80 million m^3 per year) which increases the delta margin by up to 100 m each year (Schzeglova, 1957).

The delta region of the Me Cong, or the Nam Bo lowland, extends over 40,500 square km, covering much of the lowlands of southern Vietnam. This uniform, flat, swampy plain has a mean elevation of only 2 m. Maximal yearly growth of the coastal territory (about 60–80 m each year) is observed in Ca Mau Cape (Schzeglova, 1957).

Rich alluvial soils in each of the delta regions support large population centres based on agricultural productivity. More than 80% of Vietnamese live on the coastal and delta plains, while mountain regions still have a sparse population.

CLIMATE ZONES

Winter and summer monsoons are the main influences on the climate of Southeast Asia and Indochina. North and northeast winds predominate in the region from October to March because of the high pressure that regularly occurs during the winter in the centre of Asia over the plains of Mongolia. These winds are cold and dry and mostly bring fresh, dry weather to Vietnam. Winds from the south and southwest predominate from April to September and come from the Indian and Pacific Oceans where high-pressure regions are formed during the summer. The summer monsoon winds are warm and wet and bring hot and rainy weather to the continent (Schzeglova, 1957; Fridland, 1961; Pham Ngoc Toan & Phan Tat Dac, 1993). At the same time the complicated orography of the region and specific configuration of the coastline of Vietnam locally shift the direction of monsoon winds, changing the timing of the dry and rainy seasons and influencing the amount of precipitation. This results in the formation of incredibly diverse climatic patterns in the country.

Nguyen Khanh Van *et al.* (2000) recognise seven main climate zones in Vietnam (based on earlier investigations of the region by Walter & Leith, 1960, 1964, 1967; Gaussen, Legris & Blasco, 1967; Schmid, 1974; Thai Van Trung, 1978; Hydro-Meteorological Data, 1989, 1997; Pham Ngoc Toan & Phan Tat Dac, 1993). They are as follows:

1. Monsoon tropical climate with cold winter and summer rains;
2. Monsoon tropical climate with cold winter and summer-autumn rains;
3. Monsoon tropical climate with warm winter and winter-autumn-winter rains;
4. Monsoon tropical climate with warm winter and autumn-winter rains;
5. Monsoon tropical climate with warm winter and summer rains;
6. Monsoon sub-equatorial climate with summer rains;
7. Monsoon tropical climate associated with mountains.

The climatological regions of Vietnam are outlined in Figure 3 which also provides diagrams of the typical climate for each region. A rainfall map of Vietnam compiled from all available data (Bruzon & Carton, 1930; Chevey & Carton, 1935; Schzeglova, 1957; Fridland, 1961; Walter & Leith, 1960, 1964, 1967; Gaussen, Legris & Blasco, 1967; Pham Ngoc Toan & Phan Tat Dac, 1993; Nguyen Khanh Van *et al.*, 2000) is presented in Figure 4.

Monsoon tropical climate with cold winter and summer rains

This zone extends over nearly all of northern Vietnam (Fig. 3). Dry winters with a drought period of 2–5 months, extending from December to February or March, are typical in the plains and low mountain regions of northern Vietnam. At the same time in a number of mountain regions, like the Tam Dao Ridge (21°27'N, 105°38'E), Sin Ho (22°21'N, 103°15'E) and the town of Mong Cai (21°31'N, 107°58') no dry period is observed. In other places the dry period may be very short (Nguyen Khanh Van *et al.*, 2000).

Summers in the region are wet, with the peak rainfall in July and August. Hanoi has a mean annual rainfall of 1674 mm that is very near to the mean annual rainfall of about 1700 mm for the region. Mean annual precipitation increases on the south and southeast slopes of the large mountain systems up to 2500–2800 mm, for example, on the slopes of Tam Dao Ridge and near Tam Duong (22°25'N, 103°29'E), Mong Cai and Sin Ho villages (Fig. 4). In Bac Quang (22°29'N, 104°52'E) the mean rainfall exceptionally reaches 4802 mm (Nguyen Khanh Van *et al.*, 2000). In coastal regions and on north- and northwest-facing mountain slopes minimal rates of rainfall are observed. The mean annual precipitation here is often 1200–1400 mm, for example, in Song Ma (21°04'N, 103°44'E), Yen Chau (21°03'N, 104°17'E), Bao Lac (22°57'N, 105°40'E) or Truong Duong (19°17'N, 104°26'E) towns. A minimum annual rainfall of 1126 mm is reported for Bach Long Vi island (20°08'N, 107°43'E) in the centre of Bac Bo (Tonkin) Gulf (Nguyen Khanh Van *et al.*, 2000).

Temperature regimes in the north of the country have a strong seasonality. Winter conditions with cool temperatures in Hanoi extend from November to April, usually with lows of 14–16°C (absolute minimum 2.7°C). Persistent drizzling rains, so called *craschen*, are very

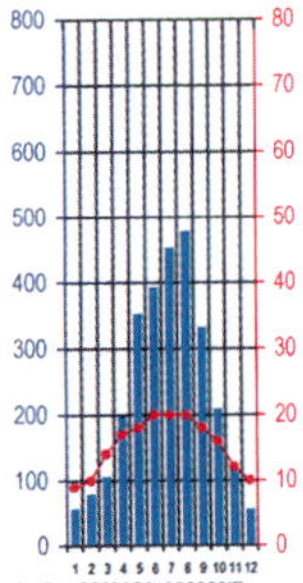

Sa Pa, 22°20'N, 103°50'E,
elev. 1570 m a.s.l.
Mean annual rainfall 2833 mm

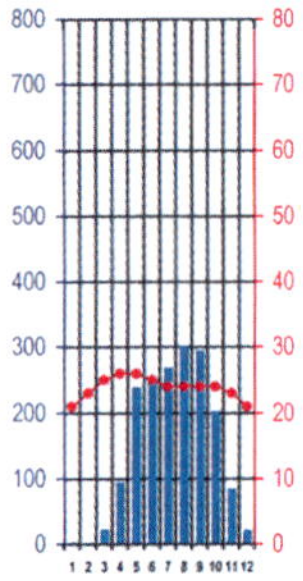

Buan Ma Thuot, 12°41'N, 108°03'E,
elev. 490 m a.s.l.
Mean annual rainfall 1789 mm

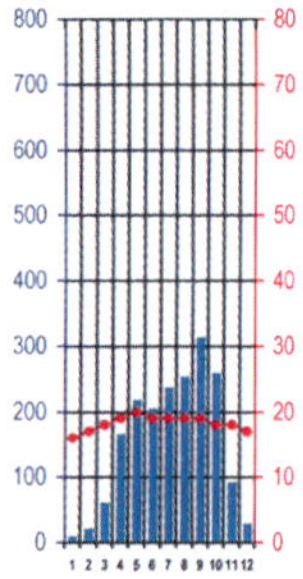

Dalat, 11°57'N, 108°26'E,
elev. 1513 m a.s.l.
Mean annual rainfall 1865 mm

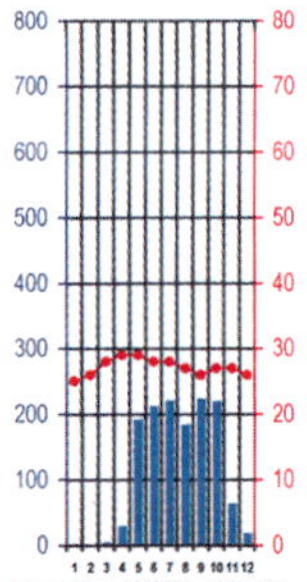

Vung Tau, 10°20'N, 107°05'E,
elev. 4 m a.s.l.
Mean annual rainfall 1369 mm

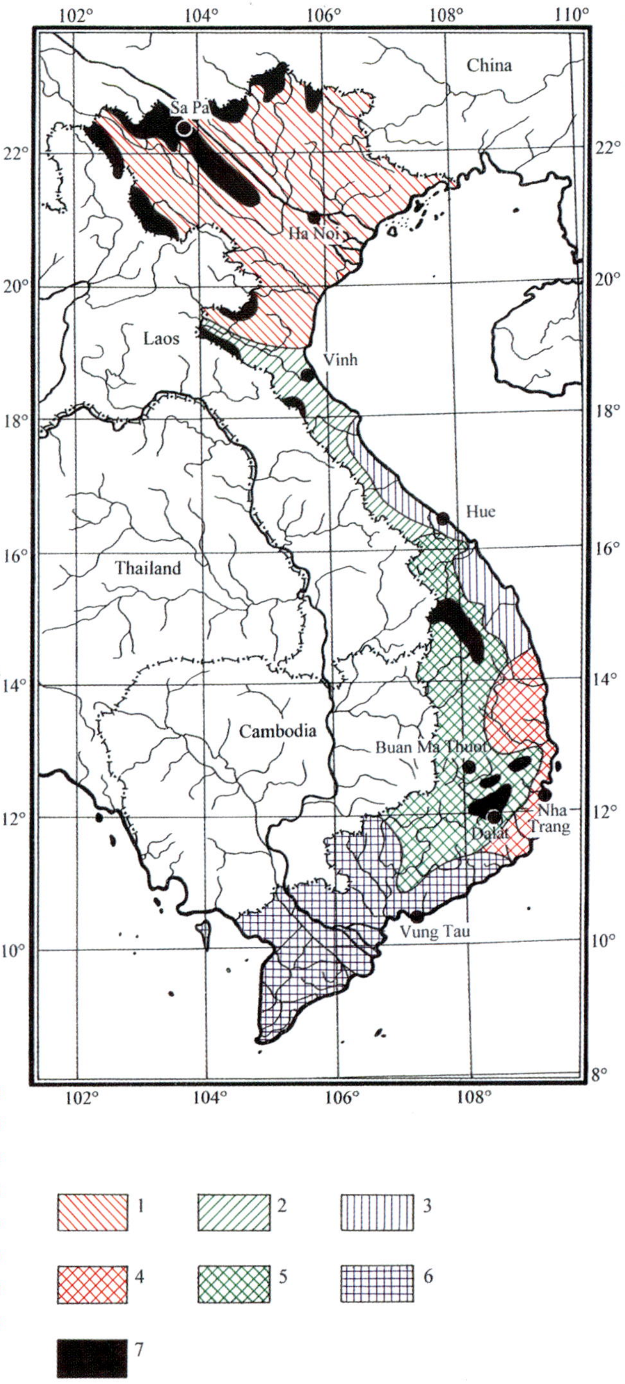

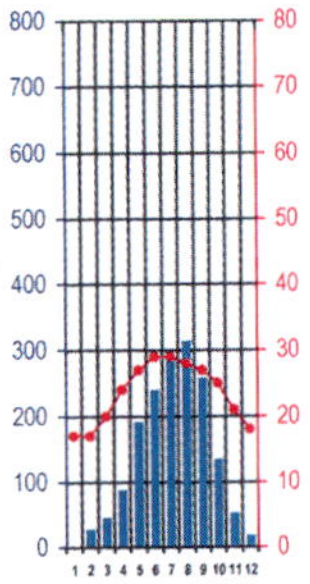

Ha Noi, 21°01'N, 105°51'E,
elev. 5 m a.s.l.
Mean annual rainfall 1674 mm

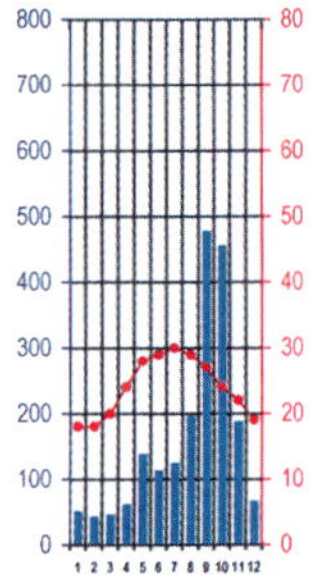

Vinh, 18°40'N, 105°40'E,
elev. 6 m a.s.l.
Mean annual rainfall 1968 mm

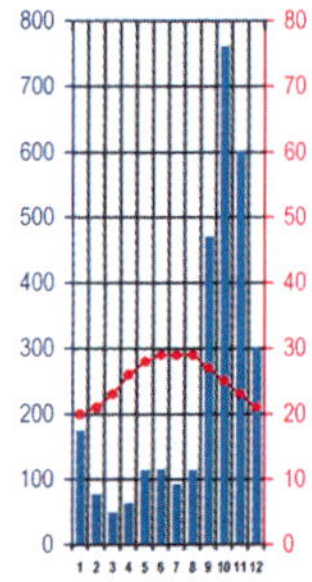

Hue, 16°24'N, 107°41'E,
elev. 17 m a.s.l.
Mean annual rainfall 2936 mm

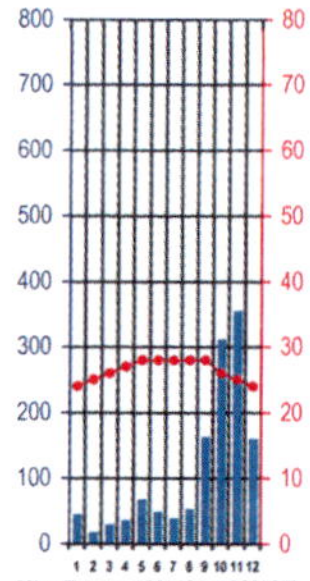

Nha Trang, 12°15'N, 109°12'E,
elev. 5 m a.s.l.
Mean annual rainfall 1324 mm

typical in the lowland plains from early February until the end of March. Summers are hot and humid, with normal maximum temperatures in Hanoi reaching 33°C (absolute maximum 42.8°C). The mean annual temperature at Hanoi is 23.5°C (Gaussen, Legris & Blasco, 1967; Rundel, 1999; Nguyen Khanh Van *et al.*, 2000). The high mountains of northern Vietnam experience cooler temperatures and higher levels of mean annual rainfall with increasing elevation. At elevations above 1000 m, hoar-frost at night is sometimes observed in open places during winter months.

Monsoon tropical climate with cold winter and summer-autumn rains

This climate type is found, for example, in the northern part of central Vietnam (Fig. 3), where temperatures and rainfall are similar to those mentioned above. Vinh town (18°40'N, 105°40'E), for example, has a mean annual rainfall similar to that of Hanoi, with 1968 mm. However, the mean annual precipitation for the region is much higher at 2342 mm. The annual rainfall varies in different places between 1707 and 3205 mm (Fig. 4). At the same time the peak rainfall in this region is in September and October. In every month here at least 50 mm of rain falls. Only a few regions, like Khe Sanh (16°38'N, 106°50'E) or Do Luong (18°54'N, 105°18'E) towns, have a distinct dry winter season lasting 2–4 months.

Summer temperatures in the region often reach 30–35°C with a maximum of 41–42°C. Winter is fresh with usual temperature lows of 14–16°C and an absolute minimum of 3–4°C (0.7°C reported from Huong Son, 18°27'N, 105°16'E).

Monsoon tropical climate with warm winter and summer-autumn-winter rains

This region spreads along the coast of central Vietnam (Fig. 3). Two rainfall peaks during summer and later autumn are typical in this region. The pronounced wet season here extends from September until January. Total rainfall continues to increase moving southwards along the coastline with 2936 mm at Hue (16°24'N, 107°41'E), but then drops to 2041 mm at Da Nang (16°02'N, 108°11'E) and 2003 mm at Hoai Nhon (14°32'N, 109°01'E). At the same time, the annual precipitation can be up to 3340 mm in Ba To (14°46'N, 108°43'E) and 3973 mm near Tra My (15°21'N, 108°13'E) close to the eastern slopes of the Truong Son Range (Fig. 4).

In this region usually there is a semi-dry season of 2–4 months from February to April or May. However, in some places, like Hue, Tra My and Dong Hoi (17°28'N, 106°37'E), significant rain of 20–50 mm falls in all of these months and a dry season is absent (Gaussen *et al.*, 1967; Nguyen Khanh Van *et al.*, 2000).

Summer in the region is hot with usual temperatures reaching 30–34°C (with an absolute maximum 40–42°C), winter median temperatures are about 18–22°C. Winter temperatures very rarely fall here below 18°C, however an absolute minimum temperature of 4.5°C has been reported from Con Co island (17°10'N, 107°22'E).

Fig. 3 (opposite). Climatalogical regions of Vietnam.
(Drawn by L. Averyanov based on Nguyen Khanh Van *et al.*, 2000).

Regions with: 1 — Monsoon tropical climate with cold winter and summer rains; **2** — Monsoon tropical climate with cold winter and summer-autumn rains; **3** — Monsoon tropical climate with warm winter and summer-autumn-winter rains; **4** — Monsoon tropical climate with warm winter and autumn-winter rains; **5** — Monsoon tropical climate with warm winter and summer rains; **6** — Monsoon sub-equatorial climate with summer rains; **7** — Monsoon tropical climate associated with mountains.

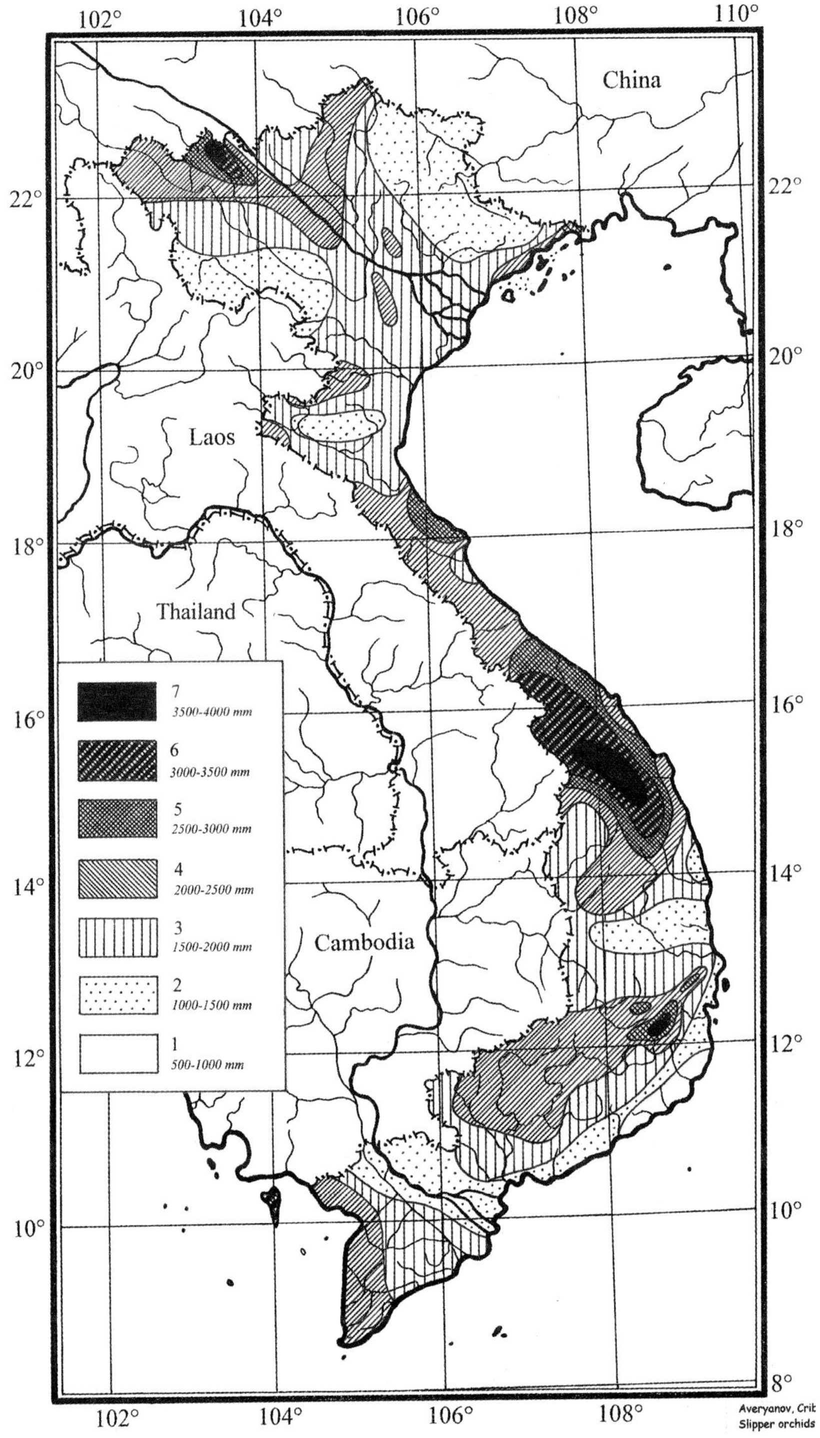
102°
104°
106°
108°
110°
China
Laos
Thailand
Cambodia
22°
20°
18°
16°
14°
12°
10°
8°
7
3500-4000 mm
6
3000-3500 mm
5
2500-3000 mm
4
2000-2500 mm
3
1500-2000 mm
2
1000-1500 mm
1
500-1000 mm
Averyanov, Crib
Slipper orchids

Monsoon tropical climate with warm winter and autumn-winter rains

This climate occurs along coastal areas in the north portion of southern Vietnam (Fig. 3). The peak months for rainfall in the region are typically October and November under the influence of northeastern monsoon winds. This is the most arid area of Indochina with a tropical dryland climate, which characterizes the coastal zone from Quy Nhon (13°46'N, 109°13'E) southwards to Phan Thiet (10°56'N, 108°06'E). Most of this region, lying in a rain shadow of the southern Truong Son Range, has a mean annual rainfall of less than 1500 mm (Fig. 4), and a dry season up to nine months (Bruzon & Carton, 1930; Schzeglova, 1957).

The driest part of the region lies along the coast from Phan Rang to Phan Thiet (Fig. 4). Thus, Ca Na (Padaran) Cape (11°18'N, 108°55'E) south of Phan Rang has only 757 mm of annual rainfall with a mean annual minimum of 409 mm (Bruzon and Carton, 1930; Schzeglova, 1957; Schmid, 1974). Three-year observations at Phan Ri (11°08'N, 108°33'E), Mui Ne (10°53'N, 108°17'E) and Phan Rang (11°33'N, 108°59'E) gave respectively mean annual rainfalls of 622 mm, 596 mm and 413 mm (Bruzon and Carton, 1930; Schzeglova, 1957). These stations experience a second summer drought in July and August when flows from the monsoon rains are blocked. Rainfall in the coastal belt is highly variable from year to year, and monthly totals are less than 100 mm for 6–9 months in every year.

In the northern areas, the Song Ba Valley and Cho Reo Depression, that spread inland from the coast, the rainfall increases and reaches a mean annual rate of 1697 mm in Quy Nhon (13°46'N, 109°13'E) and 1715 mm in Son Hoa (13°03'N, 108°59'E). These semi-arid areas have dry periods of 2–5 months.

The temperature regime of this region approaches that of sub-equatorial areas. Mean annual temperatures here are 25–26°C. Normal winter temperatures in the region lie between 20–25°C, however absolute temperature minimums of 12–14°C are reported for some stations. Common summer temperatures are 28–29°C., rarely increasing up to 38–42°C.

Monsoon tropical climate with warm winter and summer rains

This climatic zone is hot and semi-humid, and is characterized by moderate levels of rainfall, ranging from 1500 mm at Yok Don Nature Reserve (13°00'N, 107°30'E) to 2339 mm at Dac Nong (12°00'N, 107°41'E) and a strong predominance of summer precipitation (Fig. 4). This semi-humid area dominates more than half of southern Vietnam, covering the majority of lowland and middle elevation areas below 800–1000 m (Fig. 3). Typically, there are 3–4 months in winter with less than 50 mm of rainfall. However, in some places, like Bao Loc (11°28'N, 107°48'E) and Nam Dong (16°09'N, 107°43'E), the dry period may be very short. Dew is frequent in the dry season, but mist is rare.

Mean annual temperatures in the region are 22–25°C with usual summer temperatures about 22–27°C and common winter temperatures of 19–25°C. An absolute temperature maximum of 41°C is reported in Nam Dong. The absolute minimum of 3.4°C was reported in Dac To (14°42'N, 107°49'E) at an elevation of 620 m.

Fig. 4 (opposite). Rainfall map of Vietnam.

(Drawn by L. Averyanov based data from Bruzon & Carton, 1930; Chevey & Carton, 1935; Schzeglova, 1957; Fridland, 1961; Walter & Leith, 1960, 1964, 1967; Gaussen, Legris & Blasco, 1967; Pham Ngoc Toan & Phan Tat Dac, 1993; Nguyen Khanh Van *et al.*, 2000).

The rainfall peak in montane areas of this region is usually observed from August to October under the influence of the southward return of the monsoon with northeastern winds. The humid submontane climate of this zone comprises upland areas at elevations below about 1100 m that are exposed to humid winds. Mean annual rainfall here is generally above 2500 mm and quite regular, with a brief dry season lasting three months or less. Dew is abundant in the brief dry season and fog is common. This climatic zone covers mainly hilly and mountain regions of the Kon Tum, Dac Lac and Mo Nong plateaus.

Monsoon sub-equatorial climate with summer rains

The southern lowland areas of Vietnam have a hot monsoon sub-equatorial climate with distinct wet and dry periods, but temperatures that are higher than elsewhere in the country (Fig. 3).

Minimal amounts of annual precipitation are observed in the area along the east coast and on inland plains. Mean annual rainfall ranges here from 1086 mm in Phan Thiet (10°56'N, 108°06'E) to 1297 mm in Chau Doc (10°46'N, 105°08'E), 1359 mm in Cao Lanh (10°28'N, 105°38'E) and 1369 mm in Vung Tau (10°20'N, 107°05'E) (Fig. 4). The dry season in these semi-arid areas occurs in winter and lasts from four to six months.

Annual rainfall considerably increases in areas of southern Vietnam situated along the west coast facing the Gulf of Thailand and in districts near to the southwest slopes of the southern branches of the Truong Son Range. In such places annual precipitation may reach 2109 mm in Rach Gia (10°00'N, 105°05'), 2344 mm in Ca Mau (9°10'N, 105°10'E) and 2469 mm in Dong Phu (11°32'N, 106°54'E) (Fig. 4). At the same time the dry period here is very pronounced.

Considerable rainfall and a distinct dry period are typical for some islands off the southern Vietnam coast, such as Phu Quoc (10°13'N, 103°58'E) and Con Dao (8°41'N, 106°36'E), and the Truong Sa (8°39'N, 111°55'E) archipelago. Their mean annual rainfalls are, respectively, 3024 mm, 2069 mm and 2388 mm. In contrast, Cu Lao island (10°31'N, 108°56'E) situated near the south-east seashore, gets only 1244 mm of rain per annum.

Mean monthly temperatures throughout the year vary between from 25–29°C. Winter and summer temperatures usually differ by less than 2–3°C, very typical for a sub-equatorial climate. An absolute temperature minimum of 11.9°C is reported from Dong Phu town at 89 m elevation, a maximum temperature of 40°C was observed in Tan Son Nhat (10°49'N, 106°40'E) and also at Can Tho (10°02'N, 105°47'E).

Monsoon tropical climate associated with mountains

A tropical monsoon climate is typically associated with mountains above 1400 m, including the largest mountain ranges of northern Vietnam such as Hoang Lien Son, Si Lung, Den Dinh and Sam Sao. It is also observed in highland areas of some isolated mountain ranges, such as Tay Con Linh, Pu Tha Ca, Pia Da and Pia Oac. In the south of the country, this climate is found in the highlands of the Kon Tum and Lam Vien (Dalat) plateaus (Fig. 3).

Mean annual precipitation rates of 2833 mm in Sa Pa town (22°20'N, 103°50'E, elev. 1570 m) and 3552 mm in Hoang Lien Son station (22°21'N, 103°46'E, elev. 2170 m) have been recorded in northern Vietnam (Fig. 4). Rainfall peaks in the summer months. A dry period is absent for this region.

In Dalat (11°57'N, 108°26'E, elev. 1513 m) in southern Vietnam, the mean annual rainfall is 1865 mm, with a summer peak and 1–2 dry winter months of less than 50 mm rainfall. At the same time on the eastern margin of the Lam Vien Plateau, at Hon Ba station (12°05'N, 108°45'E, elev. 1480 m), very heavy rains of 3851 mm occur with no dry period (Schmid 1974). These peak very late in November under the influence of the northeastern monsoon winds.

Morning dew is very common throughout the montane zone, as well as heavy mist. Humidity and the amount of rainfall rise with elevation in all the mountain areas of Vietnam.

The temperature regime usually depends on elevation. The mean annual temperature at Sa Pa is 15.2°C at an elevation of 1570 m, while at Hoang Lien Son it is 12.8°C at an elevation of 2170 m. Summer temperatures in these regions are usually between 16–20°C and the winter mean temperature regularly ranges from 7–10°C. The absolute minimum temperature reported at 2170 m was -5.7°C. Snow and hail are not uncommon here during winter months. The maximum temperature recorded at this elevation is 24.9°C.

Dalat and Hon Ba stations in southern Vietnam have mean annual temperatures of 18.2°C and 17.4°C respectively. Summer mean temperatures are about 19–20°C, but winter temperatures lie between 14–18°C. The maximum temperature recorded in Dalat is 31.5°C, the minimum -0.1°C. In the highland areas of Ngok Linh on the Kon Tum Massif at above 1700–1800 m, local people are familiar with ice and winter night frost .

FLORA OF VIETNAM

The flora of Vietnam is still very poorly documented. However, we do have a reasonable idea about the richness and diversity of Vietnamese flora, mainly thanks to the following taxonomic accounts: *Flora Cochinchinensis* (Loureiro, 1790, 1794), *Flore Generale de l'Indochine* (ed. M.H. Lecomte & H. Humbert, 1907–1951), *Flore du Cambodge, du Laos et du Viet-Nam* by J.E. Vidal (ed. A.Aubreville, J.F. Leroy, & Ph.Morat, 1960–1997), *An illustrated flora of South Viet-Nam* and *An illustrated flora of Vietnam* (Pham Hoang Ho, 1970–1972, 1991–1993), *Vascular plant synopsis of Vietnamese flora* (ed. Averyanov *et al.*, 1990, 1996) and a number of recent floristic and taxonomic monographs (Schmid, 1974, 1989; Averyanov, Nguyen Tien Ban & Kudriavtzeva, 1988; Vu Van Dung, 1996; Nguyen Nghia Thin & Harder, 1996; Nguyen Tien Ban, 1997; Nguyen Nghia Thin, 1997a, 1998, 1999; Nguyen Nghia Thin & Nguyen Thi Thoi, 1998 and many others *). A number of publications are specifically devoted to Vietnamese orchids (Seidenfaden, 1975, 1992; Averyanov, 1988a-f, 1989, 1990c,d, 1991e, 1994, 1996a-c, 1997a-c, 1998a-b, 1999a-c, 2000a-c; Averyanov & Duong Duc Huyen, 1989, 1990, 1993; Averyanov & Vu Ngoc Long, 1989, 1990, 1991; Duong Duc Huyen & Averyanov, 1989, 1991, 1992; Averyanov *et al.*, 1996a,b, 1997a,b, 2000; Averyanov & Christenson, 1998a-e; Tran Hop, 1998; Averyanov & Averyanova, 2000; Cribb, Averyanov & Nguyen Tien Hiep, 2000).

Preliminary estimates (Tolmachev, 1974; Nguyen Nghia Thin, 1997b; Phan Ke Loc, 1998) suggest that about 9600 species of vascular plants are native to Vietnam. In addition about 750 naturalized, introduced and cultivated species have been recorded as occurring in Vietnam. Therefore, the total number of vascular plant species known from Vietnam at the present time is about 10350 species in 2256 genera and 305 families (Phan Ke Loc, 1998). However, it is estimated that at least 2400 additional species have yet to be listed, giving a

* Bibliographies of accounts of the Indochina flora have been published, each covering sequential periods of time by Petelot (1955), Vidal (1972) and Vidal *et al.* (1988, 1994).

total indigenous vascular plant flora of about 12000 species in nearly 3010 genera. This is a remarkably diverse flora for a relatively small country like Vietnam. A high degree of environmental heterogeneity within Vietnam with respect to climate, soil, landscape and topography is certainly responsible for this high diversity.

The flora of Vietnam is not only large, but also rich in endemic species. Some preliminary studies (Vo Quy, 1995) suggest that about 10% of its species and 3% of its genera are endemic to Vietnam. Among the 305 vascular plant families reported for Vietnam, the highest levels of endemism are observed in the Acanthaceae, Anacardiaceae, Annonaceae, Apocynaceae, Araceae, Arecaceae, Asclepiadaceae, Celastraceae, Ericaceae, Euphorbiaceae, Fagaceae, Myrsinaceae, Rubiaceae, Sapindaceae, Sapotaceae, Symplocaceae, Theaceae and Zingiberaceae (Schmid, 1974; Takhtajan, 1986; Rundel, 1999). About 35% of the orchid species in Indochina are endemic while about 19% are endemic in Vietnam (Averyanov, 1991e).

The Vietnamese flora is typical of Palaeotropic floras in its character (Tolmachev, 1974; Phan Ke Loc, 1998). The largest families of flowering plants in the flora are the Orchidaceae, Fabaceae, Poaceae, Euphorbiaceae, Rubiaceae, Cyperaceae, Asteraceae, Lauraceae, Fagaceae and Acanthaceae (Tab. 1). The Orchidaceae is the largest family in the flora of Vietnam with about 753 species and 135 genera reported to date (Averyanov, 1990a,b,d, 1991a,b,e, 1994; Phan Ke Loc, 1998).

TABLE 1

TEN LARGEST FAMILIES OF VASCULAR PLANTS IN THE FLORA OF VIETNAM (PHAN KE LOC, 1998).

Family	Number of species	Percentage of the total flora
1. Orchidaceae	753	7.8%
2. Fabaceae	557	5.8%
3. Poaceae	467	4.9%
4. Euphorbiaceae	416	4.3%
5. Rubiaceae	395	4.1%
6. Cyperaceae	304	3.2%
7. Asteraceae	291	3.0%
8. Lauraceae	245	2.5%
9. Fagaceae	211	2.2%
10. Acanthaceae	177	1.8%

About 3.1% of Vietnamese orchids have a pantropical or palaeoptropical distribution, about 3.7% are widely distributed in tropical Asia and the Malay archipelago, 8.8% are eastern Himalayan, Southeast Asian and West Malesian, 11% are Indochinese and from the Greater Sunda islands, and 73.4% have a continental distribution in mainland Southeast Asia. The biogeographic representation of orchids in the Vietnamese orchid flora is given in Table 2. The affinities of the orchid flora of Vietnam are closest with those of tropical mainland Asia. This is also the case for the flora of the country as a whole.

TABLE 2

TYPES OF DISTRIBUTION AND PROPORTION OF ORCHID GROUPS WITH VARIOUS DISTRIBUTION PATTERNS IN THE ORCHID FLORA OF VIETNAM (Averyanov, 1991e).

Distribution pattern	Approximate proportion in Vietnamese orchid flora
1. Pantropical	**0.3%**
2. Palaeotropical	**0.4%**
3. Asian-Australian	**2.4%**
4. Asian	**96.9%**
4.1. South Asian group	**3.7%**
4.1.1. South Asian-Malesian	1.3%
4.1.2. South Asian-Sundaic	2.4%
4.2. E. Himalayan-SE Asian group	**8.8%**
4.2.1. E. Himalayan-Malesian	2.0%
4.2.2. E. Himalayan-Sundaic	2.5%
4.2.3. E. Himalayan-Sumatran-Bornean-Philippines	0.6%
4.2.4. E. Himalayan-Sumatran-Javanese-Philippines	1.0%
4.2.5. E. Himalayan-Bornean-Philippines	0.1%
4.2.6. E. Himalayan-Sumatran-Javanese	0.8%
4.2.7. E. Himalayan-Sumatran	0.4%
4.2.8. E. Himalayan-Bornean	0.1%
4.2.9. E. Himalayan-Philippines	1.3%
4.3. Indochinese-Malesian group	**11.0%**
4.3.1. Indochinese-Malesian	1.8%
4.3.2. Indochinese-Sundaic-Philippines	0.8%
4.3.3. Indochinese-Sundaic	1.8%
4.3.4. Indochinese-Javanese-Borneo-Philippines	0.3%
4.3.5. Indochinese-Javanese-Sumatran	1.4%
4.3.6. Indochinese-Borneo-Sumatranian	0.8%
4.3.7. Indochinese-Borneo-Philippines	0.1%
4.3.8. Indochinese-Sumatran	1.1%
4.3.9. Indochinese-Javanese	0.8%
4.3.10. Indochinese-Bornean	0.8%
4.3.11. Indochinese-Philippines	1.0%
4.4. SE. Asia mainland group	**73.4%**
4.4.1. SE. Asia mainland	2.8%
4.4.2. Hindustan-Indochinese	1.4%
4.4.3. E. Himalayan-SE. Asian (mainland)	1.1%
4.4.4. E. Himalayan-S. Chinese	5.3%
4.4.5. S.Chinese	4.2%
4.4.6. E. Himalaya-Indochinese	23.5%
4.4.7. Malay-Indochinese (mainland)	4.9%
4.4.8. Indochinese (except Malay Peninsula)	11.0%
4.4.9. Vietnamese	19.2%

The ten largest genera in the orchid flora of Vietnam are listed in Table 3. They represent a typical spectrum of the largest orchid genera in tropical Asia. However, the appearance of *Paphiopedilum* amongst them is unique and emphasises that Vietnam has the richest diversity of *Paphiopedilum* species in the world, matched only by that of neighbouring southern China.

TABLE 3

TEN LARGEST ORCHID GENERA IN THE FLORA OF VIETNAM (Averyanov, 1991E).

Genus	Number of species	Percentage of the total orchid flora
Dendrobium	97	12.9%
Bulbophyllum	78	10.4%
Eria	37	4.9%
Habenaria	27	3.6%
Coelogyne	25	3.3%
Liparis	25	3.3%
Oberonia	24	3.2%
Cleisostoma	20	2.7%
Calanthe	18	2.4%
Paphiopedilum	18	2.4%

The genera with the most numerous endemic orchid species in the Vietnamese flora are *Anoectochilus, Bulbophyllum, Calanthe, Cheirostylis, Cleisostoma, Coelogyne, Cymbidium, Dendrobium, Eria, Liparis, Oberonia, Pholidota* and *Taeniophyllum* (Averyanov, 1991e). The four monotypic orchid genera, *Christensonia, Deceptor, Cleisostomopsis* and *Eparmatostigma*, that are endemic to Vietnam are closely related respectively to the larger allied genera *Aerides, Acampe, Cleisostoma* and *Malleola*.

Nine *Paphiopedilum* taxa are restricted in their distribution to Vietnam, namely, *Paphiopedilum* × *aspersum, P.* × *dalatense, P. delenatii, P. hangianum, P. helenae, P. malipoense* var. *hiepii, P. tranlienianum, P. vietnamense* and *P. villosum* var. *annamense*. Another twelve taxa, *Paphiopedilum* × *affine, P. barbigerum, P. dianthum, P. emersonii, P. gratrixianum, P. henryanum, P.* × *herrmannii, P. malipoense* var. *malipoense, P. malipoense* var. *jackii, P. micranthum, P. purpuratum* and *P. villosum* var. *boxallii*, may be regarded as Vietnamese subendemics, which occur outside Vietnam in very limited areas adjacent to the borders of the country.

MAIN PATTERNS OF VEGETATION

Monsoon tropical forests are the most typical vegetation in Vietnam. Wet, closed, evergreen, broad-leaved, mixed or coniferous woods represent them in mountain regions. However, in some hilly lowland and flat areas with more arid conditions these forests give way to open semi-deciduous and deciduous broad-leaved woodlands and semi-savannas. Along dry coastal plains and on drought-prone coastal hills in southern Vietnam, such woodlands are replaced by their shrubby derivatives, which often appear as a more or less dense sclerophytic scrub. Swampy grass-sedge communities and mangrove thickets are very common on flat coastal plains especially in the deltas of large rivers. Regrettably, much of the native primary vegetation has already been destroyed. Secondary forests, scrub and grassland, bamboo thicket, wood plantations, pastures, crop and vegetable fields, different weed communities (usually lacking all interesting native plants) nowadays replace native vegetation over extensive parts of Vietnam. A simplified vegetation map of Vietnam is presented in figure 5 in which the eight main forms of plant cover listed below are roughly distinguished.

Evergreen broad-leaved plateau forests on alkaline soils

Originally these forests covered the undulating limestone plateaus and the flat bottoms of alluvial valleys on alkaline soils associated with basement rocks rich in calcite or magnesite. In prehistoric times they covered large areas of Vietnam from sea level up to 1000 m elevation. Presently, they have been destroyed over most of their primary range. The last small remnants are to be found in a few areas on plains and slopes of limestone plateaus or terraces of river valleys. Sometimes remnants survive along rivers and appear as fragments of gallery woods in flat deforested river valleys (Fig. 5).

These broad-leaved forests thrived in a climate with a humid wet summer from May to October and a dry winter with moderate temperatures from October to April. Heavy mist and drizzle are common in these regions in March and April. The climate is are similar to that of the Hanoi area (Fig. 3).

Primary evergreen broad-leaved forests at low elevations on alkaline soils in valleys are usually wet and have a closed canopy (Fig. 6). The tallest trees regularly reach 30–35 m high. The most important canopy trees in these forests are mostly representatives of tropical families (Nguyen Nghia Thin, 1994, 1997a) such as:

Anacardiaceae (*Choerospondias, Dracontomelum*),
Clusiaceae (*Garcinia*),
Combretaceae (*Anogeissus, Terminalia*),
Datiscaceae (*Tetrameles*),
Dipterocarpaceae (*Parashorea, Vatica*),
Ebenaceae (*Diospyros*),
Fabaceae (*Erythrophleum, Paralbizzia, Saraca*),
Lauraceae (*Caryodaphnopsis, Cinnamomum, Phoebe*),
Meliaceae (*Aglaia, Chisocheton*),
Moraceae (*Streblus, Ficus*).
Sapindaceae (*Pometia*),
Sapotaceae (*Madhuca*),
Simaroubaceae (*Canarium*),
Sterculiaceae (*Pterospermum, Sterculia*),
Styracaceae (*Rehderodendron*),
Tiliaceae (*Burretiodendron*).

Warmth-loving trees, such as *Symingtonia tonkinensis* and *Liquidambar formosana* (Hamamelidaceae), and tropical oaks such as *Castanopsis indica, Lithocarpus bacgiangensis* and *L. licentii* (Fagaceae), are also common dominants in these forests (Rundel, 1999). Some typical plants of these rich woods are presented in figure 7.

The orchid flora of these lowland forests is very rich and includes numerous terrestrial species. Orchids found here are:

Calanthe alismaefolia,
C. angusta,
C. herbacea,
C. triplicata,
Corymborkis veratrifolia,
Cymbidium ensifolium,
C. lancifolium,
C. sinense,
Erythrodes blumei,
Geodorum attenuatum,
Goodyera foliosa,
G. fumata,
G. hispida,
Habenaria ciliolaris,
Hetaeria rubens,
Liparis nervosa,
Malaxis acuminata,
Nervilia fordii,
Tropidia angulosa,
T. curculigoides,
Vrydagzynea albida,
Zeuxine nervosa.

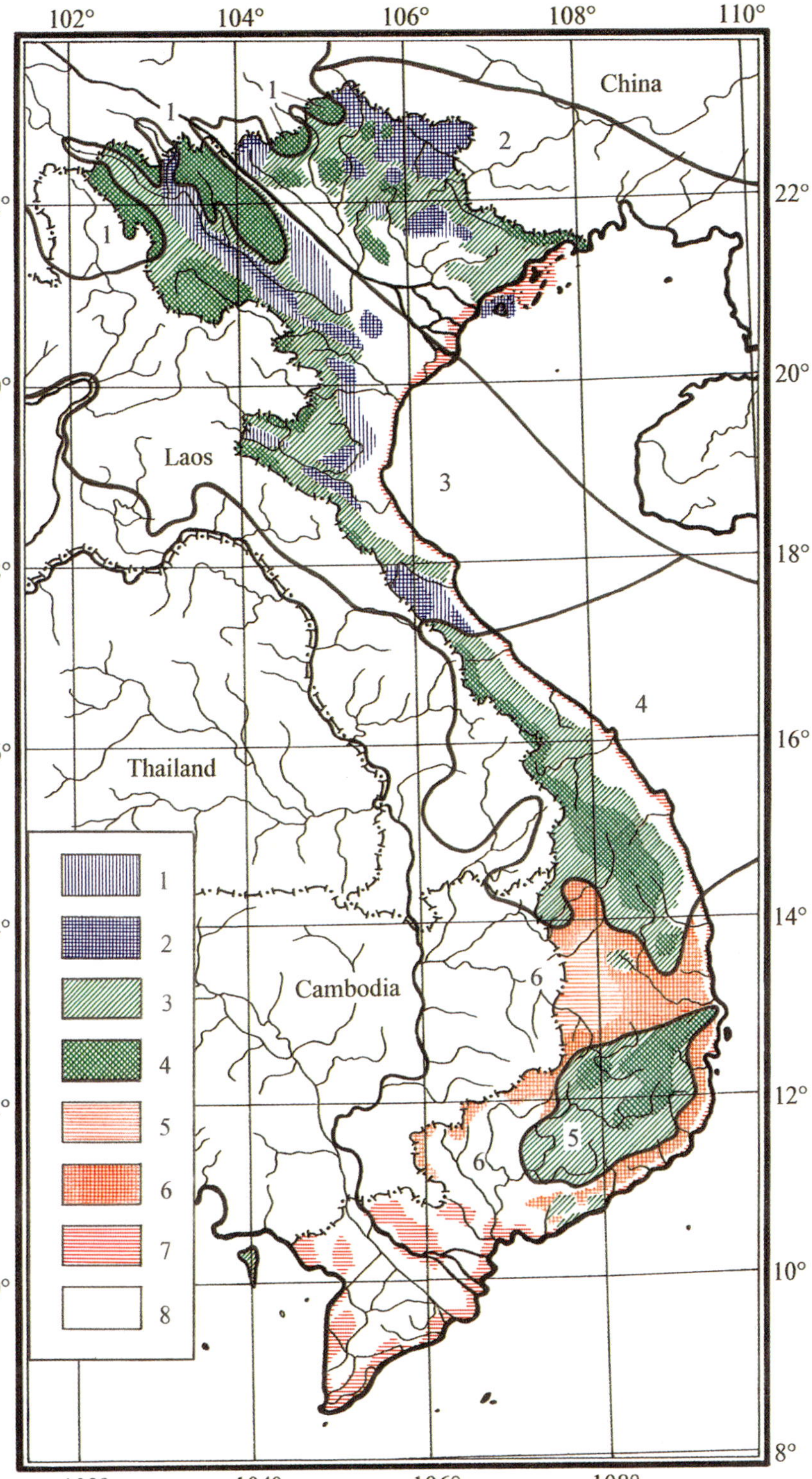
102°
104°
106°
108°
110°
22°
20°
18°
16°
14°
12°
10°
8°
China
Laos
Thailand
Cambodia
1
2
3
4
5
6
7
8

Primitive orchids, such as *Neuwiedia griffithii* and *N. inae*, the rare local endemic *Zeuxine vietnamica*, and some rare and unusual mycotrophic species, such as *Cyrtosia javanica, Didymoplexiopsis khiriwongensis, Didymoplexis pallens, Epipogium roseum* and *Erythrorchis ochobiensis*, occur in these forests. The creeping epiphytic liana, *Vanilla annamica*, with a fleshy thick stem reaching 10 and more m long, can also be found.

The canopies of large trees in these forests support a rich and diverse epiphytic orchid flora. The most usual epiphytic orchids found in these forests are:

Aerides odorata,
Callostylis rigida,
Ceratostylis himalaica,
Cleisostoma aspersum,
C. fuerstenbergianum,
C. rostratum,
C. williamsonii,
Coelogyne fimbriata,
Cymbidium aloifolium,
C. bicolor,
Dendrobium acinaciforme,
D. aduncum,
D. anosmum,
D. fauhaberianum,
D. hercoglossum,
D. lindleyi,
D. linguella,
D. nobile,
D. parishii,
D. wardianum,
Eria pannea,
Liparis viridiflora,
Malleola seidenfadenii,
Ornithochilus difformis,
Pelatantheria insectifera,
Phalaenopsis lobbii,
Pholidota rubra,
Robiquetia succisa,
Thrixspermum centipeda,
Trichotosia dasyphylla.

Among the epiphytic species are found some rare local Vietnamese endemics, such as *Biermannia calcarata, Dendrobium nobile* var. *alboluteum, Phalaenopsis gibbosa* and *Pteroceras simondianum.*

In the past *Paphiopedilum concolor* has occasionally been found growing on limestone outcrops in lowland forests. Rarely, *Paphiopedilum malipoense* var. *malipoense* and var. *hiepii* have also been observed, growing in the ground litter on shady steep slopes. However, *Paphiopedilum* species are not typically found in lowland forests.

Evergreen and semi-deciduous forests on limestone mountains

Spectacular karst landscapes formed by narrow rocky ridges and crowded numerous column-like mesas with very steep slopes and numerous sheer cliffs are widely distributed in northern Vietnam. The highly eroded, marble-like limestone of such mountains creates very specific edaphic conditions, which has resulted in the formation of a unique vegetation. Tropical evergreen and semi-deciduous, broad-leaved and mixed forests grow in these areas, mainly at elevations from sea level up to 1600 m. However, the major part of this forest type lies between 300–900 m (Fig. 8). In a number of places remnants of deciduous broad-leaved or coniferous forest spread along the highest ridges, usually at elevations between 1400–1700 m (Fig. 5).

The primary vegetation in the limestone mountains of Vietnam usually exhibits a clear zonation. Evergreen, broad-leaved, highly closed and wet forests are usually found in the foothills and on the steep slopes of highly eroded limestone ridges. Dominants, such as *Burretiodendron hsienmu, Gleditchia* sp. and *Symingtonia populnea*, reach 25–30 m tall and

Fig. 5 (opposite). Main vegetation patterns and floristic regions in Vietnam.
(Drawn by L. Averyanov based on original data and available literature).

Vegetation types: 1 — Evergreen broad-leaved plain forests on alkaline soils; **2** — Evergreen and semideciduous broad-leaved, mixed and coniferous limestone mountain forests; **3** — Evergreen lowland forests on silicate rocks at 0–1000 m a.s.l.; **4** — Evergreen montane and highland forests on silicate rocks at 1000–3000 m a.s.l.; **5** — Semideciduous dry lowland forests; **6 —** Deciduous dry lowland forests and savanna-like woodlands; **7 —** Coastal vegetation, lowland wetlands and mangrove thickets; **8** — Secondary, weed and agricultural plant communities, forestry and industrial plantations.
Floristic regions: 1 — Sikang-Yunnan Province; **2** — South Chinese Province; **3** — North Indochinese Province; **4** — Central Annamese Province; **5** — South Annamese Province; **6** — South Indochinese Province.

Fig. 6. Primary evergreen broad-leaved lowland forest on limestone hills at 350 m.
N. Vietnam, Cao Bang Prov., Ba Be National Park. (Photo.: L. Averyanov).

Fig. 7. Typical plants of primary evergreen broad-leaved lowland limestone forests. (Photos: L. Averyanov).

a — *Saraca indica* L. (Fabaceae);
b — *Amorphophallus paeonifolius* (Denst.) Nicols. (Araceae);
c — *Poikilospermum suaveolens* (Blume) Merr. (Cecropiaceae);
d — *Stixis balansae* DC. (Capparidaceae);
e — *Cycas balansae* Warb. (Cycadaceae);
f — *Glyptopetalum tonkinensis* Pitard (Celastraceae);
g — *Tacca plantaginea* (Hance) Drenth (Taccaceae);
h — *Chirita* sp. (Gesneriaceae);
i — *Balanophora* sp. (Balanophoraceae);
j — *Aspidistra* sp. (Convallariaceae).

Fig. 8. Primary evergreen and semi-deciduous broad-leaved and mixed forests on remnant limestone ridges and mesas at 800–900 m.

N. Vietnam, Cao Bang Prov., Thang Hen Lake area. (Photo.: L. Averyanov).

up to 2 m in diam. near the base (Averyanov *et al.*, 2000). The lower subcanopy and understory strata in these forests are well developed and composed of numerous tree and shrub species from a number of families (Nguyen Nghia Thin, 1994, 1997a) such as:

Annonaceae (*Mitrephora*),
Clusiaceae (*Calophyllum*),
Ebenaceae (*Diospyros*),
Euphorbiaceae (*Trigonostemon*),
Moraceae (*Dimerocarpus, Ficus, Streblus, Teonongia*),
Oleaceae (*Jasminum*),
Rosaceae (*Eriobotrya*),
Rubiaceae (*Canthium*),
Rutaceae (*Murraya*),
Sapindaceae (*Xerospermum*),
Sapotaceae (*Sinosideroxylon*),
Sterculiaceae (*Heritiera, Pterospermum*).

On the upper slopes and along the ridge-tops, broad-leaved forests are replaced by a rather open, light, semi-deciduous mixed forest in which conifers are dominant, notably *Pinus fenzeliana* (syn. *P. kwangtungensis*), *Pseudotsuga sinensis* var. *brevifolia* and *Tsuga chinensis* (Nguyen Tien Hiep, 1998; Phan Ke Loc, Nguyen Tien Hiep & Averyanov, 1999a). Many other gymnosperms, such as *Amentotaxus argotaenia*, *Calocedrus macrolepis*, *Cephalotaxus mannii*, *Cupressus torulosa*, *Dacrycarpus imbricatus*, *Dacrydium elatum*, *Fokienia hodginsii*, *Keteleeria evelyniana*, *Nageia fleuryi*, *Podocarpus neriifolius*, *P. pilgeri*, *Taxus chinensis*, and *Xanthocyparis vietnamensis* (Fig. 9b) are also frequently found in these forest stands (Farjon *et al.*, 2002). Cycads, such as *Cycas balansae*, *C. hainanensis*, *C. miquelii*, *C. multifrondis* and *C. taiwaniana*, are also sometimes found here at elevations from sea level up to 600 m. The commonest dominant deciduous trees are species of *Acer*, *Alangium*, *Carpinus*, *Lithocarpus*, *Quercus*, *Ulmus* and *Platycarya strobilacea*. *Illicium jiadifengpi*, *Pistacea weinmannifolia*, *Schefflera pesavis*, *Tirpitzia sinensis*, species of *Euonymus*, *Photinia*, *Madhuca* and *Sorbus* are characteristic of the subcanopy of these forests. The most typical shrubs and small trees here are representatives of *Celastrus*, *Lespedeza*, *Pittosporum*, *Sophora*, *Wikstroemia*, *Rhododendron*, *Vaccinium*, and also some species of dwarf palms, for example, species of *Guihaia*, *Livistona*, *Luculia* and *Rhapis* (Phan Ke Loc, Nguyen Tien Hiep & Averyanov, 1999a; Averyanov *et al.*, 2000).

On limestone mountains above 1200 m, especially on isolated peaks and along narrow ridges, open mossy coniferous woods are often formed, with more or less significant participation of broad-leaved, semi-deciduous trees only in undercanopy. Numerous herbaceous plants, other than orchids, are commonly found in the mixed and coniferous forests on limestone, including some typical species of *Ainsleya* (Asteraceae), *Alpinia*, *Hedychium* (Zingiberaceae), *Arisaema* (Araceae), *Laportea* (Urticaceae), *Lysimachia* (Primulaceae), *Selaginella* and some species of *Begonia* with brightly variegated leaves. Species of Convallariaceae (including numerous species of *Aspidistra*, *Disporum*, *Ophiopogon*, *Peliosanthes*, *Polygonatum* and *Tupistra*) and Urticaceae (*Elatostema*, *Pellea* and *Pellionia*) are also well represented. Succulent species of *Impatiens*, many kinds of sedges (*Carex*, *Scleria*), numerous ferns and herbs from the Acanthaceae, Gesneriaceae and Melastomataceae also occur here (Phan Ke Loc, Nguyen Tien Hiep & Averyanov, 1999a; Averyanov *et al.*, 2000). Typical species from these forests are presented in figure 9.

Orchids reach an extraordinary abundance and diversity in the woods that spread along the ridges of karst limestone. Plentiful light under the canopy of the trees defines their abundance on the forest floor.

Some orchids grow on open rocks and bluffs and form understory thickets. Among these are *Acampe rigida*, *Aerides rosea*, *Cleisostoma paniculatum*, *C. striatum*, *Hygrochilus parishii*, *Flickingeria fimbriata*, *Renanthera coccinea*, *Thrixspermum calceolus*, *Vanda fuscoviridis* and *Vandopsis gigantea*, large specimens of which reach more than a metre in height.

Fig. 9. Typical plants of primary evergreen and semi-deciduous broad-leaved, mixed and coniferous highland limestone forests. (Photos: L. Averyanov).

a — *Pinus fenzeliana* Hand.-Mazz. (Pinaceae);
b — *Xanthocyparis vietnamensis* Farjon et al. (Cupressaceae);
c — *Pseudotsuga brevifolia* W.C.Cheng & L.K.Fu (Pinaceae);
d — *Nageia wallichiana* (Presl) O.Kuntze (Podocarpaceae);
e — *Amentotaxus hatuyenensis* N.T.Hiep (Taxaceae);
f — *Clematis* sp. (Ranunculaceae);
g — *Paris polyphylla* Smith (Trilliaceae);
h — *Arisaema franchetianum* Engl. (Araceae);
i — *Trevesia sphaerocarpa* Grusw. & N.Skvorts. (Araliaceae);
j — *Tsuga chinensis* (Franch.) Pritzel (Pinaceae).

Other orchids with creeping rhizomes, sometimes with vine-like tendencies cover mossy rocks and the bases of large trees with an unbroken cover. Species that are commonest in this group include:

Bulbophyllum ambrosia,
B. andersonii,
B. macraei,
B. purpureifolium,
B. reptans,
Callostylis rigida,
Coelogyne fimbriata,
C. lockii,
C. malipoensis,
C. schultesii,
Epigeneium amplum,
Eria coronaria,
E. rhombodalis
E. pubescens,
Flickingeria fimbriata,
Gastrochilus pseudodistichus,
Monomeria barbata,
Otochilus pseudoporrectus,
Pelatantheria insectifera,
Pholidota articulata,
P. leveilleana,
P. recurva,
P. roseans,
P. rubra,
P. yunnanensis.

The majority of orchids in the humid, mossy, mixed and coniferous forests have no strong preference for an epiphytic, lithophytic or terrestrial mode of life. Many are regularly observed on mossy limestone outcrops as lithophytes, as terrestrials in shallow soils, or as epiphytes at the base of gnarled mossy trees on mountain summits. The commonest among them are:

Appendicula reflexa,
Cleisostoma rostratum,
Cymbidium aloifolium,
C. bicolor,
C. floribundum,
C. suavissimum,
Dendrobium chrysanthum,
D. fimbriatum,
D. loddigesii,
D. nobile,
D. ochreatum,
D. thyrsiflorum,
Eria apertiflora,
E. carinata,
E. boniana,
E. clausa,
E. corneri,
E. coronaria,
E. paniculata,
E. pusilla,
E. siamensis,
E. sutepensis,
Liparis averyanoviana,
L. balansae,
L. bootanensis,
L. distans,
L. latilabris,
L. mannii,
L. petelotii,
L. stricklandiana,
L. viridiflora,
Oberonia cavaleriei,
Pelatantheria insectifera,
Podochilus microphyllus.

At the same time a number of obligate epiphytic orchids are also found in these forests, usually growing on the mossy branches of old, dwarf, gnarled trees found along the ridge-tops. Typical species of this group are:

Bulbophyllum hirtum,
B. retusiusculum,
B. umbellatum,
B. xylophyllum,
Ceratostylis himalaica,
Chiloschista trudelii,
Cleisostoma birmanicum,
C. crochetii,
C. fuerstenbergianum,
C. williamsonii,
Cryptochilus luteus,
Dendrobium acinaciforme,
D. cariniferum,
D. lindleyi,
D. longicornu,
D. nobile,
D. truncatum,
Epigeneium chapaense,
Eria pannea,
Gastrochilus minutiflorus,
Holcoglossum wangii,
Liparis delicatula,
Luisia appressifolia,
L. morsei,
Malleola insectifera,
Ornithochilus difformis,
Panisea yunnanensis,
Phalaenopsis lobbii,
Pholidota missionariorum,
Porpax lanii,
Renanthera citrina,
Schoenorchis gemmata,
Thelasis pygmaea,
T. khasiana,
Trichotosia dasyphylla,
T. velutina.

Cheirostylis bipunctata, C. chinensis, C. cochinchinensis, C. marmorifolia, C. spathulata, C. yunnanensis, C. takeoi, Hemipilia calophylla, Liparis cordifolia, Oberonia ensiformis and *Thunia alba* are obligate lithophytic orchids typical of these forests.

Most *Paphiopedilum* species occurring in Vietnam are found in these primary forests. Among these are *P. × aspersum, P. barbigerum, P. concolor, P. dianthum, P. emersonii, P. hangianum, P. helenae, P. henryanum, P. × herrmannii, P. hirsutissimum, P. malipoense, P. micranthum, P. purpuratum, P. tranlienianum* and *P. vietnamense.* All grow almost exclusively on steep rocky cliffs as a lithophytic plants and only *Paphiopedilum dianthum* may rarely be observed as a humus epiphyte near the base of old mossy trees.

Broad-leaved, evergreen, closed and humid primary forest is found in the middle and lower reaches of limestone ridges. A numerous of obligate terrestrial orchids may be found there, often growing in damp crevices of the karst limestone filled by well-structured soil rich in humus. Typical species in such places are:

Calanthe alleizetti,
C. alismaefolia,
C. argenteostriata,
C. triplicata,
Cephalantheropsis longipes,
Cryptostylis arachnites,
Cymbidium ensifolium,
C. lancifolium,
C. sinense,
Goodyera schlechtendaliana,
G. viridiflora,
Liparis nervosa,
L. paradoxa,
Nervilia punctata,
N. plicata,
Phaius flavus,
P. longicruris,
Tropidia angulosa,
T. curculigoides.

A number of other terrestrial orchids are also found here. These include tiny jewel orchids (*Anoectochilus calcareus*, *A. roxburghii*, *Cheirostylis marmorifolia*, *Goodyera hispida* and *Rhomboda petelotii*), mycotrophic species (*Cyrtosia javanica*, *Didymoplexiopsis khiriwongensis*, *Didymoplexis pallens* and *Epipogium roseum*), the creeping vine *Vanilla albida*, and some rhizomatous and tuber-bearing terrestrials (*Bletilla ochracea*, *Habenaria ciliolaris*, *Habenaria dentata*, *Phaius tankervilleae*). *Spiranthes sinensis* and *Arundina graminifolia* are the commonest species in open secondary grasslands all over the area (Averyanov *et al.*, 2000).

Evergreen lowland forests on acidic substrates between sea level and 1000 m

In the past, evergreen broad-leaved tropical forests developed on acidic rocks covered vast plains, hills and low mountain areas all over Vietnam (Fig. 5). Granite, gneiss, rhyolite, dacite, sandstone, schist, shale and other rocks rich in silicate minerals are the commonest basement rocks underlying soils in these regions. The elevation, annual rainfall and seasonality are other important factors that determine the distribution and differentiation of these forests. They formed a more or less pronounced belt at about 800 m in northern Vietnam and below about 1000 m in southern Vietnam. The annual rainfall ranges mostly from 1500–2000 up to 3000 mm. In drier areas of southern Vietnam that have less than 1500 mm annual rainfall, evergreen forests give way to semi-deciduous and deciduous dry lowland forests, open woodlands and savannas.

Usually broad-leaved lowland forests on acidic soils are very wet and closed. They have a very rich species composition with numerous typical representatives of tropical tree genera in each forest layer (Fig. 10). The largest trees may reach 40–50 m high and more than 2 m in diameter. They mainly belong to genuine tropical families (Nguyen Nghia Thin, 1994, 1997a; Rundel, 1999), such as:

Aceraceae (*Acer*),
Anacardiaceae (*Swintonia*),
Apocynaceae (*Alstonia*),
Burseraceae (*Canarium, Dacryodes*),
Clusiaceae (*Calophyllum*),
Dipterocarpaceae (*Dipterocarpus, Hopea*),
Elaeocarpaceae (*Elaeocarpus*),
Euphorbiaceae (*Macaranga, Endospermum, Bischofia, Sapium*),
Fagaceae (*Castanopsis, Lithocarpus, Quercus*),
Ixonanthaceae (*Ixonanthes*),
Lauraceae (*Cinnamomum, Litsea, Machilus, Notaphoebe, Phoebe*),
Meliaceae (*Aglaia, Amoora, Chisocheton, Dysoxylon, Heynea, Sandoricum*),
Moraceae (*Artocarpus, Ficus*),
Rhizophoraceae (*Anisophyllea*),
Ulmaceae (*Celtis*),
Sapindaceae (*Sapindus, Xerospermum*),
Sterculiaceae (*Scaphium*),
Theaceae (*Schima*).

Fig. 10. Primary, evergreen, broad-leaved, lowland forest on granite at 250 m.
N. Vietnam, Ha Tinh Prov., Rao An river. (Photo.: L. Averyanov).

A great number of woody species, mainly 5–15 m tall trees, can be seen in the understory of these forests. The commonest belong to the following families and genera (Rundel, 1999):

Aquifoliaceae (*Ilex*),
Clusiaceae (*Calophyllum, Cratoxylon, Garcinia*),
Elaeocarpaceae (*Elaeocarpus*),
Euphorbiaceae (*Baccaurea, Chaetocarpus*),
Fagaceae (*Castanopsis, Lithocarpus*),
Icacinaceae (*Apodytes*),
Juglandaceae (*Engelhardtia*),
Lauraceae (*Cryptocarya, Lindera, Litsea, Machilus, Neolitsea*),
Magnoliaceae (*Manglietia*),
Meliaceae (*Chisocheton*),
Myristicaceae (*Horsfieldia*),
Myrtaceae (*Cinnamomum*),
Rosaceae (*Pygeum*),
Rubiaceae (*Adina*),
Sabiaceae (*Meliosma*),
Sapindaceae (*Mischocarpus*),
Symplocaceae (*Symplocos*),
Theaceae (*Pyrenaria*),
Xanthophyllaceae (*Xanthophyllum*).

In the northern part of the country Fabaceae, Fagaceae, Lauraceae, Magnoliaceae and Myrtaceae dominate the canopy layer, although in the south species from the Anacardiaceae, Burseraceae, Clusiaceae, Combretaceae, Dipterocarpaceae, Lauraceae, Meliaceae, Moraceae, Myrtaceae, Rubiaceae, Sapindaceae, Sterculiaceae and Theaceae predominate.

Several canopy strata are always observed in these forests with numerous species in each tier. The understorey of these forest formations is rich and diverse, with a notably large number of shrubby Rubiaceae, Melastomataceae, Lauraceae and Acanthaceae, which sometimes form a continuous cover 1.2–1.5 m in height. Numerous palms, lianas and diverse epiphytes form an important component of this forest community (Schmid, 1974). Some typical common plants may be seen in figure 11.

In wetter conditions on granite, gneiss or dacite at elevations above 600 m some gymnosperms, notably *Podocarpus neriifolius*, *Dacrycarpus imbricatus* and *Dacrydium elatum*, can be found in the canopy. Lianas, such as *Freycinetia* (Pandanaceae), *Pseudostachyum polymorphum* (Poaceae), *Ancistrocladus cochinchinensis* (Ancistrocladaceae), and diverse numbers of rattans, *Calamus tetradactylus*, *C. pseudoscutellaris*, *C. bousigonii*, are typical in these habitats (Rundel, 1999). Interesting herbaceous communities rich in terrestrial orchids develop in such shady humid habitats on rocky substrates here. Typical terrestrial orchids are:

Acanthephippium odoratum,
Anoectochilus roxburghii,
A. siamensis,
Calanthe lyroglossa,
Cephalantheropsis longipes,
C. obcordata,
Corymborkis veratrifolia,
Cymbidium ensifolium,
C. lancifolium,
C. sinense,
Geodorum attenuatum,
G. densiflorum,
G. pulchellum,
G. recurvum,
G. siamense,
Habenaria pantlingiana,
Hetaeria oblongifolia,
H. pauciflora,
Liparis nervosa,
L. nigra,
Ludisia discolor,
Malaxis octodentata,
Phaius flavus,
P. indigoferus,
P. mishmensis,
P. tankervilleae,
Tainia angustifolia,
T. hongkongensis,
T. hookeriana,
T. latifolia,
T. pauciflora,
T. ruybarrettoi,
T. viridifusca,
Tropidia angulosa,
T. curculigoides.

A number of remarkable mycotrophic orchids, such as *Aphyllorchis annamensis*, *A. evrardii*, *A. montana*, *Didymoplexiopsis khiriwongensis*, *Epipogium roseum*, *Lecanorchis javanica* and *Stereosandra javanica*, also inhabit these forests. Among them the large leafless liana, *Galeola nudifolia*, may reach more than 10 m long.

Other terrestrial orchids of this area prefer more open habitats and easily survive the dry periods by developing underground tubers. Among these are:

Fig. 11. Typical plants of primary evergreen broad-leaved lowland forests on silicate rocks. (Photos: L. Averyanov).

a — *Erythrophleum fordii* Oliv. (Fabaceae);
b — *Sterculia hyposticta* Miq. (Sterculiaceae);
c — *Terminalia catappa* L. (Combretaceae);
d — *Dysoxylum rubrocostatum* Pierre (Meliaceae);
e — *Leea rubra* Blume (Leeaceae);
f — *Calamus rudentum* Lour. (Arecaceae);
g — *Oxyspora* sp. (Melastomataceae);
h — *Helixanthera parasitica* Lour. (Loranthaceae);
i — *Gloriosa superba* L. (Liliaceae);
j — *Alpinia blepharocalyx* K. Schum. (Zingiberaceae).

Brachycorythis laotica, Cryptostylis arachnites, Cymbidium cyperifolium, Eulophia andamanensis, E. flava, E. graminea, E. macrobulbon, E. macrostachya, E. spectabilis, Habenaria tonkinensis, Malaxis latifolia, Pachystoma pubescens, Spathoglottis pubescens.

Open wet rocks along mountain streams and rivers are common habitats of some orchids, such as *Epipactis atromarginata*, *Arundina chinensis* and *Phaius longicornu.*

However, the largest group of orchids in evergreen lowland forests are epiphytes. Most are small and shrubby, short creeping vines or tufted plants growing in the canopies of large trees. Typical species of this group are:

Acampe ochracea, Acriopsis indica, A. liliifolia, Aerides falcata, A. houlletiana, A. odorata, Agrostophyllum brevipes, A. callosum, A. planicaule, Appendicula hexandra, Bulbophyllum affine, B. astelidum, B. averyanovii, B. blepharistes, B. careyanum, B. devangiriense, B. elassonotum, B. frostii, B. hiepii, B. hirtum, B. odoratissimum, B. pecten-veneris, B. retusiusculum, B. sigaldiae, B. umbellatum, Ceratostylis himalaica, C. radiata, C. siamensis, C. subulata, Cleisostoma birmanicum, C. duplicilobum, C. fuerstenbergianum, C. williamsonii, Coelogyne lentiginosa, C. viscosa, Cymbidium aloifolium, C. atropurpureum, C. bicolor, C. dayanum, Dendrobium acinaciforme, D. amabile, D. anosmum, D. aphyllum, D. crystallinum, D. ellipsophyllum, D. farmeri, D. fimbriatum, D. gratiosissimum, D. heterocarpum, D. lindleyi, D. nobile, D. ochraceum, D. oligophyllum, D. parishii, D. porphyrophyllum, D. salaccense, D. stuartii, D. terminale, D. thyrsiflorum, D. truncatum, D. unicum, D. uniflorum, Eparmatostigma dives, Eria acervata, E. albidotomentosa, E. amica, E. apertiflora, E. clausa, E. corneri, E. eriopsidobulbon, E. obscura, E. paniculata, E. pannea, E. pubescens, E. thao, Flickingeria angustifolia, F. fimbriata, F. forcipata, Gastrochilus calceolaris, G. intermedius, Liparis caespitosa, L. elliptica, Luisia psyche, Micropera poilanei, Ornithochilus difformis, Pholidota articulata, P. chinensis, P. convallariae, P. guibertiae, P. imbricata, P. leveilleana, P. pallida, P. recurva, P. rubra, Polystachya concreta, Pomatocalpa spicata, Pteroceras leopardinum, Robiquetia spathulata, R. succisa, Sarcoglyphis mirabilis, Thecopus maingayi, Thecostele alata, Thrixspermum pauciflorum, Trias disciflora, T. nasuta, Trichotosia dalatensis, T. microphylla, T. pulvinata.

Numerous highly specialized epiphytes inhabit the terminal branches and twigs of tree canopies. Amongst them are:

Biermannia calcarata, Bulbophyllum bryoides, B. catenarium, B. crassiusculifolium, B. sessile, Eria muscicola, E. perpusilla, E. pusilla, Malleola insectifera, Pennilabium angraecum, Porpax elwesii, P. reticulata, Pteroceras semiteretifolium, Taeniophyllum fasciculatum, Thelasis pygmaea, Thrixspermum formosanum, Trichotosia dasyphylla.

More or less large epiphytic lianas or scrambling orchids are also an important component of these lowland forests. Among them the most characteristic orchid species are:

Arachnis annamensis, *Callostylis rigida*, *Epigeneium amplum*, *Renanthera annamensis*, *R. imschootiana*, *Staurochilus fasciatus*, *Thrixspermum calceolus*, *T. centipeda*.

Few lithophytic species are found in lowland evergreen forests and can be found on the rocky outcrops that usually occur along canyons of montane streams and rivers. These include:

Doritis pulcherrima, *Eria spirodela*, *Flickingeria fimbriata*, *Habenaria rhodocheila*, *Liparis elliptica*, *Ludisia discolor*, *Pholidota chinensis*, *P. imbricata*, *P. pallida*, *Podochilus microphyllus*.

Four *Paphiopedilum* species have been reported as inhabitants of these lowland forests at elevations of 700 m and higher. Among them, *Paphiopedilum appletonianum* and *P. delenatii* occur equally often as terrestrials and lithophytes, *Paphiopedilum callosum* grows as terrestrial and *P. villosum* is usually found as a typical humus epiphyte, very rarely occurring as a lithophyte on granite cliffs. It may be that, in the past, all of these species and other orchids also inhabited primary evergreen broad-leaved forests at lower elevations. However, this cannot now be verified, because primary evergreen lowland forests below 500 m are more or less destroyed in most parts of the country and many species have been lost.

Evergreen montane and highland forests on acidic substrates between 1000 and 3000 m

Evergreen, montane and highland forests are widespread in the extensive mountain areas above 900–1000 m in the north and above 1200–1400 m in the south of the country (Fig. 5). The great majority of these mountain systems are composed of acidic silicate rocks. The humid and cool climatic conditions of these mountains produce a relatively sharp change in vegetation structure between the lowland and the montane forest communities. This change is marked by a decline in representatives of tropical families, especially Anacardiaceae, Dipterocarpaceae, Euphorbiaceae, Meliaceae and Simaroubaceae, but an increase in temperate families such as Fagaceae, Theaceae and Magnoliaceae and a diverse assemblage of conifers. Most dipterocarp trees, for example, drop out at 600–900 m, while the upper limit of this family occurs commonly at 1100 m. Similarly, the lower elevational limits of many characteristic montane tree species are at 900–1200 m (Rundel, 1999).

Montane forests are generally shorter in stature than lowland ones, and are characterized by a dense evergreen upper canopy. Lianas are much less abundant and diverse than in lowland evergreen forests, while epiphytes are particularly abundant and diverse (Rundel, 1999). Montane forest habitats in Vietnam have annual rainfall totals of 2000 to more than 3000 mm, and a relatively short or absent dry season.

Primary montane habitats include a range of intergrading communities dominated by evergreen broad-leaved trees, conifers and mixed forest. Broad-leaved forests regularly are situated on mountain slopes and include a great number of tree species, including:

Aceraceae (*Acer*),
Clusiaceae (*Garcinia*),
Elaeocarpaceae (*Elaeocarpus*),
Fabaceae (*Cylindrokelupha*),
Fagaceae (*Castanopsis, Lithocarpus, Quercus*),
Hamamelidaceae (*Rhodoleia, Symingtonia*),
Illiciaceae (*Illicium*),
Juglandaceae (*Engelhardtia*),
Lauraceae (*Actinodaphne, Cinnamomum, Cryptocarya, Lindera, Litsea, Machilus, Neolitsea, Notaphoebe, Phoebe*),
Magnoliaceae (*Michelia, Talauma*),
Myrtaceae (*Syzygium*),
Rosaceae (*Pyrus, Sorbus, Pygeum*),
Sapotaceae (*Madhuca*),
Theaceae (*Eurya, Pyrenaria, Schima, Gordonia*),
Ulmaceae (*Gironniera*)

Woody and semi-woody species of Acanthaceae, Melastomataceae, Rubiaceae and Urticaceae predominate in the understory. Also common here are some woody ferns (*Cyathea, Ctenitopsis, Cybotium*) which may reach 10 and more metres in height in humid river canyons. White, pink or bright red root parasites (*Balanophora, Sapria, Rhopalocnemis*) are not uncommon in these habitats. Along rocky mountain streams in deep canyons may be seen rich communities dominated by Aceraceae (*Acer*), Betulaceae (*Carpinus*), Rosaceae (*Pyrus, Pygeum*), Elaeocarpaceae (*Sloanea*), Myrsinaceae (*Ardisia*), Oleaceae (*Fraxinus*), Araliaceae (*Schefflera, Trevesia, Tupidanthus, Macropanax*). A number of trees and shrubs, including species of Lauraceae (*Cinnamomum, Eugenia*), Fagaceae (*Castanopsis*), Salicaceae (*Salix*), Sterculiaceae (*Reevesia*), Nyssaceae (*Nyssa*), Elaeocarpaceae (*Elaeocarpus*), Pandanaceae (*Pandanus, Freycinetia*). Moraceae (*Ficus*), Rubiaceae (*Lasianthus*), Chloranthaceae (*Chloranthus*), form the typical riparian community on better-drained rocky soils. Palms, such as *Caryota sympetala* and *Pinanga banaensis*, and large understory herbs, including ferns (*Cyathea, Marattia, Angipteris, Cibotium, Rumohra, Polystichum, Cyclosorus*), *Begonia* (Begoniaceae), *Impatiens* (Balsamiaceae), *Musa* (Musaceae), and *Curculigo* (Hypoxidaceae) are also typical of these habitats. Soils kept constantly wet by mists along steep water courses often support diverse assemblages of terrestrial orchids and other herbs. Typical terrestrial orchids here are:

Acanthephippium simplex,
A. striatum,
Anoectochilus acalcaratus,
A. chapaensis,
A. elwesii,
A. lanceolatus,
A. lylei,
A. roxburghii,
A. siamensis,
A. sikkimensis,
A. tridentatus,
Apostasia nuda,
A. odorata,
A. wallichii,
Calanthe clavata,
C. densiflora,
C. herbacea,
C. lyroglossa,
C. sylvatica,
C. velutina,
Cephalantheropsis longipes,
C. obcordata,
Collabium assamicum,
C. chinense,
C. chloranthum,
Collabiopsis formosanum,
Cymbidium ensifolium,
C. lancifolium,
C. sinense,
C. wilsonii,
Diglyphosa evrardii,
Eria gagnepainii,
Goodyera foliosa,
G. fumata,
Habenaria pantlingiana,
Mischobulbon cordifolium,
M. macranthum,
Nephelaphyllum tenuiflorum,
Phaius flavus,
P. tankervilleae,
Pristiglottis umbrosa,
Tainia hookeriana,
Zeuxine nervosa,
Z. parvifolia.

Some rare leafless mycotrophic orchids, *Aphyllorchis annamensis, A. evrardii, A. montana, Chamaegastrodia poilanei, Lecanorchis javanica* and *Yoania prainii*, occur in these conditions. The large yellowish vine, *Galeola nudifolia*, whose stems reach 6–8 m long, belongs also to this group. In open sunny places may be found some terrestrial and climbing orchids, such as *Cryptostylis arachnites, Eulophia spectabilis, Peristylus chapaensis, P. constrictus, P. prainii, P. candidus, P. tentaculatus, P. calcaratus, P. densus, Vanilla albida, V. annamica, V. aphylla*, and *V. pierrei.* A number of orchids grow on sheer rocky cliffs along mountain streams and rivers in these forests, including:

Appendicula cornuta,
A. hexandra,
A. reflexa,
Bulbophyllum tixieri,
Dendrobium alleizettii,
D. fimbriatum,
D. sociale,
Flickingeria fimbriata,
Liparis luteola,
Monomeria barbata,
Pholidota leveilleana,
P. rubra,
Podochilus microphyllus.

However, epiphytic orchids represent the largest and most diverse group in primary mountain forests, the most frequent at between 1000–2500 m being:

Agrostophyllum callosum,
Appendicula gracilis,
Bulbophyllum crassiusculifolium,
B. delitescens,
B. elassonotum,
B. frostii,
B. hymenanthum,
B. longiflorum,
B. luanii,
B. monanthum,
B. odoratissimum,
B. orientale,
B. pecten-veneris,
B. picturatum,
B. putidum,
B. retusiusculum,
B. sigaldiae,
Callostylis rigida,
Ceratostylis himalaica,
C. radiata,
C. siamensis,
C. subulata,
Cleisostoma birmanicum,
C. discolor,
C. duplicilobum,
C. striatum,
C. williamsonii,
Coelogyne assamica,
C. brachyptera,
C. cycnoches,
C. eberhardtii,
C. fimbriata,
C. lawrenceana,
C. sanderae,
C. stricta,
C. viscosa,
Cymbidium schroederi,
Dendrobium aduncum,
D. amabile,
D. aphyllum,
D. bellatulum,
D. cariniferum,
D. crystallinum,
D. dalatense,
D. ellipsophyllum,
D. farmeri,
D. faulhaberianum,
D. gratiosissimum,
D. hercoglossum,
D. heterocarpum,
D. khanhoaense,
D. langbianense,
D. linguella,
D. lomatochilum,
D. longicornu,
D. nobile,
D. oligophyllum,
D. pachyglossum,
D. parishii,
D. uniflorum,
Epigeneium amplum,
E. cacuminis,
Eria acervata,
E. bidupensis,
E. biflora,
E. carunculosa,
E. clausa,
E. floribunda,
E. foetida,
E. globulifera,
E. lactiflora,
E. lanigera,
E. lasiopetala,
E. paniculata,
E. pannea,
E. tomentosa,
E. truncata,
Flickingeria angustifolia,
F. ritaeana,
Gastrochilus calceolaris,
G. hainanensis,
G. intermedius,
G. pseudodistichus,
G. simplicilabius,
Holcoglossum amesianum,
H. lingulatum,
H. subulifolium,
Hygrochilus parishii,
Liparis elliptica,
L. latilabris,
L. luteola,
L. stricklandiana,
Malleola seidenfadenii,
Oberonia acaulis,
O. anthropophora,
O. caulescens,
O. dalatensis,
O. dolichocaulis,
O. ensiformis,
O. evrardii,
O. langbianensis,
O. quadridentata,
Panisea tricallosa,
Pennilabium angraecum,
Pholidota articulata,
P. chinensis,
P. convallariae,
P. leveilleana,
P. pachyglossa,
P. recurva,
Podochilus microphyllus,
Polystachya concreta,
Pomatocalpa spicata,
Pteroceras leopardinum,
Robiquetia spathulata,
R. succisa,
Schoenorchis eberhardtii,
Sunipia andersonii,
S. pallida,
S. scariosa,
Taeniophyllum glandulosum,
T. minimum,
T. obtusum,
T. pahangense,
Trichotosia dalatensis,
T. dasyphylla,
T. microphylla,
T. pulvinata,
T. velutina,
Vanda bidupensis,
V. concolor,
V. denisoniana,
V. pumila.

A number of conifers are usually a more or less important element of forests on mountain slopes toward the ridges and summits (Fig. 12). Among them are such species as *Cephalotaxus mannii*, *Dacrycarpus imbricatus*, *Dacrydium elatum*, *Nageia wallichiana*,

Fig. 12. Highland montane primary coniferous forest (with *Pinus dalatensis* and *P. krempfii*) on granite at 1700–1800 m.

S. Vietnam, Lam Dong Prov., Chu Dang Sin ridge (Photo.: L. Averyanov).

Podocarpus neriifolius and *Taxus chinensis*. Some conifers such as *Fokienia hodginsii*, *Keteleeria evelyniana*, *Pinus krempfii* and *P. dalatensis*, may reach 40–45 m in height and 2 m in diameter. These are generally emergents above a forest dominated by Fagaceae, Lauraceae, and Magnoliaceae. Important dominant species of these woods are presented in figure 13.

On mountain summits and along ridge-tops, broad-leaved evergreen closed wet forest gives way to mixed and coniferous forest. Drier montane habitats up to 2000 m often support open forests of *Keteleeria evelyniana*, *Pinus kesiya* and *P. merkusii*, three montane conifers widespread throughout Vietnam (Nguyen Tien Hiep, Phan Ke Loc & Averyanov, 1999).

Humid montane forests from 1200 to 2400 m throughout Vietnam may support a rich forest dominated by *Fokienia hodginsii*, *Dacrycarpus imbricatus*, *Dacrydium elatum*, *Cunninghamia konishii*, *Nageia wallichiana* and *Podocarpus neriifolius* (Nguyen Tien Hiep, Phan Ke Loc & Averyanov, 1999; Rundel, 1999).

A coniferous community of distinctly temperate affinities enters northern Vietnam at higher elevations from 2400 to 2900 m in the north-western part of the country, with occasional dominance of *Tsuga dumosa* and *Abies delavayi* (Nguyen Tien Hiep, Phan Ke Loc & Averyanov, 1999; Rundel, 1999).

Pinus fenzeliana and *P. wangii* are typical conifers in a few scattered montane regions in northern and central Vietnam. The humid montane coniferous forests of southern and central Vietnam are typically composed of two local endemic species of pine, *Pinus krempfii* and *P. dalatensis* (Nguyen Tien Hiep, Phan Ke Loc & Averyanov, 1999; Rundel, 1999).

The subcanopy tree and shrub layer of open pine forests usually includes representatives of the following families and genera (Schmid, 1974):

Ericaceae (*Agapetes, Craibiodendron*),
Euphorbiaceae (*Breynia, Phyllanthus*),
Fabaceae (*Albizia, Archidendron*),
Fagaceae (*Lithocarpus, Quercus*),
Myricaceae (*Myrica*),
Myrtaceae (*Tristania*),
Proteaceae (*Helicia*),
Rhamnaceae (*Rhamnus*),
Rubiaceae (*Gardenia*),
Theaceae (*Adinandra, Anneslea, Eurya, Ternstroemia*).

Epiphytic orchids and ferns are abundant and diverse in montane pine forests. However, they reach their maximum abundance in coniferous forests on mountain peaks and along the highest ridges, which are commonly bathed in clouds, heavy dew and fog all through the year. Annual rainfall here may reach 4000 mm and more without any pronounced dry season. However, cold winters with night temperatures falling below freezing result in highland vegetation and the appearance of a number of semi-deciduous and deciduous trees and shrubs in such habitats. Soils here are shallow, but rich in organic matter and highly acid.

Very old gnarled conifers, especially *Fokienia hodginsii*, *Pinus krempfii*, *P. wangii*, *P. kesiya*, *P. dalatensis*, *Podocarpus neriifolius* and *Nageia wallichiana*, and ericaceous trees and dwarf shrubs, such as *Agapetes*, *Lyonia*, *Rhododendron* and *Vaccinium*, are the main dominants in such highland communities (Fig. 14). Trees here rarely reach more than 10 m in height with a mean canopy cover about 75% (Rundel, 1999). Often low woody ericaceous thickets 3–5 m in height cover isolated mountain peaks and narrow ridges. Many interesting plants occur only in these highland conditions (Fig. 15). In the extreme northwestern highland regions *Tsuga dumosa* and *Abies delavayi* may appear as occasional emergents in cloud forests. Lush epiphytic mosses, lichens, orchids and ferns cover all tree stems, branches, ground and rock outcrops in such habitats with thick moist unbroken continuous cover. The commonest orchid species here are:

Fig. 13. Typical plants of highland montane primary forests on silicate rocks. (Photos: L. Averyanov).

a — *Fokienia hodginsii* (Dunn) A.Henry & Thomas (Cupressaceae);
b — *Pinus* (*Ducampopinus*) *krempfii* Lecomte (Pinaceae);
c — *Pinus dalatensis* Ferre (Pinaceae);
d — *Podocarpus neriifolius* D.Don (Podocarpaceae);
e — *Schima wallichii* (DC.) Korth. (Theaceae);
f — *Amentotaxus poilanei* (Ferre & Rouane) Ferg. (Taxaceae);
g — *Symingtonia populnea* (R.Br.) van Steenis (Hamamelidaceae);
h — *Magnolia* sp. (Magnoliaceae);
i — *Kadsura coccinea* (Lem.) A.C.Smith (Schizandraceae);
j — *Rhodoleia championii* Hook.f. (Hamamelidaceae).

Bulbophyllum pinicolum, B. pumilio, B. semiteretifolium, B. thaiorum, Ceratostylis himalaica, Coelogyne calcicola, C. filipeda, C. mooreana, C. ridleyi, C. rigida, C. schultesii, Cymbidium devonianum, C. eburneum, C. lowianum, Dendrobium christyanum, D. virgineum, D. wattii, Epigeneium cacuminis, E. chapaense, Eria amica, E. bipunctata, E. globifera, E. paniculata, E. pannea, E. siamensis, E. thao, Liparis bootanensis, L. cespitosa, L. chapaensis, L. compressa, Monomeria dichroma, Otochilus albus, O. fuscus, Panisea albiflora, Pholidota articulata, P. convallariae, P. leveilleana, P. recurva, Schoenorchis gemmata, Sunipia annamensis, Trichotosia dasyphylla, T. microphylla.

Cymbidium insigne, *Goodyera schlechtendaliana*, *Bulbophyllum ngoclinhensis* and *B. reptans* are common on the highest peaks of southern Vietnam.

Numerous temperate floristic elements are found here because of the connection between the highlands of Vietnam and the highland areas of southern China (Nguyen Nghia Thin, 1998; Nguyen Nghia Thin & Nguyen Thi Thoi, 1998), including above 2000 m a number of terrestrial and lithophytic orchids:

Fig. 14. Highland montane cloud thickets with emergent trees of *Pinus dalatensis* along a mountain ridge at 2600 m.

S. Vietnam, Kontum Prov., Ngoc Linh mountain system. (Photo.: L. Averyanov).

Fig. 15. Typical plants of highland montane cloud forests on silicate rocks. (Photos: L. Averyanov).

a — *Rhododendron irroratum* Fr. subsp. *kontumense* (Sleum.) Chamb. (Ericaceae);
b — *Rhododendron moulmainense* Hook. (Ericaceae);
c — *Lyonia* sp. (Ericaceae);
d — *Vaccinium* sp. (Ericaceae);
e — *Aeschynanthus moningeriae* (Merr.) Chun (Gesneriaceae);
f — *Phyllagathis megalocentra* C.Hans. (Melastomataceae);
g — *Rhopalocnemis phalloides* Junghuhn (Balanophoraceae);
h — *Balanophora laxiflora* Hemsl. (Balanophoraceae);
i — *Monotropastrum pumilum* H.Andres (Ericaceae);
j — *Sciaphila clemensiae* Hemsl. (Triuridaceae).

Amitostigma bidupense, *A. keiskeoides*, *Gastrodia taiensis*, *Goodyera foliosa*, *G. schlechtendaliana*, *Hancockia uniflora*, *Platanthera angustata*, *Pleione grandiflora*, *P. maculata*, *P. praecox*, *P. vietnamensis*.

Five *Paphiopedilum* species, *P. appletonianum*, *P. callosum*, *P. delenatii*, *P. gratrixianum* and *P. villosum*, inhabit these highland forests in Vietnam. The distribution of all *Paphiopedilum* species in montane areas of Vietnam is closely connected with the more or less closed and wet coniferous and mixed forests. *Paphiopedilum appletonianum* and *P. callosum* typically grow here as terrestrials or lithophytes between (700) 1000 and 2000 m. In the same habitats, *Paphiopedilum villosum* is sometimes quite frequent, usually growing as a humus epiphyte on tall broad-leaved trees. *Paphiopedilum delenatii* and *P. gratrixianum* occur typically as lithophytes on granite, gneiss and rhyolites. However, both species inhabit very restricted areas at elevations around 1000 m.

Semi-deciduous and deciduous dry lowland forests from sea level up to 500 (700) m

In Vietnam, semi-deciduous and deciduous, dry, lowland forests, and open, savanna-like woodlands represent the eastern extension of a vegetation type that extends across mainland Southeast Asia in habitats with less than 2000 mm annual rainfall and with 5–7 months of drought. These plant communities are distributed in semi-arid southern Vietnam from sea level to 500 (700) m and mainly form on shallow soils over basalt and granite (Fig. 5). Most of these open woodlands have been highly impacted by human activities and represent one of the most endangered vegetation types. In intact habitats emergent trees, such as *Tetrameles nudiflora* or *Lagerstroemia angustifolia*, grow along edges of these forests and may reach 40–50 m in height. However, in most cases these woods have dominants no taller than 15–20 m (Fig. 16).

Semi-deciduous trees Dipterocarpaceae are typical of the upper canopy of the dry lowland forests of Vietnam. Other dominants of this forest layer belong to many families, typical among them are (Schmid, 1974):

Anacardiaceae (*Buchanania*, *Dracontomelum*),
Annonaceae (*Miliusa*),
Bignoniaceae (*Stereospermum*),
Bombacaceae (*Bombax*),
Burseraceae (*Canarium*),
Combretaceae (*Anogeissus*, *Terminalia*),
Datiscaceae (*Tetrameles*),
Dilleniaceae (*Dillenia*),
Dipterocarpaceae (*Dipterocarpus*, *Hopea*, *Shorea*),
Ebenaceae (*Diospyros*),
Fabaceae (*Afzelia*, *Albizia*, *Cassia*, *Dalbergia*, *Pterocarpus*),
Fagaceae (*Lithocarpus*, *Quercus*),
Lythraceae (*Lagerstroemia*),
Meliaceae (*Aglaia*, *Amoora*, *Sandoricum*, *Toona*),
Moraceae (*Artocarpus*, *Ficus*),
Myrtaceae (*Eugenia*),
Rubiaceae (*Adina*),
Simaroubaceae (*Irvingia*),
Sapindaceae (*Erioglossum*, *Pometia*, *Sapindus*),
Sapotaceae (*Palaquium obovatum*),
Sterculiaceae (*Sterculia*, *Pterospermum*),
Ulmaceae (*Celtis*, *Holoptelea*),
Verbenaceae (*Vitex*).

Many smaller trees and shrubs are observed in the understory of these open forests and a large diversity and abundance of grasses, bamboos and cycads. Some typical plants of dry lowland forests may be seen in figure 17. Notable amongst the grasses are *Arundinella*, *Pseudopogonantherum*, *Imperata*, *Heteropogon*, *Themeda* and *Alloteropsis*, some reaching as much as 1–1.5 m in height. Lianas and epiphytes are common, but nowhere abundant. Small creeping succulent vines, like *Dischidia* and *Hoya*, and large epiphytic ferns, such as *Drynaria*, *Pseudodrynaria* and *Platycerium*, are typical of these forests.

Fig. 16. Deciduous dry lowland forest and savanna-like woodland at 200–300 m.
S. Vietnam, Dac Lac Prov. (Photo.: L. Averyanov).

Despite the dry conditions, the diversity of epiphytic orchids in lowland *Dipterocarpus* forests in the south of Vietnam is very high with notable local endemism, many of them having obvious xeromorphic features. These include:

Acampe papillosa,
Acriopsis indica,
A. liliifolia,
Adenoncos vesiculosa,
Aerides crassifolia,
A. falcata,
A. houlletiana,
A. multiflora,
Arachnis annamensis,
A. cannaeformis,
A. hookeriana,
A. labrosa,
A. maingayi,
Ascocentrum christensonianum,
A. curvifolium,
A. garayi,
Ascolabium pusillum,
Brachypeza laotica,
Bulbophyllum averyanovii,
B. bariense,
B. blepharistes,
B. dayanum,
B. elassonotum,
B. hirtum,
B. lemniscatoides,
B. orientale,
B. refractum,
B. tripudians,
Chiloschista exuperei,
C. parishii,
Christensonia vietnamica,
Cleisomeria lanatum,
Cleisostoma equestre,
C. inflatum,
Coelogyne lentiginosa,
Cymbidium aloifolium,
C. atropurpureum,
C. bicolor,
Dendrobium crumenatum,
D. crystallinum,
D. exile,
D. farmeri,
D. gratiosissimum,
D. indivisum,
D. lindleyi,
D. lomatochilum,
D. podagraria,
D. porphyrophyllum,
D. pseudotenellum,
D. secundum,
D. truncatum,
D. unicum,
Eparmatostigma dives,
Eria acervata,
E. cochinchinensis,
E. obscura,
E. pannea,
Luisia brachystachys,
L. filiformis,
L. psyche,
Malaxis ophridis,
Nervilia plicata,
Oberonia acaulis,
O. caulescens,
O. emarginata,
O. gammiei,
O. longibracteata,
O. lycopodioides,
O. rasmussenii,
O. rufilabris,
O. tixieri,
O. trochopetala,
Papilionanthe hookeriana,
P. pedunculata,
Pelatantheria ctenoglossum,
P. eakroensis,
Peristylus densus,
Pholidota leveilleana,
P. rubra,
Polystachya concreta,
Rhynchostylis coelestis,
R. gigantea,
R. retusa,
Staurochilus fasciatus,
Thecopus maingayi,
Thecostele alata,
Thrixspermum centipeda,
Trichoglottis retusa,
T. seidenfadenii,
Trichotosia dasyphylla,
Vanda pumila.

The diversity of herbaceous perennials is also remarkably large. Most develop underground tubers or thick rhizomes and their leaves usually die down during the dry season and appear again with the onset of the rains. Such species usually bloom before or during the first days of the rainy season. Many terrestrial orchids inhabit these woods, including:

Anthogonium gracile,
Brachycorythis galeandra,
B. laotica,
Eulophia andamanensis,
E. flava,
E. graminea,
E. macrobulbon,
E. pauciflora,
E. pulchra,
E. spectabilis,
Geodorum attenuatum,
G. densiflorum,
G. pulchellum,
G. recurvum,
G. siamense,
Habenaria apetala,
H. commelinifolia,
H. dentata,
H. erostrata,
H. lindleyana,
H. lucida,
H. mandersii,
H. medioflexa,
H. myriotricha,
H. poilanei,
H. rostellifera,
H. rostrata,
H. rumphii,
H. viridiflora,
Pachystoma pubescens,
Pecteilis cochinchinensis,
P. henryi,
P. susannae,
Spathoglottis affinis,
S. aurea,
S. eburnea,
S. plicata,
S. pubescens.

Though the orchid flora of lowland dry forests in southern Vietnam is rich and diverse no *Paphiopedilum* species have yet been reported from this interesting area.

Fig. 17. Typical plants of dry deciduous lowland forests and savanna-like woodlands. (Photos: L. Averyanov).

a — *Dipterocarpus alatus* Roxb. ex G.Don (Dipterocarpaceae);
b — *Erythrina fusca* Lour. (Fabaceae);
c — *Dillenia hookeri* Pierre (Dilleniaceae);
d — *Cycas pectinata* Buch.-Ham. (Cycadaceae);
e — *Anthocephalus chinensis* (Lam.) Walp. (Rubiaceae);
f — *Tetracera scandens* (L.) Merr. (Dilleniaceae);
g — *Kaempferia rotunda* L. (Zingiberaceae);
h — *Strophanthus caudatus* (Apocynaceae);
i — *Nepenthes thorelii* Lecomte (Nepenthaceae);
j — *Uvaria purpurea* Blume (Annonaceae).

Coastal vegetation, lowland wetlands and mangrove

The main plant communities along the coastline of Vietnam are coastal scrub, coastal lowland wetlands and mangrove thickets (Fig. 5). Nearly all have been heavily impacted by human activities and greatly altered from their natural condition. Many of these areas are the most arid in Vietnam with a typical mean annual rainfall of less than 1700 mm.

Shrubs form more or less dense thickets along the beaches, on dunes and on sea slopes (Fig. 18). Common species here are *Scaevola koenigii* (Goodeniaceae), *Cerbera manghas* (Apocynaceae), *Pandanus tectorius* (Pandanaceae), *Cycas rumphii*, *C. inermis* (Cycadaceae), *Caesalpinia godefroyana* (Fabaceae), *Clerodendron inerme* (Verbenaceae), *Litchi longana* (Sapotaceae), *Pleiospermum littorale* (Rutaceae), *Sindora siamensis* (Fabaceae), *Vitex trifolia* (Verbenaceae) and others (Schmid, 1974; Averyanov, Nguyen Tien Ban & Kudriavtzeva, 1988). Some of these are shown in figure 19.

Moist depressions and coastal wetlands amongst young low dunes are occupied by *Baeckia frutescens* (Myrtaceae), *Melaleuca leucadendron* (Myrtaceae), *Acronychia laurifolia* (Rutaceae), *Canthium didymum* (Rubiaceae), *Alstonia spathulata* (Apocynaceae) and numerous grasses and sedges.

Extensive stands of mangroves are common in the deltas of large rivers and on swampy salty coastal lowlands in Vietnam (Fig. 20). Mangrove diversity is relatively high, the commonest species being *Avicennia alba*, *A. officinalis* (Verbenaceae), *Lumnitzera racemosa*, *L. littorea* (Combretaceae), *Bruguiera gymnorrhiza*, *B. sexangula*, *Ceriops tagal*, *Kandelia candel*, *Rhizophora apiculata*, *R. mucronata*, *R. stylosa* (Rhizophoraceae), and *Sonneratia caseolaris* (Sonneratiaceae), and the typical mangrove palms *Nypa fruticans* and *Phoenix paludosa.* Common plants associated with this type of vegetation are presented in figure 21.

Few orchids have been found in these areas and none is typical of lowland coastal habitats.

Secondary and agricultural plant communities

Various secondary forests, thickets, scrublands and grasslands have developed under the long history of human impact on native vegetation. Such plant communities along with farms, wood and industrial plantations now cover vast areas of the country (Fig. 5). A great number of weed and exotic species are found in these areas, but usually all the native ones have been lost.

A few orchids, such as *Arundina graminifolia*, *Phaius tankervilleae*, *Spiranthes sinensis*, *Zeuxine strateumatica*, and some species of *Geodorum*, *Habenaria* and *Peristylus*, are capable of surviving in these areas. These represent a very small group in comparison with the orchid flora that once inhabited these areas.

Fig. 18. Coastal slopes with typical scrub vegetation.
S. Vietnam, sea coast of Con Dao island. (Photo.: L. Averyanov).

Fig. 19. Typical plants of coastal scrub vegetation.
(Photos: L. Averyanov).

a — *Pandanus tectorius* Parkinson (Pandanaceae);
b — *Sindora siamensis* Teysmann ex Miquel (Fabaceae);
c — *Phoenix humilis* Royle (Arecaceae);
d — *Barringtonia asiatica* (L.) Kurz. (Lecythidaceae);
e — *Ipomoea stolonifera* (Cyr.) Gmel. (Convulvulaceae).

Fig. 20. Mangrove thickets dominated by *Nypa fruticans*.
S. Vietnam, Con Dao island. (Photo.: L. Averyanov).

Fig. 21. Typical plants of the coast and mangrove thickets. (Photos: L. Averyanov).

a — *Rhizophora mucronata* Lam. (Rhizophoraceae);
b — *Avicennia marina* (Forsk.) Vierh. (Verbenaceae);
c — *Scaevola taccada* (Gaertn.) Roxb. (Goodeniaceae);
d — *Flagellaria indica* L. Flagellariaceae);
e — *Crinum asiaticum* L. (Amaryllidaceae).

Fig. 22. Typical and sub-endemic orchids of the Vietnamese part of the Sikang-Yunnan floristic province. (Photos: L. Averyanov).

a — *Calanthe alleizettei* Gagnep;
b — *Galeola nudifolia* Lour.;
c — *Liparis balansae* Gagnep.;
d — *Epigeneium chapaense* Gagnep.;
e — *Goodyera schlechtendaliana* Rchb.f.;
f — *Eria gagnepainii* Hawkes & Heller;
g — *Sarcoglyphis smithianus* (Kerr) Seidenf.;
h — *Cryptochilus luteus* Lindl.;
i — *Mischobulbum cordifolium* (Hook.f.) Schltr.;
j — *Dendrobium longicornu* Lindl.

Fig. 23. Endemic and sub-endemic orchids of the South Chinese floristic province.
(Photos **a–f**, **h–j**: L. Averyanov; **g**: P.J. Cribb).

a — *Renanthera citrina* Aver.;
b — *Liparis emarginata* Aver.;
c — *Dendrobium guangxiense* Tsi & Chen;
d — *Holcoglossum wangii* E.A.Christenson;
e — *Vanda fuscoviridis* Lindl.;
f — *Pteroceras simondianum* (Gagnep.) Aver.;
g — *Coelogyne malipoensis* Tsi;
h — *Hemipilia calophylla* Par. & Rchb.f.;
i — *Bulbophyllum purpureifolium* Aver.;
j — *Cymbidium floribundum* Lindl.

Fig. 24. Typical, endemic and sub-endemic orchids of the North Indochinese floristic province. (Photos: L. Averyanov).

a — *Pholidota guibertiae* Finet.;
b — *Biermannia calcarata* Aver.;
c — *Micropera poilanei* (Guillaum.) Garay;
d — *Didymoplexiopsis khiriwongensis* Seidenf.;
e — *Dendrobium nobile* var. *alboluteum* Huyen & Aver.;
f — *Eria sutepensis* Rolfe ex Downie;
g — *Bulbophyllum andersonii* (Hook.f.) J.J.Smith;
h — *Eria coronaria* (Lindl.) Rchb.f.;
i — *Malleola seidenfadenii* E.A.Christenson;
j — *Bulbophyllum delitescens* Hance.

Fig. 25. Endemic and sub-endemic orchids of the Central Annamese floristic province.
(Photos: **a–h**, **j**: L. Averyanov; **i**: E.Mokhov).

a — *Cleisocentron klossii* (Ridl.) Garay;
b — *Pristiglottis umbrosa* Aver.;
c — *Coelogyne sanderae* Kraenzl.;
d — *Aphyllorchis annamensis* Aver.;
e — *Schoenorchis eberhardtii* (Finet) Aver.;
f — *Eria thao* Gagnep.;
g — *Bulbophyllum ngoclinhensis* Aver.;
h — *Monomeria dichroma* (Rolfe) Schltr.;
i — *Pleione vietnamensis* Aver. & Cribb;
j — *Bulbophyllum averyanovii* Seidenf.

Fig. 26. Endemic and sub-endemic orchids of the South Annamese floristic province.
(Photos: L. Averyanov).

a — *Arachnis annamensis* (Rolfe) J.J.Smith;
b — *Bulbophyllum frostii* Summerhayes;
c — *Calanthe velutina* Ridl.;
d — *Trichotosia dalatensis* (Gagnep.) Seidenf.;
e — *Vanda bidupensis* Aver. & E.A.Christenson;
f — *Bulbophyllum sigaldiae* Guillaum.;
g — *Sunipia pallida* Aver.;
h — *Liparis nigra* Seidenf.;
i — *Panisea albiflora* (Ridl.) Seidenf.;
j — *Papilionanthe pedunculata* (Kerr) Garay.

Fig. 27. Typical, endemic and sub-endemic orchids of the South Indochinese floristic province. (Photos: L. Averyanov).

a — *Ascolabium pusillum* (Aver.) Aver.;
b — *Trichoglottis seidenfadenii* Aver.;
c — *Pteroceras semiteretifolium* Pedersen;
d — *Brachycorythis laotica* (Gagnep.) Summerh.;
e — *Dendrobium ochraceum* De Wild.;
f — *Ascocentrum garayi* E.A.Christenson;
g — *Bulbophyllum blepharistes* Rchb.f.;
h — *Dendrobium unicum* Seidenf.;
i — *Malaxis octodentata* Seidenf.;
j — *Dendrobium secundum* (Blume) Lindl.

PART II. SLIPPER ORCHIDS

WHAT IS A SLIPPER ORCHID?

Slipper orchids are a distinctive group. They can be easily recognized by their unusual flowers that have a slipper-shaped, deeply saccate median petal (technically called a lip or labellum) lowermost in the flower, which gives it a curious appearance. Hence, the common name of these plants. Similarly, the Latin name of largest slipper orchid genus, *Paphiopedilum*, can be translated literally as "the slipper of Paphos", Paphos being the birthplace of Aphrodite, the Greek goddess of love and beauty (known as Venus to the Romans), and *pedilon* a Greek word for a slipper or shoe. The curious lip and unusual form of the flower distinguish slipper orchids immediately from all other orchids.

Botanically, slipper orchids belong to five genera: *Cypripedium* L., *Mexipedium* V.A. Albert & M.W. Chase, *Paphiopedilum* Pfitzer, *Phragmipedium* Rolfe and *Selenipedium* Rchb.f. (Fig. 28). About 50 species of *Cypripedium*, which are often called Lady's slippers, inhabit temperate and montane areas of the Northern Hemisphere. Tropical America is the home of the 25 or so species of *Mexipedium, Phragmipedium* and *Selenipedium*. Approximately 75 species of *Paphiopedilum* occur in tropical Asia from southern India and the eastern Himalayas in the west to the Philippines, New Guinea and the Solomon Islands in the east. The tropical Asiatic species of *Paphiopedilum* have been widely hybridized in cultivation for about 140 years. Species of *Phragmipedium* have also been hybridized but to a lesser extent.

Great numbers of slipper orchid species and artificial hybrids bred from them are currently cultivated and they are amongst the commonest ornamental plants in many countries. Despite this the majority of species are rare in the wild. Most occur in inaccessible habitats, such as vertical bluffs of rocky mountains, cliffs near waterfalls, or in the canopies of high trees. With rare exceptions, slipper orchids are an integral component of primary forests and quickly and irreversibly become extinct when native plant communities are damaged by man. As a result, very few people have seen these wonderful plants in their natural habitats, which still remain in the last surviving tracts of virgin forest.

Mountain regions of Vietnam, where the last small pockets of some of the most diverse primary forests still survive, are very rich in slipper orchids. In spite of their rarity, the diversity of slipper orchids in Vietnam is higher than in any other region. Many of the Vietnamese species are not only rare but are narrowly endemic, national treasures of worldwide significance.

The increasingly awareness of the need to protect and save Vietnamese slipper orchids in the wild is the main goal of this review of these endangered and spectacular plants which are rapidly disappearing from their native haunts.

ORIGIN AND AFFINITIES OF SLIPPER ORCHIDS

Slipper orchids are a small but remarkable offshoot of the main line of orchid evolution (Fig. 29). The orchid family (Orchidaceae) demonstrates one of the most specialized lines of flowering plant evolution. It is likely that all orchids derived from *Hypoxis*-like ancestors

a
b
c
d

Fig. 28 (opposite and above). Slipper orchid diversity. (Photos **a, c, d, g**: P.J. Cribb; photos **b, e, f, h**: L. Averyanov).

a–d — Representatives of the genus *Cypripedium*:

a — *C. calceolus* L. from Europe,

b — *C.* × *ventricosum* Sw. from Russian Far East,

c — *C. formosanum* Hayata from Taiwan,

d — *C. macranthos* Sw. from SE Siberia.

e — *Paphiopedilum callosum* (Rchb.f.) Stein from Thailand.

f — *Paphiopedilum spicerianum* (Rchb.f.) Pfitz. from Burma.

Species of the genus *Phragmipedium*:

g — *P. longifolium* (Warsc. ex Rchb.f.) Rolfe

h — *P. besseae* C.H. Dodson & J. Kühn from Tropical America.

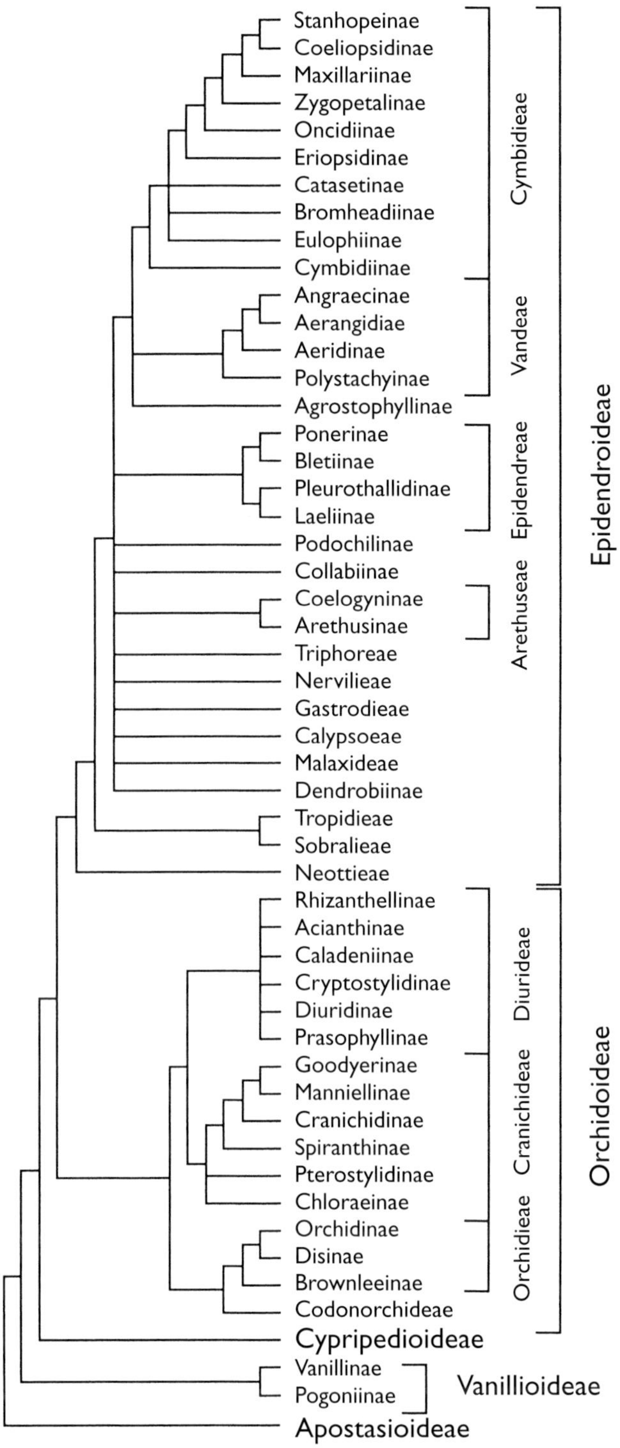

Fig. 29. A consensus tree based upon total data available, showing the five major clades of the Orchidaceae, corresponding to the subfamilies Apostasioideae, Vanilloideae, Cypripedioideae, Orchidoideae and Epidendroideae. (Courtesy of Mark Chase *et al.*, in press).

that had six tepals (three tepals of an outer whorl called sepals and three tepals of an inner whorl called petals) and six stamens (again three in an outer whorl and three in an inner one). Several trends of morphological specialization can be observed in the evolution and formation of different orchid groups. However, a reduction of the number of stamens and fusion of the remaining fertile stamen(s) with the pistil is the main general floral transformation that led to the evolution of the family. Successive reduction in the number of stamens has led to formation of groups of orchids with three, two and one stamen remaining in flower (Fig. 30).

Representatives of the most primitive group of orchids have three or two stamens in their flowers. These plants, called the Apostasioid orchids, are united into subfamily Apostasioideae. It is small group of terrestrial orchids confined to tropical Asia and northern Australia and comprises two genera, *Apostasia* and *Neuwiedia*, with about 16 species. Their flowers are almost regular and resemble somewhat those of *Hypoxis* itself.

A second group, that also retains two stamens in its flower, is the slipper orchids (subfamily Cypripedioideae) and it also represents a distinct lineage. This subfamily includes five genera (*Cypripedium, Mexipedium, Paphiopedilum, Phragmipedium* and *Selenipedium*) and about 150 species widely distributed in Eurasia, North America and the South American tropics.

The apostasioids and cypripedioids (slipper orchids) are sometimes placed in separate families, the Apostasiaceae Lindl. and Cypripediaceae Lindl. (Rasmussen, 1985; Szlachetko, 1995), from the orchids. However, nowadays most authorities include them in the Orchidaceae which is phylogenetically a natural group. The Orchidaceae, with an estimated 25,000 species in about 800 genera, is the largest family among flowering plants, matched only by the Aster family (Asteraceae, or Compositae).

More than 99 % of all orchid species have only a single stamen in the flower and this is one of the main features of the orchid family (Fig. 29). In the most recent taxonomic treatments, the monandrous orchids (with one stamen) are usually separated into three subfamilies: Vanilloideae, Orchidoideae and Epidendroideae (Kores *et al.*, 1997; Pridgeon *et al.* 1999, 2000) (Fig. 29). They are distributed all over the world except in polar areas and arid deserts.

Progressive adaptation to different modes of insect pollination has been the main factor in the evolution of the orchid flower. In the unspecialized flower of Apostasioid orchids, strong bilateral symmetry is absent. In this case, as in other regular flowers, all the stamens play a nearly equal role in pollination. This is not so in other orchids.

Pollination of bilaterally symmetrical flowers has resulted in the loss of anthers so that only one or two anthers remain functional. This evolutionary line led to the evolution of the slipper orchids and the monandrous orchids of subfamilies Vanilloideae, Orchidoideae and Epidendroideae (Fig. 29, 30).

In the slipper orchids, two fertile anthers are placed on either side of the column (Fig. 30). The central stamen is sterile and is borne at the apex of the column. It is curiously modified into large shield-like organ, the staminode, that prevents direct access by the pollinator from the front to the centre of flower. The two other stamens are hidden behind the staminode, one on each side of the column. The saccate lip of slipper orchids has evolved as a trap for pollinators. The inside walls of the lip also very slippery but a ladder of hairs lies on the interior dorsal wall. This leads under the stalked ventral stigma to one of two exits at the base of the lip on either side of the column.

The median sepal, lateral petals, staminode and lip play the main roles in attracting pollinators to the flower of slipper orchids. These parts of the flower are usually brightly coloured. The lateral sepals that lie behind the lip are usually connate to form a synsepal.

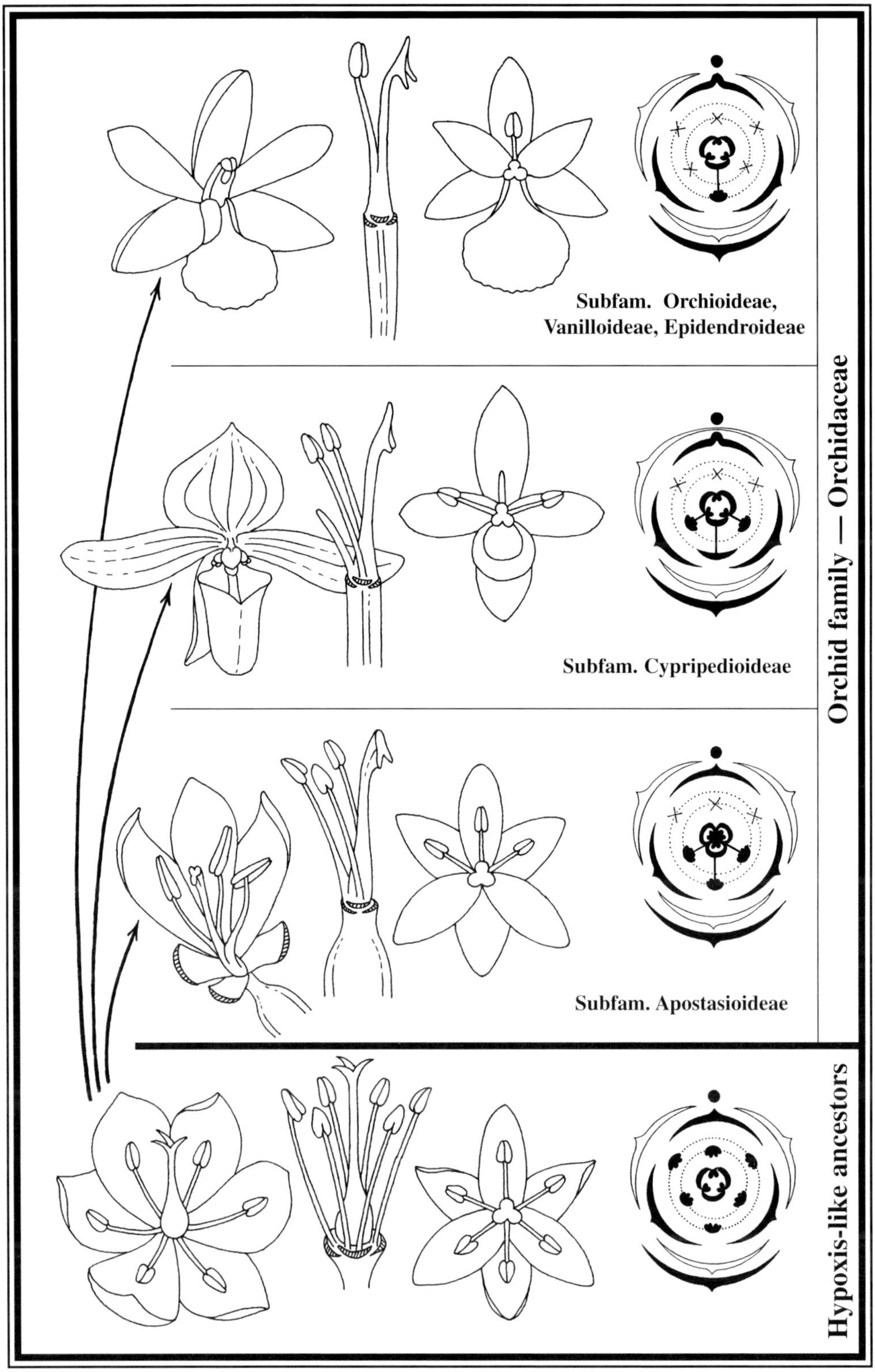
Subfam. Orchioideae, Vanilloideae, Epidendroideae
Subfam. Cypripedioideae
Subfam. Apostasioideae
Orchid family — Orchidaceae
Hypoxis-like ancestors

They partially lose their attractive function and, as a result, they are usually reduced in size. Usually, the synsepal has a bilobed apex and two distinct main nerves reflecting its origin as two lateral sepals.

The median petal (lip) in the monandrous subfamilies usually plays a role as a landing platform for pollinators and can provide also attraction for pollinators, such as nectar, and a convenient path to the anther. The vanilloid orchids are a small group that includes *Vanilla*, a genus of about 70 species of lianas. The orchidoid orchids are mostly terrestrials with tubers or fleshy rhizomes and include most European, North American, terrestrial African and temperate Australian orchids. The epidendroid orchids, the largest group, are predominantly epiphytes or lithophytes and include all the showy tropical genera, such as *Cattleya, Oncidium, Odontoglossum, Phalaenopsis* and *Vanda*.

MORPHOLOGY

Slipper orchids may be very diverse in appearance (Fig. 28). Some temperate representatives of the genus *Cypripedium* are small dwarf herbs, *Selenipedium* species and *Cypripedium irapeanum* and *C. subtropicum* are large bamboo-like plants with stems reaching 1–3 m tall. We are only concerned here, however, with the tropical Asiatic Slipper orchids occurring in Vietnam. All belong to the genus *Paphiopedilum*.

Habit. *Paphiopedilum* species are medium-sized herbs with a very short abbreviated stem bearing a number of distichously arranged leaves borne in a fan (Fig. 31). In all species rhizomes are present but they are usually very short. However, some species, such as *P. malipoense* and *P. micranthum* produce elongated rhizomes that connect numerous growths in an underground network. In these species a single clone might cover an area of several square metres in favourable conditions. In other species, e.g. *P. hirsutissimum* or *P. dianthum*, plants may have one to twenty or more growths which densely cluster.

Leaves. The leaves of *Paphiopedilum* species are usually strap-like, obovate or elliptic-oblong and spreading (Fig. 31). Each is articulated to a channelled sheath which is closely overlapping and imbricate on the stem. In some species, such as *P. dianthum* or *P. hirsutissimum*, the leaves can measure more than 50 cm in length, but in species such as *P. helenae* mature specimens sometime have leaves no longer than 3 cm. The upper surface of the leaves may be uniformly green, or tessellated with dark and light green irregular marks and well defined dark green longitudinal veins. Such mottling is typical for leaves of representatives of subgenera *Parvisepalum* and *Brachypetalum* and in sect. *Barbata* of the typical subgenus. The lower surface of the leaves in most species of subgenera *Parvisepalum*

Fig. 30 (opposite). Simplified scheme of flower transformation during historical derivation of main groups of orchids from *Hypoxis*-like ancestors. (Drawing: L. Averyanov).

Simplified drawings of flowers show main features of flowers, reduction stamens and formation of column (gynostemium). Floral diagrams (schematic vertical projection of all flower parts on one plane) represent flower structure (in diagrams the following designations are used: black round — inflorescence axis; outer black brace — flower bract; outer white braces — tepals of outer whorl, or sepals; inner black braces — tepals of inner whorl, or petals; fertile stamens arranged in two whorls are marked by black roundish bodies, staminode by small black bracket and places of reduced stamens designated by crosses; central roundish figures present section of one- or three-locular ovary; lines which connect stamens, staminode and ovary signify connection with formation of column, or gynostemium).

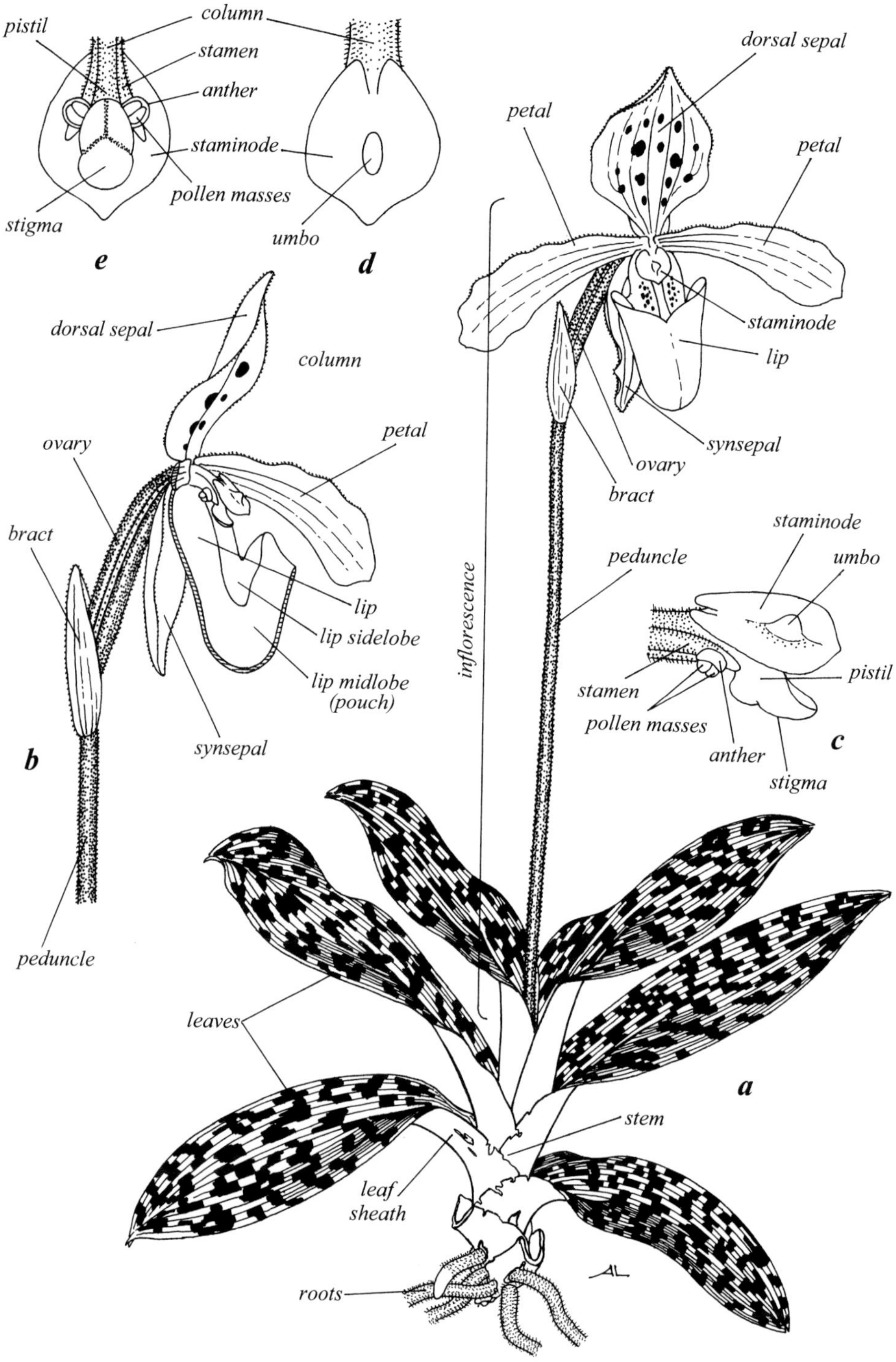

Fig. 31. Plant and floral morphology of *Paphiopedilum*. (Drawing: L. Averyanov).

a — flowering plant; **b** — flower with cut petal and longitudinal section of the lip; **c–e** — column (gynostemium), side view (**c**), view from above (**d**) and from below (**e**).

and *Brachypetalum* is densely purple-spotted. In other species dull purple markings are pronounced only near the base of leaves. Few species have uniformly green lower surfaces of the leaves. Purple pigmentation on the leaves is also lacking in albinistic forms. It is remarkable that the character and intensity of tessellation is a very variable feature in most species. The margin of young leaves in some species, for example, in *P. malipoense* may be ciliate. Leaves of species typical of dry habitats, such as *P. helenae* and *P. dianthum*, are thick, fleshy and rigid.

Inflorescence. The inflorescence of paphiopedilums is usually erect or arching (Fig. 31). *Paphiopedilum helenae, P. henryanum* and *P. dianthum* which grow on vertical rocky cliffs or *P. villosum* that grows epiphytically may have a more or less horizontal or drooping peduncle, but in most species the peduncle is more or less erect as in *P. malipoense, P. micranthum* and *P. appletonianum*. Some species such as *P. dianthum* have an multi-flowered inflorescence usually bearing from two to five flowers. A few species that usually have one-flowered inflorescences, such as *P. delenatii* and *P. concolor*, develop two-flowered inflorescences in favourable conditions. Other species usually produce a single flower, although in populations of *P. appletonianum* and *P. malipoense* two-flowered spikes are very rarely observed. The inflorescence axis of most species is shortly densely pubescent, while in *P. villosum, P. micranthum* and *P. malipoense* it is villose. On the other hand, in *P. dianthum* it is nearly glabrous. Flower bracts are conduplicate and vary in shape from lanceolate or ovate and acute to elliptic and rounded depending on the species. The bracts are usually less pubescent than the other parts of the inflorescence but are generally ciliate and hairy along the mid-vein on the outer side. Glabrous bracts are found in *P. dianthum*.

Flower. The *Paphiopedilum* flower has two tepals in outer whorl: a dorsal sepal and a synsepal, and three petals in the inner whorl (Fig. 31, 32).

The median or dorsal sepal is large, upwardly directed and usually showy, marked with stripes or spots on plane background. The dorsal sepal is sometimes held above the saccate lip, partially sheltering the lip opening from rain-water. In other species, however, it stands erect and acts as a flag. It lies opposite to lower downwardly directed synsepal that is formed by the completely united lateral sepals. The origin of the synsepal is apparent in its two main nerves and the twin longitudinal keels on its outer surface. It lies behind the lip and usually has a much more modest and dull colouration than dorsal sepal. Both the dorsal sepal and synsepal are usually densely pubescent on their outer surfaces and in several species, such as *P. micranthum*, within as well.

The two lateral petals are found each side of the upper sepal and usually spread slightly below the horizontal. They may be spathulate, elliptic, broadly ovate or circular. Narrowly lanceolate, spirally twisted petals that taper from base to apex are typical of *P. dianthum*. Petals of this species have glandular pubescence at the apex and carry small bundles of glandular hairs along their ciliated basal margins. Species such as *P. callosum* and *P. appletonianum*, besides ciliae, have prominent marginal warts on their upper and sometimes on the lower petal margins. Several species have petals with a markedly undulate upper margin, perhaps best seen in *P. hirsutissimum*. The petals of all species have prominent long ciliae near the base.

The third median petal in the flower of *Paphiopedilum* is highly modified to form a sack-like or slipper-shaped lip which serve as a trap for potential pollinating insects. The lip is three-lobed with a deeply saccate mid-lobe and basal in-turned side-lobes. An inflated, thin-textured lip is characteristic of subgenus *Parvisepalum*. In subgenus

a
b
c
d
e
g
f

Brachypetalum the lip is smaller, ovoid, and more solid-textured. In both these subgenera the apical margin is involute. In other Vietnamese species the margin of the mid-lobe is not involute but the warty and glossy side-lobes are well developed and incurved to form a tube at the base of the lip. The lip of *Paphiopedilum* species acts as a trap for potential pollinators and is equipped with margins that prevent direct escape of insects trapped therein but has a ladder of hairs leading from the pouch up to the base of the lip for insects to climb to enable them to escape through a gap on either side of the column, passing in succession beneath the stigma and an anther.

The remaining stamens and pistil of the slipper orchid are united into stalk-like column or gynostemium in the center of the flower (Fig. 31, 32).

The main feature of the column is a prominent apical shield-like staminode, which derives from a sterile median stamen of the outer whorl. Though varying greatly within a species, the shape, colour and hairiness of the staminode still have significance in the taxonomy of the genus.

Behind and below the staminode lies a stalked ovate or circular stigma (Fig. 31, 32). The convex, finely papillose disk of the stigma is divided into three parts by shallow lines that reflects origin of pistil from three united carpels.

Two stamens with short filaments and broadly elliptic anthers lie behind the stigma and on the sides of the column stalk (Fig. 31, 32). They represent lateral stamens of the inner whorl. The filament of most species is extended into a prominent point at the apex. The anther loculi are usually proximate but occasionally separated from each other. The pollen masses are usually sticky but, in such species as *P. delenatii* and *P. micranthum*, they are powdery.

An inferior unilocular ovary with parietal placentation is typical of *Paphiopedilum* (Fig. 32). Pubescent ovaries are found in all the Vietnamese species except *P. dianthum*. In *P. micranthum*, *P. malipoense* and *P. villosum* the ovary is villose. Most species have a more or less cylindrical ovary but in *P. villosum* the ovary is triangular in cross-section. The ovary is commonly green or flushed with dull purple.

Fruit. The slipper orchid fruit is a unilocular, elongate, dry capsule with three wide and three narrow valves. The capsule opens from near the apex by six splits. Fruits commonly ripen in natural conditions after six to ten months after fertilisation.

LIFE HISTORY

The seeds of *Paphiopedilum* species vary from ellipsoidal or shortly spindle-shaped to elongate and narrowly spindle-shaped, mainly 0.4–1.1 mm long. A small embryo, 0.3–0.4 mm long, is surrounded in seeds of slipper orchids by a papery seed coat derived from the inner and outer integuments of the ovule. Ripe seeds are very light and are dispersed by wind. Orchid seeds do not have any endosperm and rely for nutrition upon an intimate association with a mycorrhizal fungus.

Fig. 32 (opposite). Morphology of a *Paphiopedilum* flower. (Photos: L. Averyanov).

Paphiopedilum gratrixianum: **a** — flower, frontal view; **b** — flower with cut petal and sagittal section of the lip, side view; **c** — flower, view from behind; **d–f** — column (gynostemium), side view (**d**), view from above (**e**) and from below (**f**); **g** — cross section of ovary.

The seeds mature between six and ten months after fertilization and disperse through slits in the capsule that develop as it turns brown and dries out. Germination occurs in the dark in the soil, on rocks or on trees soon after dispersal when the suspensor of the embryo is infected by a suitable mycorrhizal fungus. The protocorm produces long rhizoids that also become infected with hyphae of mycorrhizal fungus. A root soon emerges behind the shoot apex. The rhizomatous protocorm elongates and produces further roots. The first leaf may appear a few months after germination. The first root, which contains little xylem but copious phloem, is covered in rhizoids and is mycotrophic from the start. The cortex is densely infected with fungal pelotons at this stage. However, later roots are progressively less infected and by the fifth, the roots are characteristically slender with well-developed xylem and lack pelotons.

In mature plants the bud containing the leaves and inflorescence for the following year develops during the previous growing season. The new growth begins to develop usually before flowering has finished. Most Vietnamese slipper orchids flower in nature during spring or early summer (Tab. 10). Among them are *P. appletonianum, P. callosum, P. concolor, P.*

TABLE 4.

BLOOMING SEASONS OF VIETNAMESE SLIPPER ORCHIDS.

Flowering samples of the listed species were observed in natural habitats during months which are marked with blue. Months when flowering was observed under cultivation in the Northern Hemisphere (according to personal observations and published data of Braem, 1988a; Braem, C.Baker & M.Baker, 1998, 1999; Gruss & Perner, 1998, 1999; Cavestro & Chiron, 1999; Senghas & Schettler, 1999) are marked with asterisk.

	1	2	3	4	5	6	7	8	9	10	11	12
P. appletonianum	*	*	*	*	*	*	*		*	*	*	*
P. barbigerum										*	*	*
P. callosum					*	*						
P. concolor	*	*	*	*	*	*	*	*	*	*	*	*
P. delenatii	*	*	*	*	*	*					*	*
P. dianthum						*						
P. emersonii					*	*						
P. gratrixianum	*		*	*				*	*	*	*	*
P. hangianum				*	*	*						
P. helenae							*	*	*			
P. henryanum	*	*	*	*			*	*			*	*
P. hirsutissimum	*	*	*	*	*	*	*	*		*	*	*
P. malipoense			*	*		*						
P. micranthum		*	*	*								
P. purpuratum	*	*	*				*	*	*	*	*	*
P. tranlienianum									*	*	*	
P. vietnamense	*			*								*
P. villosum	*	*	*	*	*	*	*		*	*	*	*
P. × affine												
P. × dalatense						*	*					
P. × herrmannii						*	*	*	*			

emersonii, P. hangianum, P. hirsutissimum, P. malipoense, P. micranthum, P. × herrmannii, P. vietnamense and *P. villosum*. The other species, *P. barbigerum, P. dianthum, P. gratrixianum, P. helenae, P. henryanum, P. purpuratum*, and *P. tranlienianum* flower in nature in the autumn while *P. delenatii* and *P. villosum* flower during the winter. All species of Vietnamese slipper orchids, except *P. delenatii*, flower during the dry season. Flowers last between four to eight weeks. Variation in the flowering time is common for all species that have been removed into cultivation. Available data on blooming time of Vietnamese slipper orchids in natural populations and in cultivation are presented in Table 4.

Slipper orchids can live for many years, a new sympodial growth appearing annually. Old portions of rhizome with bases of old erect shoots successively decay. In species growing in dry habitats, however, dead portions of rhizomes may remain for a very long time. In such cases, the number of dead (but not yet decayed) shoot bases will be equal to the minimum age of each plant. According to our observations, the minimum age of mature flowering samples of *P. helenae* in wild populations was at least 30–40 years. In nature, the actual age of studied plants will be greater. Hybrid slipper orchids that were made by growers in Europe in the last three decades of the 19th Century are still going strong after more than a century in cultivation.

POLLINATION

Cross-pollination by insects is certainly a well-founded hypothesis for all species of *Paphiopedilum*. However, little data on pollinators of these plants are actually available.

Some evidence exists that hover-flies (Syrphidae family) pollinate the flowers of *P. villosum* (Bänziger, 1996; Cribb, 1998). Hover-flies are lured mainly by food deception. In *P. villosum* long-distance attraction is supposedly by the urine-like scent of the flower. The fruity or honey fragrance of *P. concolor, P. delenatii* or *P. malipoense* probably plays a similar role . However, most species of the genus have odourless flowers and the bright coloration of the flower may be the main factor in long-distance attraction of pollinators.

In some cases food deception based on the obvious mimicry by the flowers of nectarless orchids of nectar-bearing flowers of other plants may be supposed. This may be seen in *P. micranthum* flowers that certainly resemble in colour flowers of some *Rhododendron* species (Fig. 33) that produce abundant nectar to attract a great number of different pollinators. In limestone areas of the north of Vietnam and in southern China, numerous examples of *Rhododendron* species with large white or pink flowers have been observed near populations of flowering *P. micranthum*. Most visitors to *Rhododendron* flowers also occasionally visit flowers of *P. micranthum*. Obviously only a few of these can act as pollinators of *P. micranthum*. Insects that are too small often drown inside the lip where some rainwater usually accumulates. We have seen numerous small dead flies and moths in the lips of *P. micranthum*. Larger insects that cannot escape the lip trap by the narrow exits due to their large size also die and sometimes are found in the lip. Some dead beetles (Cetoniinae) and large flies (Syrphidae) have been found in this way. Without doubt, other large insects trapped in the lip are able to gnaw through the lip wall to escape. Some flowers with nibbled holes in the lip wall observed in populations of *P. micranthum* certainly support this supposition.

The water remaining inside the lip of *P. micranthum* after rainfall may theoretically be an additional attraction for some insects as the weather during flowering period of this species is often very dry. Some of these insects entering the lip-trap may act as unspecific pollinators.

Fig. 33. Flowers of *Rhododendron* sp. (**a**) and *Paphiopedilum micranthum* (**b**) growing together along limestone ridges in Cao Bang Prov., northern Vietnam, Distr. at 1500 m. (Photos: L. Averyanov).

Occasional pollination of nectarless orchid flowers may be connected with food deception based on mimicry with nectar bearing *Rhododendron* flowers that attract numerous pollinators.

Despite numerous kinds of insects having been seen inside the lip, hover-flies are still the only insects that have been reliably observed as pollinators of slipper orchid flowers (Bänziger, 1996; Cribb, 1998).

The flowers of *P. delenatii, P. malipoense, P. micranthum* and their allies strongly resemble flowers of *Cypripedium* species that are regularly bee-pollinated (Nilsson, 1979). These species of *Paphiopedilum* may be also bee- rather than fly-pollinated.

In species of subgenus *Paphiopedilum* close-range attraction of insects may be effected by the glittering staminode which mimic droplets of honeydew or moisture. The warts on the side lobes of the lip appear to provide an attractive potential perch for the hover-flies or other insects but they are slippery and the flies fall into the slipper-shaped lip, the incurved margins of which prevent wing action so that they cannot fly out of the lip's opening. Escape is along a ladder of hairs and beneath the stigma and anthers in turn to eventually emerge at the base of the column. The pollen, which is very sticky adheres to the back of the insect. If the pollinator visits another flower and is trapped, the pollen can be transferred to the stigma of that flower as the insect squeezes beneath the stigma, the flexibility of the lip pressing the insect against the papillose stigma surface.

Efficiency of slipper orchid pollination varies from year to year, from species to species and from population to population throughout the range. According to studies of Vietnamese populations of *P. appletonianum, P. hirsutissimum* and *P. gratrixianum*, successful fertilization levels may reach 90%. On the other hand, in 1998 we found less than 1% success in populations of *P. micranthum*. The same picture was observed in numerous populations of *P. helenae* between 1997 and 2000. Species such as *P. concolor, P. dianthum* and *P. henryanum* achieve moderate fertilization success, ranging from 20–60% in Vietnamese populations .

NATURAL HYBRIDIZATION

Many observations suggest that the flowers of *Paphiopedilum* species have no obligate specific pollinators and are visited by numerous kinds of insects. Obviously, some of these that can act as pollinators may be equally attracted by flowers of different species of slipper orchids. Thus, it is not too surprising that natural hybrids are observed from time to time in sympatric populations of some *Paphiopedilum* species.

The first time that natural hybridization was suspected in *Paphiopedilum* was in 1890 when Rolfe described, as *P.* × *siamense*, a putative hybrid of *P. callosum* and *P. appletonianum* that had been received in a collection of *P. callosum* from Thailand. A number of other natural hybrids have been reported in *Paphiopedilum* (Cribb, 1998). Some of them were found in Vietnam.

Thus, De Wildeman (1906) described *P. affine* from North Vietnam. On the basis of study of the description and cultivated specimens flowered in 1912 in Bruxelles, it is likely that *P. affine* is of hybrid origin (Cribb, 1998). The flower is similar in size and shape to *P. villosum*, but the staminode is shorter and broader and lacks a central umbo. The tessellated leaves and staminode shape suggest that the widespread *P. appletonianum* is the other parent.

Plants that display intermediate morphological features between *P. callosum* and *P. villosum* are sometimes collected in Lam Dong province of southern Vietnam. From time to time, they may be seen in cultivation in Dalat, the orchid capital of Vietnam, among collections of other wild orchids. There is little doubt that these plants, called *P.* × *dalatense*, have hybrid origin (Averyanov, 2001).

Paphiopedilum × *herrmannii*, another natural hybrid, was described as a distinct species by Fuchs and Reisinger (1995) based on plants imported from northern Vietnam but lacking exact provenance. According to molecular analysis of leaf material, it was suspected that this plant may be a natural hybrid of *P. hirsutissimum* var. *esquirolei* and *P. barbigerum* (Cribb, 1998). However, a number of *P.* × *herrmannii* specimens have recently been found growing between adjacent populations of *P. hirsutissimum* var. *esquirolei* and *P. helenae* (Averyanov, 1999). Hence, it is more likely that its second parent is *P. helenae* rather than *P. barbigerum*.

Other recently described problematic taxa are *P. tranlienianum* (Gruss & Perner, 1998) and *P. caobangense* (Nguyen Thien Tich, 1999) which are undoubtedly conspecific. As often happens, they were described on the basis of commercial collections lacking reliable provenance. The authors mentioned regions of Cao Bang and Bac Thai (now divided into Bac Can and Thai Nguyen provinces) Provinces as possible areas of origin. Obviously, this information may be misleading and brings certain difficulties for understanding of the nature of these plants. From observations of this orchid in private collections and in nature it is possible that it may also have a hybrid origin.

Natural hybrids of *P. malipoense* and *P. micranthum* have been reported from China and called *P* × *fanaticum* Koop. & Hasegawa. The appearance of such hybrids is not surprising because both parental species bloom at the same time and often occur together in North Vietnam along the northern border of the country. However, the hybrid has yet to be reported from Vietnam.

Some reports also exist of the occurrence in Indo-china of plants that are intermediate in their morphology between *P. bellatulum* and *P. concolor*. Among orchid growers these plants have been given the name "*P. conco-bellatulum*", a name that has not been formally or validly described to date. They were supposedly introduced into cultivation from Guangxi and Yunnan, and may possibly represent some introgression between the two species. The occurrence of such hybrid plants is possible in limestone areas of the far north-west regions of Vietnam.

In conclusion, it appears likely that hybridization may be of some importance in the origin of some slipper orchid species. For example, the Malayan species *P. godefroyae*, erroneously reported in the past for Vietnam, may well have originated through ancient hybrid introgression between *P. bellatulum* and *P. niveum* but has subsequently stabilized as a distinct species.

ECOLOGY

Slipper orchids in Vietnam may be divided into two distinct groups. Species of the first group exclusively inhabit limestone regions in the north of the country. But those of the second group are found in regions where acidic silicate rocks, sandstones or schists are prevalent.

Paphiopedilum × *aspersum, P. barbigerum, P. concolor, P. dianthum, P. emersonii, P. hangianum, P. helenae, P. henryanum, P.* × *herrmannii, P. hirsutissimum, P. malipoense, P. micranthum, P. purpuratum, P. tranlienianum* and *P. vietnamense* are obligate calcicoles. They are found in northern Vietnam where highly eroded limestone mountains are widely distributed. Shady shelves and deep crevices on steep rocky cliffs and bluffs of solid marble-like limestone or rocky outcrops are the most typical habitats of these species.

From the sea level up to 400–500 m, *Paphiopedilum concolor* may be commonly observed in the dry evergreen broad-leaved forest that usually develops in limestone areas of northern Vietnam. The rare Vietnamese endemics, *Paphiopedilum tranlienianum* and *P. vietnamense* also occur at these elevations.

Between 500–900 m, semi-deciduous, broad-leaved forest replace evergreen forest. Some conifers, such as *Pseudotsuga brevifolia* and *Podocarpus neriifolius*, appear occasionally in these forests above 650–750 m. Slipper orchids that inhabit the zone between 600–900 m include *Paphiopedilum. helenae, P.* × *herrmannii, P. hirsutissimum* var. *esquirolei, P. emersonii* and *P. hangianum. Paphiopedilum dianthum, P. henryanum* and *P. malipoense* may also occasionally be found at these elevations.

Mixed and coniferous forests with a large representation of conifers, such as *Fokienia hodginsii, Pinus fenzeliana, Pseudotsuga brevifolia* and *Tsuga chinensis*, spread along limestone ridges from 900–1500 m. Some other rare gymnosperms, including *Amentotaxus, Cephalotaxus, Nageia, Podocarpus, Cupressus* and *Xanthocyparis*, are also found in these forests. The rocks in primary mixed and coniferous forests towards the tops of limestone mountains are always covered in mosses, ferns and orchids. *Paphiopedilum dianthum, P. henryanum, P. malipoense* and *P. micranthum* regularly occur here. In many undisturbed forests species such as *P. malipoense* and *P. micranthum* are important co-dominants in communities of lithophytic herbs.

At 1500–1600 m, the highest limestone regions in the north of Vietnam, *Paphiopedilum dianthum, P. micranthum, P. purpuratum* and *P. hirsutissimum* var. *chiwuanum* have been observed.

Species of the second group of slipper orchids inhabit mountains that are composed mainly of acidic rocks. Although widely distributed throughout the country, they are commoner in central and southern Vietnam. *Paphiopedilum delenatii, P. gratrixianum, P. appletonianum* and *P. callosum* occur on soils rich in silicates. Some of them also grow as lithophytes in the crevices and ledges of vertical cliffs of granite and gneiss. On the other hand, *Paphiopedilum gratrixianum* is usually found on rhyolite. *Paphiopedilum villosum* very rarely grows as a lithophyte but it can be found regularly as a humus epiphyte on tall trees. Although it rarely grows in soil, this species has always been found on mountains composed of acidic rocks. Species of this group have been found between 700–2200 m. In their habitat preferences,

Paphiopedilum delenatii, P. appletonianum, P. callosum and *P. villosum* certainly prefer mixed and coniferous woods with *Fokienia hodginsii, Keteleeria evelyniana, Pinus dalatensis, P. kesiya* and *P. krempfii.*

The orchids of Vietnam, like orchids elsewhere are found as terrestrials, lithophytes or epiphytes. Some species may change their preference in different habitats, but their main mode usually remains species' specific. About 22% of the Vietnamese orchid flora is terrestrial, 8% lithophytic and 70% epiphytic (Fig. 34). Amongst epiphytic orchids 15% are humus epiphytes, living in the forks of old stems of trees where relatively large amounts of humus and leaf litter accumulate. These are usually large orchids, for example, species of *Cymbidium* that can produce more than 50 growths. About 50% of epiphytes are trunk or branch epiphytes. About 5% of epiphytes live on small branches, twigs or even leaves in the tree canopy. Slipper orchids of Vietnam may be terrestrials, lithophytes or humus epiphytes (Fig. 35).

Terrestrial species with strongly tessellated leaves, such as *Paphiopedilum appletonianum, P. callosum* and *P. purpuratum*, usually grow in deep shade on the forest floor, mostly on steep mountain slopes, in well-drained conditions. They prefer well-developed leaf mould soils rich in humus. However, occasionally they may be also be found as lithophytes. In very wet and mossy cloud forests, *Paphiopedilum appletonianum* may sometimes be observed as a humus epiphyte at the base of old large trees.

Most Vietnamese species, including *Paphiopedilum concolor, P. delenatii, P. dianthum, P. emersonii, P. gratrixianum, P. hangianum, P. helenae, P. henryanum, P. hirsutissimum, P. malipoense, P. micranthum, P. tranlienianum* and *P. vietnamense* are lithophytes. These usually grow in dappled shade of rather open forests, often on rocky cliffs and just below ridge-tops. However, *P. gratrixianum, P. hirsutissimum, P. micranthum* and especially *P. malipoense* often grow in deep shade. Sometimes *P. micranthum* may be found in primary open herbaceous communities of the summits of mountains. Vertical shady cliffs below ridge-tops are the typical habitats of *P. concolor, P. helenae, P. henryanum* and *P. hirsutissimum.* Sometimes these species grow in narrow crevices or on ledges on dry lime dust. Occasionally, *P. helenae* can be observed on vertical limestone cliffs where its roots are strongly attached to the rocks. *Paphiopedilum dianthum* may be rarely found as a humus epiphyte in wet cloud forests between 1200–1500 m.

Only *Paphiopedilum villosum* of the Vietnamese slipper orchids grows regularly as an epiphyte. Very rarely it can also be found as lithophyte, usually on inaccessible shelves of high vertical bluffs of canyons above streams in the mountains.

In areas of Vietnam where slipper orchids grow, the rainfall and consequent humidity are high but rainfall is usually seasonal and plants in such areas often have to survive considerable dry periods. Their thick leathery leaves are ideally adapted to survive periodic droughts and plants recover rapidly when the rains return.

North, northeast and northwest slopes are certainly preferred by slipper orchids in degraded dry forests. However, in primary forest these plants are equally common on slopes of any exposure. In such places slipper orchids often reach their optimum and bloom better on warmer and lighter southern slopes. Usually slipper orchids grow in small colonies, however in the past large colonies with thousands of flowering specimens were not uncommon.

Moisture levels around the roots, soil type and pH, availability of mycorrhizal fungi and suitable pollinators, and light levels are important factors in the establishment and success of *Paphiopedilum* populations. Anthropogenic destruction of natural habitats, changing environmental conditions and commercial collecting are the main present day causes of the rapid extinction of slipper orchids all over the country.

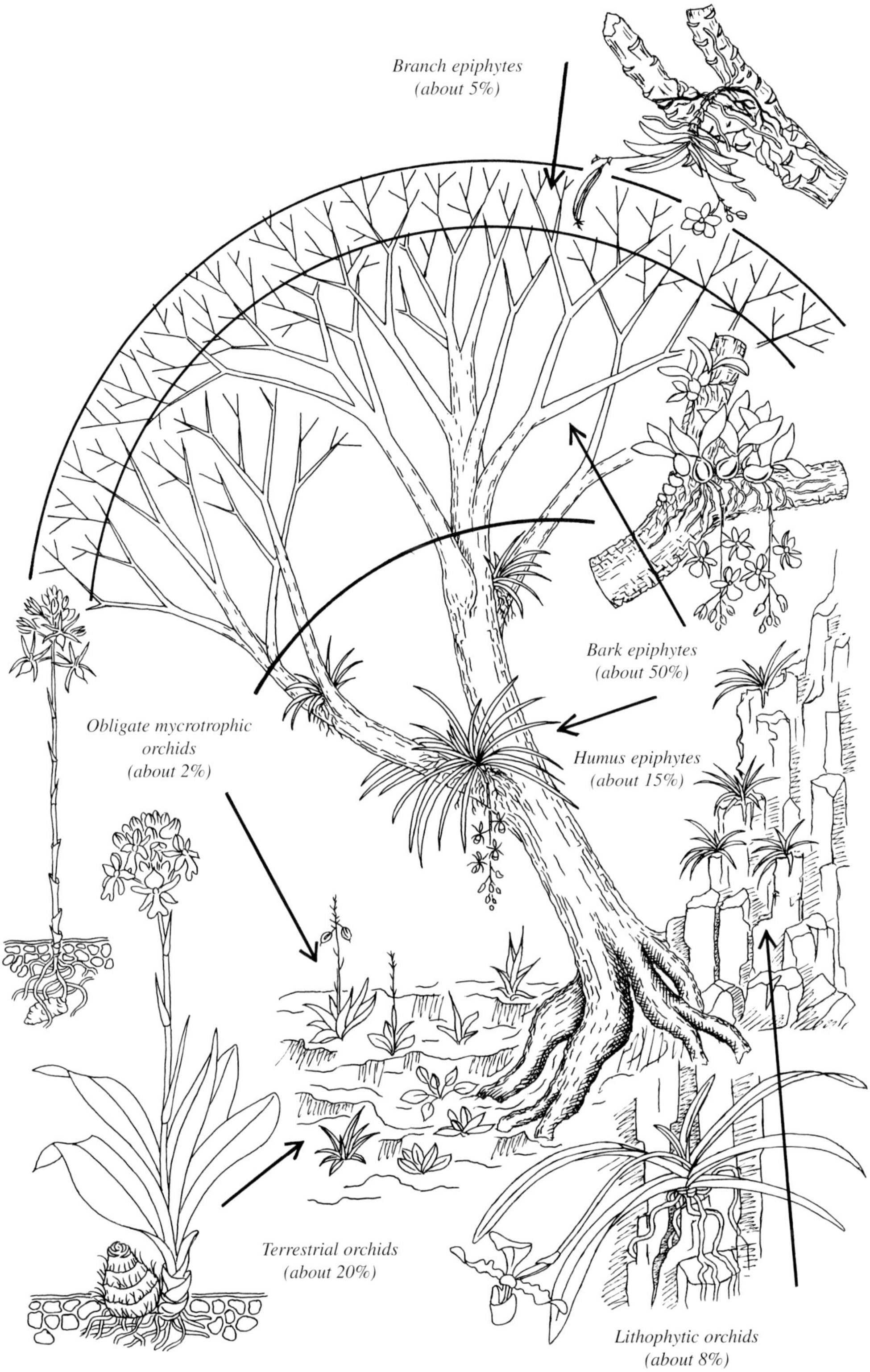
Branch epiphytes
(about 5%)
Bark epiphytes
(about 50%)
Humus epiphytes
(about 15%)
Obligate mycrotrophic
orchids
(about 2%)
Terrestrial orchids
(about 20%)
Lithophytic orchids
(about 8%)

Fig. 35. Illustration of terrestrial, lithophytic and epiphytic life modes of Vietnamese slipper orchids.

a — *Paphiopedilum appletonianum* growing as terrestrial herb on the forest floor in S. Vietnam, Prov. Kontum (Photo.: L. Averyanov);

b — Lithophytic plant of *Paphiopedilum dianthum* on vertical solid limestone cliff in N. Vietnam, Prov. Ha Giang (Photo.: P.J. Cribb);

c — *Paphiopedilum villosum* growing as a humus epiphyte on high tree in S. Vietnam, Prov. Lam Dong (Photo.: P.K.Loc).

Fig. 34 (opposite). Ecological groups of orchids and their habitats. (Drawing: L. Averyanov).

Orchids in the flora of Vietnam may be obligate mycotrophic plants (about 2%), terrestrials (about 20%), humus epiphytes (about 15%), bark epiphytes (about 50%), branch epiphytes (about 5%) and lithophytes (about 8%). *Paphiopedilum* species commonly grow as terrestrial herbs, humus epiphytes or lithophytes.

GEOGRAPHICAL DISTRIBUTION

The range of *Paphiopedilum* extends from the tropical foothills of the Himalayas eastwards across southern China to the Philippines and throughout South-east Asia and the Malay Archipelago to New Guinea and the Solomon Islands (Fig. 36–39). One species, *P. druryi*, has a very limited isolated distribution in southern India (Fig. 37).

Paphiopedilum certainly originated in mainland Southeast Asia. Its extension southwards and eastwards into Malesia and the Southwest Pacific is the result of successive migrations of ancestral species and their divergent radiation into a number of locally endemic, usually allopatric species.

The most primitive species of the genus, in subgenus *Parvisepalum*, are found in mainland Southeast Asia, mostly in southern China and northern Vietnam, with one species in the south of Vietnam (Fig. 36). Each species has a rather limited distribution, and that of the subgenus demarcates a compact region that may indicate the area of possible origin of the genus.

Subgenus *Brachypetalum* is limited to south China, southeast Myanmar, Thailand, northern Malaya and Indochina (Fig. 36). Only one species of the subgenus, *P. concolor*, is found in Vietnam. Of all the species, it has the most extensive distribution and that nearly coincides with the distribution of the subgenus as a whole (Fig. 36).

Most *Paphiopedilum* species belongs to one of the sections of subgenus *Paphiopedilum* whose distribution encompasses the entire range of the genus. Each section has its own distinct and more limited range.

Species of the typical section (sect. *Paphiopedilum*) inhabit areas of mainland Asia from India to southern China, Thailand and Indochina with the greatest concentration of species in southwestern and southern China (Fig. 37). *Paphiopedilum hirsutissimum* and *P. villosum* have the widest distribution and that coincides with main distribution of the section. Most other species of the section are more or less local endemics.

The range of sect. *Barbata* is almost as extensive as that of the genus. *Paphiopedilum appletonianum, P. callosum, P. bullenianum* and *P. javanicum* are the most widespread species in this group. The first two are widely distributed in mainland Asia, and the last two are common on the islands of West Malesia. The greatest concentration of local endemics in this section is observed in Sumatra, Borneo and the Philippines (Fig. 38).

Sect. *Pardalopetalum* has two species in mainland South-east Asia, one local endemic in Philippines and one species widely distributed in West Malaysia and the Greater Sunda Islands of the Malay archipelago (Fig. 37).

Five closely related species of the sect. *Cochlopetalum* are restricted to the islands of Sumatra and Java (Fig. 37). On the other hand related species of the sect. *Coryopetalum* inhabit Borneo but also spread to the Philippines, Sulawesi and New Guinea. All are narrow endemics, except *P. philippinense*, which is more widely distributed in the Philippines and in the northern part of Borneo (Fig. 36).

The greatest species concentration of *Paphiopedilum* is found in south China (Yunnan, Guangxi) and north Vietnam, where 18–20 species may be found. Well-defined centres of diversity also lie in northern Sumatra and in northern Borneo with nine species in each area. Less rich, but distinct areas of diversity are also found in the south-eastern foothills of Himalayas, in northern Myanmar and Thailand, the central part of the Malay Peninsula and the Philippines. In these regions from four to six slipper orchid species occur (Fig. 39).

It is notable that local endemism is typical in the genus. 72% of the known species are locally endemic with very limited geographical distribution, such as *Paphiopedilum delenatii, P. helenae, P. henryanum, P. tranlienianum, P. vietnamense* and *P. hangianum*. A number of

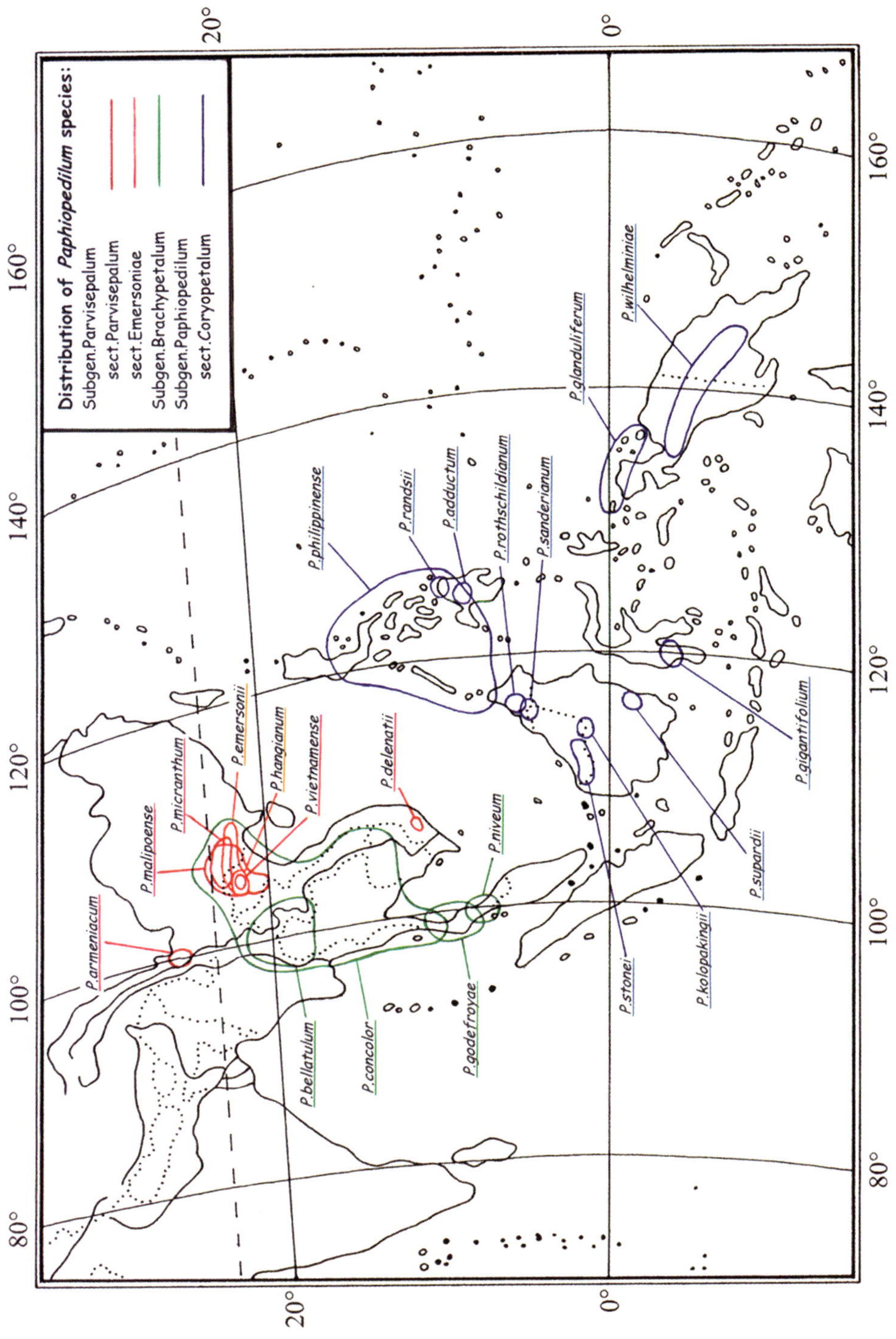

Fig. 36. Distribution of *Paphiopedilum* species from subgenera *Parvisepalum* (sections *Parvisepalum* and *Emersonianum*), *Brachypetalum* and *Paphiopedilum* (section *Coryopetalum*).

Distribution areas of sect. *Parvisepalum* are outlined with red, sect. *Emersoniae* in orange, subgen. *Brachypetalum* in green and sect. *Coryopetalum* in blue. (Drawn by L. Averyanov based on Cribb, 1998).

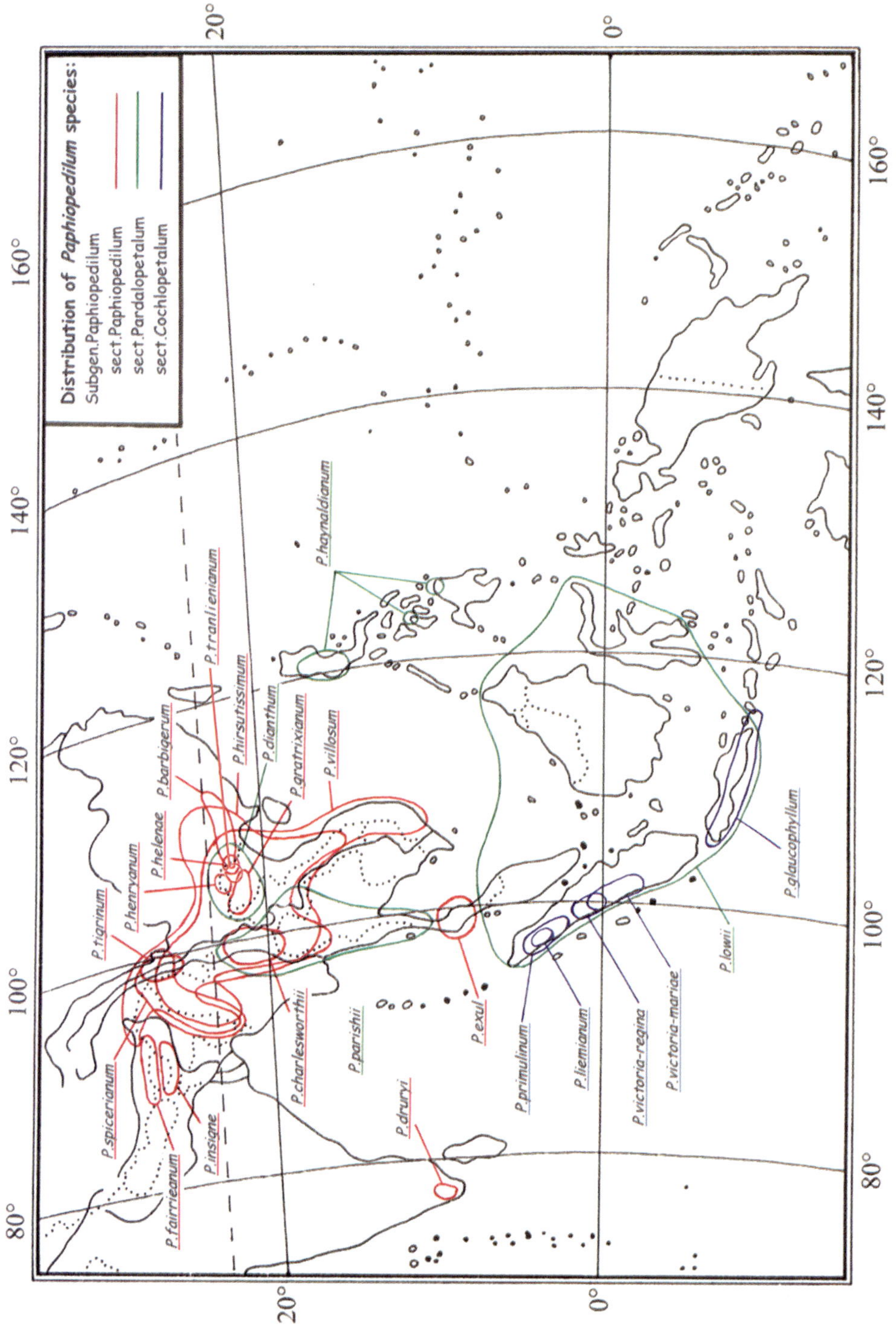

Fig. 37. Distribution of *Paphiopedilum* species from subgenus *Paphiopedilum* (sections *Paphiopedilum, Pardalopetalum* and Cochlopetalum).

Distribution areas of species of sect. *Paphiopedilum* are outlined in red, *Pardalopetalum* in green and sect. *Cochlopetalum* in blue. (Drawn by L. Averyanov based on Cribb, 1998).

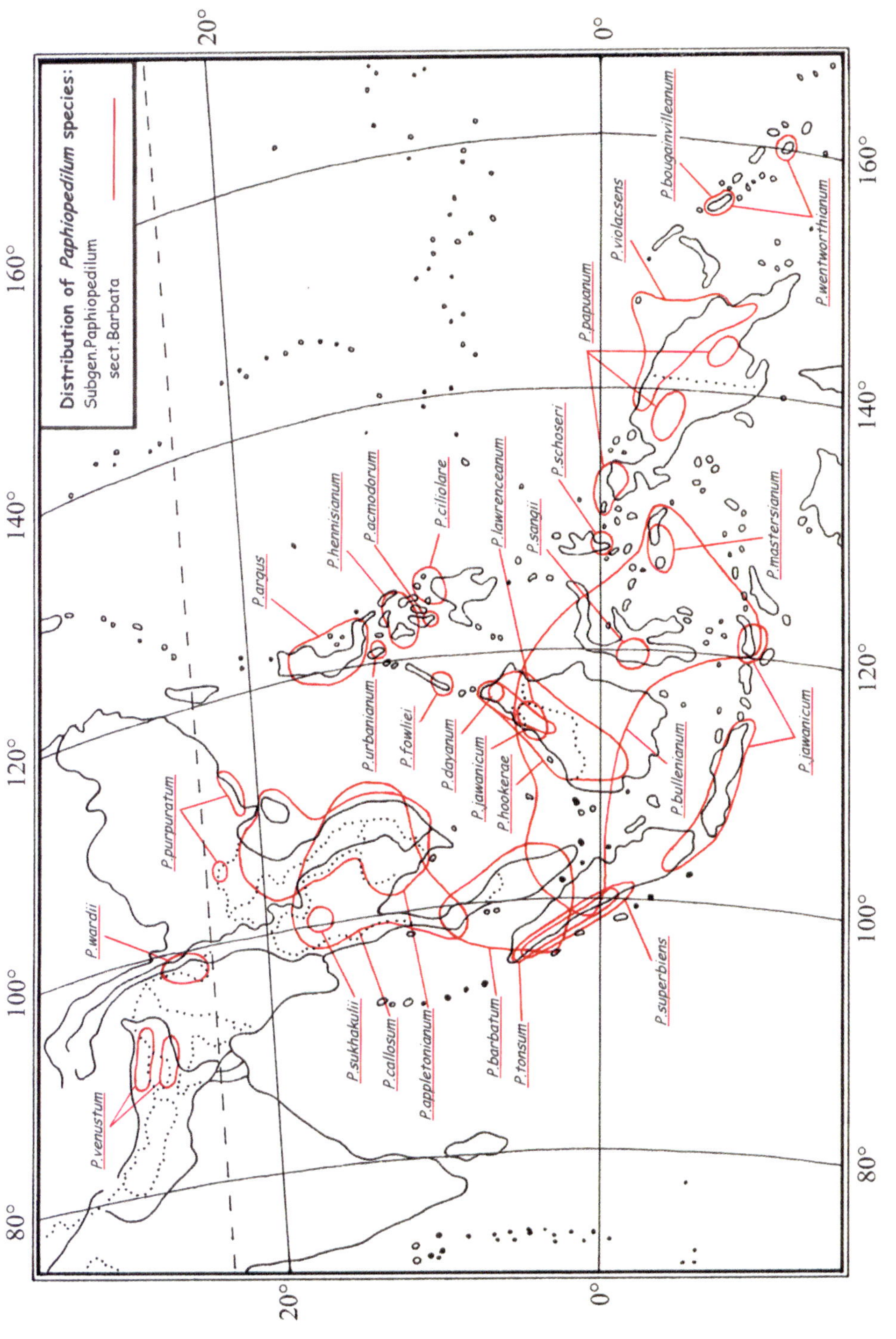

Fig. 38. Distribution of *Paphiopedilum* species from subgenus *Paphiopedilum* (section *Barbata*). Distribution areas of species are outlined in red. (Drawn by L. Averyanov based on P. Cribb, 1998).

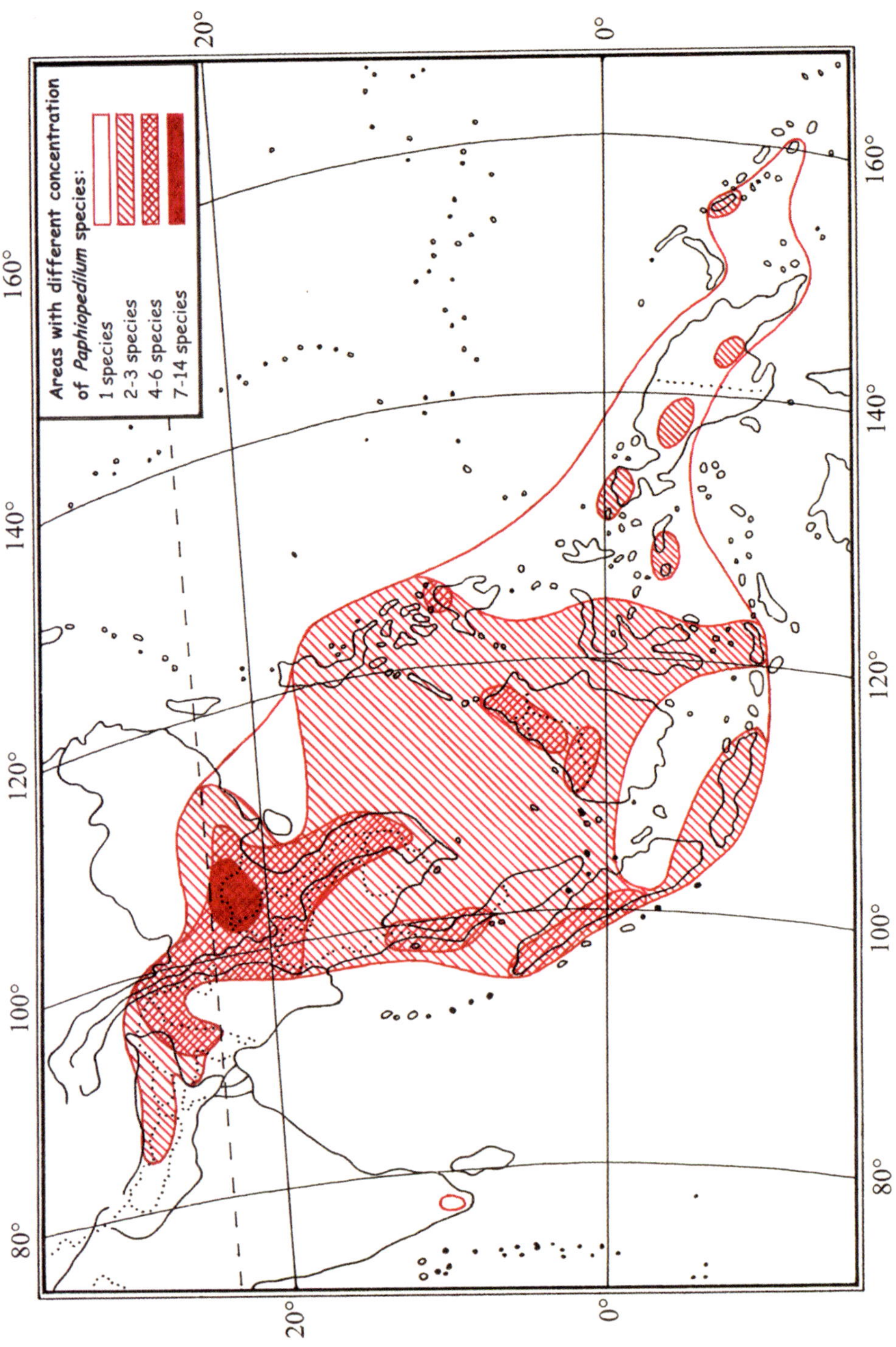

Fig. 39. Species diversity of *Paphiopedilum*.

At least one species is reported from the area outlined in red, two or three species in areas marked with simple line shading, four to six species in areas marked with cross hatching, and seven–14 species in areas marked in red. (Drawn by L. Averyanov based on P. Cribb, 1998).

such species are known in only one or few localities. For example, local endemism of *Paphiopedilum* is very typical in the Philippines flora, where 12 species occur. Nevertheless, they are not concentrated in any particular region because of their strictly allopatric endemism.

Less than 12% of *Paphiopedilum* species are relatively widely distributed. These include *Paphiopedilum villosum, P. hirsutissimum, P. appletonianum* and *P. callosum.*

INFRAGENERIC CLASSIFICATION OF PAPHIOPEDILUM

Because of its horticultural interest, the genus *Paphiopedilum* has attracted considerable attention from botanists. Since its description by Pfitzer in 1886, numerous taxonomic studies have been devoted to it and different approaches to the infrageneric classification of *Paphiopedilum* have been reviewed in recent monographs (Cribb, 1998; Braem, Baker C. & Baker M., 1998). Here, we follow the scheme outlined by Cribb (1998), which is based on earlier morphologically based treatments (Brieger, 1973; Karasawa & Saito, 1982; Atwood, 1984; Cribb, 1987) but also considers the molecular work of Cox *et al.* (1997). It is summarized in Table 5.

Based upon comparative morphological studies (Dressler, 1981, 1993; Cribb, 1987, 1998; Averyanov, 1991c, d), three clear evolutional trends define the three subgenera *Parvisepalum, Brachypetalum* and *Paphiopedilum.*

Subgenus *Parvisepalum*, comprising seven species, has an inflated, thin-textured lip, large convex or conduplicate staminode and granular pollinia. This is very isolated and primitive group of slipper orchids and has characters somewhat intermediate between those of subgenus *Paphiopedilum* and the genus *Cypripedium.*

Based on the texture and colouration of the leaves, and the form of the petals and lip, subgenus *Parvisepalum* can be divided into two distinct groups, sect. *Parvisepalum* with five species and sect. *Emersonianum* with two species.

The second subgenus *Brachypetalum* comprises four species characterized by their concolorous white or yellowish flowers, elliptic to suborbicular petals, ovoid thick-textured lip and tessellated leaves. This obviously homogeneous group represents a single clade that is morphologically somewhat intermediate between subgenera *Parvisepalum* and *Paphiopedilum.*

Subgenus *Paphiopedilum* is the most specialized and phylogenetically advanced. This large and heterogeneous group includes 60 species divided into five sections: sections *Paphiopedilum, Barbata, Pardalopetalum, Cochlopetalum* and *Coryopetalum.*

The typical section, which includes 13 species has uniformly green leaves, a single-flowered inflorescence, spathulate petals and a obcordate staminode that usually has a glossy boss or umbo in the centre. Species of this section are certainly related to those of sect. *Barbata,* which have tessellated leaves, a single-flowered inflorescence, a lip with well-developed, warty side-lobes and a broad lunate staminode lacking an umbo.

Species of other three sections: sect. *Pardalopetalum* (with four species), sect. *Cochlopetalum* (with five species) and sect. *Coryopetalum* (including 11 species) have many-flowered inflorescences and highly specialized and modified flowers that are regarded as phylogenetically advanced. This is especially true of section *Coryopetalum* which has flowers having very long, highly metamorphosed petals and a curiously evolved mechanism of pollination by Syrphid flies.

A list of the accepted species of *Paphiopedilum* classified into infrageneric categories is provided below. The taxa are arranged according to probable degree of their specialization. Species occurring in Vietnam are marked in bold type.

TABLE 5.

INFRAGENERIC TAXONOMY OF THE GENUS PAPHIOPEDILUM.

The Genus *Paphiopedilum* Pfitzer

A. subgenus *Parvisepalum* Karas. & Saito

1. sect. *Parvisepalum* Aver. & Cribb

1. *P. armeniacum* S.C.Chen & F.Y.Liu
2. *P. delenatii* Guillaumin
3. *P. vietnamense* O.Gruss & H.Perner
4. *P. malipoense* S.C.Chen & Z.H.Tsi
5. P. *micranthum* T.Tang & F.T.Wang

2. sect. *Emersonianum* Aver. & Cribb

6. *P. hangianum* H.Perner & O.Gruss
7. *P. emersonii* Koop. & P.J.Cribb

B. subgenus *Brachypetalum* (Hallier) Pfitzer

8. *P. bellatulum* (Rchb.f.) Stein
9. *P. concolor* (Bateman) Pfitzer
10. *P. godefroyae* (God.-Leb.) Stein
11. *P. niveum* (Rchb.f.) Stein

C. subgenus *Paphiopedilum*

3. sect. Paphiopedilum

12. *P. hirsutissimum* (Lindl. ex Hook.) Stein
13. *P. charlesworthii* (Rolfe) Pfitzer
14. *P. insigne* (Wall. ex Lindl.) Pfitzer
15. *P. barbigerum* T.Tang & F.T.Wang
16. *P. tranlienianum* O.Gruss & H.Perner
17. *P. helenae* Aver.
18. *P. exul* (Ridl.) Rolfe
19. *P. henryanum* Braem
20. *P. gratrixianum* (Masters) Rolfe
21. *P. villosum* (Lindl.) Stein
22. *P. tigrinum* Koop. & N.Haseg.
23. *P. druryi* (Bedd.) Stein
24. *P. spicerianum* (Rchb.f.) Pfitzer
25. *P. fairrieanum* (Lindl.) Stein

4. sect. *Barbata* (Kraenzl.) V.A.Albert & Borge Pett.

26. *P. appletonianum* (Gower) Rolfe
27. *P. bullenianum* (Rchb.f.) Pfitzer
28. *P. hookerae* (Rchb.f.) Stein
29. *P. sangii* Braem
30. *P. mastersianum* (Rchb.f.) Stein
31. *P. papuanum* (Ridl.) Ridl.
32. *P. bougainvilleanum* Fowlie
33. *P. violacsens* Schltr.
34. *P. wentworthianum* Schoser & Fowlie
35. *P. tonsum* (Rchb.f.) Stein
36. *P. argus* (Rchb.f.) Stein
37. *P. barbatum* (Lindl.) Pfitzer
38. *P. callosum* (Rchb.f.) Stein
39. *P. hennisianum* (M.W.Wood) Fowlie
40. *P. fowliei* Birk
41. *P. lawrenceanum* (Rchb.f.) Stein
42. *P. dayanum* (Lindl.) Stein
43. *P. ciliolare* (Rchb.f.) Stein
44. *P. superbiens* (Rchb.f.) Stein
45. *P. acmodorum* Schoser ex M.W.Wood
46. *P. javanicum* (Reinw. ex Lindl.) Pfitzer
47. *P. schoseri* Braem
48. *P. urbanianum* Fowlie
49. *P. purpuratum* (Lindl.) Stein
50. *P. sukhakulii* Schoser & Senghas
51. *P. wardii* Summerh.
52. *P. venustum* (Wall. ex Sims) Pfitzer ex Stein

5. sect. *Pardalopetalum* Hallier f. & Pfitzer

53. *P. dianthum* T.Tang & F.T.Wang
54.*P. parishii* (Rchb.f.) Stein
55. *P. lowii* (Lindl.) Stein
56. *P. haynaldianum* (Rchb.f.) Stein

6. sect. *Cochlopetalum* Hallier f. ex Pfitzer

57 *P. glaucophyllum* J.J.Sm.
58 *P. liemianum* (Fowlie) Karas. & Saito
59 *P. primulinum* M.W.Wood & P.Taylor
60. *P. victoria-mariae* (Rolfe) Rolfe
61. *P. victoria-regina* (Sander) M.W.Wood

7. sect. *Coryopetalum* Pfitzer

62. *P. philippinense* (Rchb.f.) Stein
63. *P. randsii* Fowlie
64. *P. sanderianum* (Rchb.f.) Stein
65. *P. gigantifolium* Braem, M. Baker & C. Baker
66. *P. kolopakingii* Fowlie
67. *P. stonei* (Hook.) Stein
68. *P. adductum* Asher
69. *P. glanduliferum* (Blume) Stein
70. *P. wilhelminiae* L.O.Williams
71. *P. rothschildianum* (Rchb.f.) Stein
72. *P. supardii* Braem & Loeb

PART III. TAXONOMY OF VIETNAMESE SLIPPER ORCHIDS

HISTORY

The Portuguese missionary João de Loureiro was the first naturalist to explore the Indochina. As mathematician and naturalist to the King of Cochinchina, he spent more than 35 years in this country (Merrill, 1946). During that time he extensively studied plants in the territories that now constitute modern Vietnam. Most of his collections and observations were made around the city of Hué, the old capital of the Kingdom. His discoveries were summarized in his *Flora Cochinchinensis* which was published in Portugal a year before his death (Loureiro, 1790). The appearance of this work, containing original descriptions of more than 185 new genera and about 630 new species caused a sensation in European botanical circles and was one of the seminal botanical events of the age. Three years later it was republished in Germany (Loureiro, 1793). Such large and well-known orchid genera as *Aerides, Galeola, Phaius, Renanthera* and *Thrixspermum* were discovered and described by him. However, he did not include any slipper orchids, because the areas where they occur were then inaccessible.

The earliest collection of slipper orchids from Vietnam was probably by the French missionary M. Ban of the Apostolic Mission to Tonkin. He collected *Paphiopedilum concolor* near Ke-so and sent plants to M. Lionet in France who flowered them. Coloured illustrations of it were published by Godefroy-Lebeuf (1886) and Linden & Rodigas (1886) as *Cypripedium concolor* var. *tonkinense*, said to differ from the typical form in its broader leaves and slightly differently marked flowers.

Commercial collecting began at around the turn of the century. By 1903, both Bronckardt and Micholitz were in Vietnam collecting orchids for European nurseries. We know most about Wilhelm Micholitz, who was the main orchid collector for Messrs Sander & Sons of St. Albans, England. He arrived in Saigon on 20th October 1903 and then travelled via Nha Trang to the mountainous area of Lang Bian and collected in the gorges and hills around the town of Dalat. In a letter to Frederick Sander (Kew Archives) he mentions that Bronckardt had already been there and describes how he collected *P. gratrixianum* and *P. purpuratum* at Dankia (Lam Dong Prov.). Collecting proved to be difficult because "the country is swarming with tigers". He remained in the region until the end of December. His collections included two other slipper orchids: *P. villosum* var. *annamense* and *P. concolor*. He visited Vietnam again in 1905 when he visited Saigon and travelled overland to Vientiane in Laos (23rd September) and then in 1911 on to the Lai Chau area in Tonkin (4th November), when a rival collector, Carl Roebelen, was already in the field in Vietnam. Micholitz was again in Vietnam in 1913.

The discovery of *P. delenatii* in Vietnam by a French army officer in 1913 or 1914 was a landmark event, identifying the country as a possible source of exotic slipper orchids. The French botanist A. Guillaumin was almost certainly mistaken in reporting that this species was found in Tonkin in northern Vietnam (but see account of *P. vietnamense*). Mr. Poilane rediscovered it in 1922 near Nha Trang in Vietnam and sent plants to M. Delenat and M.

Mornay, the head horticulturist of the 'Fleuriste Municipal de la Ville de Paris'. Mornay's plant received a 'Certificate de Merité avec felicitations' from the National Society of France in 1925.

The earliest published account of Indo-Chinese slipper orchids is that of Guillaumin (1924) who listed ten species for Indo-China of which four are recorded for Vietnam. Later, he also listed ten species from Indo-China with seven reported from Vietnam: *P. affine, P. callosum, P. concolor, P. delenatii, P. godefroyae, P. gratrixianum,* and *P. villosum* (Guillaumin, 1934). Gagnepain later described additionally two collections of watercolour paintings of Vietnamese slipper orchids by Simond and Eberhardt that had been donated to the Paris Herbarium (Gagnepain, 1951). These included six of Simond's paintings of slipper orchids: *P. callosum, P. concolor, P. godefroyae, P. hookerae, P. insigne* and *P. villosum.*

Seidenfaden (1975) provided an updated list of 11 species for Indochina but did not specify exact distributions. However, he commented that a painting by Simond, identified by Gagnepain as *Paphiopedilum hookerae*, was more likely to be *P. appletonianum* and that the inclusion of *P. godefroyae* was likely to be the result of a misidentification of *P. concolor*. Averyanov (1988f) listed nine species for Vietnam adding *Paphiopedilum amabile* (=*P. appletonianum*) and *P. hirsutissimum* to the list. He later added *Paphiopedilum henryanum* to the flora of Vietnam (Averyanov, 1990d).

Seidenfaden (1992), in his comprehensive treatment of Indochinese orchids, reported nine species in Indochina, all found in Vietnam. He included *P. henryanum, P. appletonianum* and *P. purpuratum,* but reduced *P. affine* to the synonymy of *P. gratrixianum.* In the fourth edition of the illustrated account of the Vietnamese flora by Pham Hoang Ho (2000), 11 species of *Paphiopedilum* were included, including *P. amabile* (=*P. appletonianum*), *P. godefroyae* (mentioned probably on the basis of cultivated plants) and *P. insigne* reported from Laos. Eleven species of *Paphiopedilum* were listed in the revised account of Vietnamese orchids by Averyanov (1994). *Paphiopedilum malipoense* was newly reported but *P. godefroyae* was still mistakenly listed despite the cautionary comments of Seidenfaden (1975). Tran Hop (1998), the author of *The Orchids of Vietnam*, mentioned 12 *Paphiopedilum* species, adding the first records of *Paphiopedilum parishii* and *P. emersonii.* All records in his account are based on the literature and on plants in cultivation. Nguyen Thien Tich (2001) in the first volume of his *The Orchids of Vietnam* mentions 23 *Paphiopedilum* species, adding *P. hangianum* to the known flora. Some of his records are based on plants in cultivation.

Finally, as a result of extensive botanical exploration by Institute of Ecology and Biological Resources of the National Center for Natural Sciences and Technology of Vietnam (Averyanov, 1996a, 1998a, b, 1999, 2000b; Averyanov *at al.*, 1996a, b, 1997a, b; N.T.Hiep, Averyanov & P.K.Loc, 1998; N.T.Hiep, P.K.Loc & Averyanov, 1999; P.K.Loc *et al.*, 1999; P.K.Loc, N.T.Hiep & Averyanov, 1999a, b; Cribb, Averyanov & N.T.Hiep, 2000) and of extensive commercial exploration (Braem, 1987; Lückel, 1987; Fuchs & Reisinger, 1995; Gruss & Perner, 1998, 1999; N.T.Tich, 1998a, b, 1999; Cavestro & Chiron, 1999; Perner & Gruss, 1999; Senghas & Schettler, 1999), 18 species and four natural hybrids have now been credibly reported from Vietnam. Schuiteman and de Vogel (2000), authors of *Orchid Genera of Thailand, Laos, Cambodia and Vietnam*, reported 19 *Paphiopedilum* species for Vietnam but provided no detailed references. Although the status of all wild *Paphiopedilum* species in Vietnam is critical and all are either endangered or vulnerable, only three species, *P. appletonianum, P. delenatii* and *P. hirsutissimum,* were included in the recently published official *Red Data Book of Vietnam* (1996).

THE GENUS PAPHIOPEDILUM

Paphiopedilum Pfitzer, Morph. Stud. Orch.: 11 (1886); Stein, Orchideenbuch: 448–492 (1892); Guillaumin in Bull. Soc. Bot. France. 71: 548–558 (1924); id., Fl. Gen. Indo-Chine 6: 636–646 (1934); Karasawa & Saito in Bull. Hiroshima Bot. Gard. 5: 1–69 (1982); Cribb, Gen. Paphiopedilum: 1–222 (1987); id., Gen. Paphiopedilum: 1–427 (1998); Braem, Paphiopedilum: 1–249 (1988); Seidenfaden, Orch. Indochina: 16–20 (1992); Aver., Identif. Guide. Vietnam. Orch.: 15–22 (1994); Braem, C.Baker & M.Baker, Gen. Paphiopedilum 1–2: 1–363 (1998–1999); Pham Hoang Ho, Ill. Fl. Viet. 3: 762–765 (2000), *nom. conserv.* **Type:** *Paphiopedilum insigne* (Wall. ex Lindl.) Pfitz. (*Cypripedium insigne* Wall. ex Lindl.).
Stimegas Raf. Fl. Tellur. 4: 45 (1838), *nom.rej.*
Cordula Raf. l.c.: 46 (1838), *nom.rej.*

DESCRIPTION. *Terrestrial, lithophytic* or *epiphytic* sometimes stoloniferous *herbs* with fibrous roots arising from a short rhizome. Leafy erect *shoots* clustered with 3–6 leaves. *Leaves* distichous, coriaceous, conduplicate, spreading, ligulate, elliptic to oblong, obtuse to acute, often tridenticulate at apex, pure green or tessellated with dark and light green on upper surface, light green below and sometimes finely spotted or flushed with purple, glabrous or ciliate on margins. *Inflorescence* terminal, one- to many-flowered; rachis hairy or glabrous, green or spotted with purple-violet; bracts conduplicate, elliptic, hairy; pedicel very short; ovary unilocular, 3-ribbed, glabrous or hairy. *Flowers* 1–several, showy, large. *Dorsal sepal* erect to hooded over lip, lanceolate to broadly ovate, obtuse or acute, glabrous or pubescent. *Lateral sepals* fused to form a concave synsepal that is more or less similar to the dorsal sepal. *Petals* spreading or pendent, flat, reflexed or spiraling, often with undulate margins, from linear-lanceolate to broadly obovate, obtuse or acute at apex often pubescent and ciliate in basal half. *Lip* deeply pouched and inflated, slipper- or urn-shaped with more or less pronounced incurved side lobes, hairy within and usually glabrous outside. *Column* short, stalked; anthers two, bilocular, on short filaments; pollen masses powdery or viscid; staminode terminal on column, sessile or short stalked, transversely reniform, oblong to linear, flat, convex or conduplicate, glabrous or papillose often ciliate on margin; stigma stalked, tripartite, papillose. *Capsule* erect to pendent, three-ribbed, cylindrical (Fig. 31–32).

About 72 species distributed in tropical Asia and Pacific islands from India to the Philippines, New Guinea and the Solomon Islands (Fig. 36–39).

KEY TO THE VIETNAMESE SPECIES OF PAPHIOPEDILUM

1. Lip inflated, sub-spherical to ovoid, calceolate with an involuted upper apical margin; dorsal sepal and synsepal more or less similar to petals in colour and shape, white, yellowish, dull yellow-greenish or pink, more or less of one colour; petals broadly-ovate to circular, less than twice as long as broad, very rarely petals narrowly-ovate (in this case petals and lip hooked at apex) **2**

Lip goblet-like, cylindrical, with incurved side-lobes and apical margin of the pouch straight, not involuted; dorsal sepal and synsepal distinctly differ from petals in colour and shape, petals much narrower than dorsal sepal, dorsal sepal and petals commonly variously coloured, often contrastly striped and spotted; petals tapering, narrowly-oblong or spathulate, more than twice as long as broad, never hooked at the apex **(*P.* subgen. *Paphiopedilum*)**. **8**

2. Dorsal sepal and synsepal distinctly smaller than lateral petals; lip inflated, more or less spherical, thin-textured, lightly grooved along the veins; staminode large, broad at apex, flat or longitudinally conduplicate, sometimes with involute side margins; flowers white, pale yellowish-green or pink, sometimes with brown or purple stripes along nerves; leaves tessellated or uniformly green; plant of calcareous or acidic silicate substrates **(*P.* subgen. *Parvisepalum*)** **3**

Dorsal sepal and synsepal more or less similar with petals size; lip is ovoid, rather thick in texture, plain on surface; staminode middle sized, distinctly 3- or 5-dentate at the apex, with a large narrow central tooth; flowers pale yellowish-white, sepals and petals usually with small purple spots; leaves markedly tessellated; plant of calcareous substrates **(*P.* subgen. *Brachypetalum*)** . . . . **7. *P. concolor***

3. Leaves distinctly tessellated above and heavily spotted or uniformly tinged with dirty-purple-violet beneath, 10–15(20) cm long, 1.5–4 cm wide; flowers pink, pink-purple or pale green; lip longer or hardly shorter than lateral petals; underground stolons may be present or not; plant of calcareous or acidic silicate substrates **(*P.* sect. *Parvisepalum*)** **4**

Leaves uniformly green on both sides, rare with few small purple-violet marks beneath near the base, (15)20–35 cm long, 4–6 cm wide; flowers white or pale yellowish-green; lip distinctly smaller than lateral petals; plants of calcareous substrates, never developing stolons **(*P.* sect. *Emersonianum*)** **7**

4. Staminode flat, with a broad apex; sepals and petals uniformly light pink to pink-purple; plants of calcareous or acidic silicate substrates, without stolons **5**

Staminode longitudinally conduplicate or with a deep fold along midvein at the apex; sepals apple-green or yellowish-pink with purple stripes along veins; plants of calcareous substrates, with more or less elongate underground stolons **6**

5. Flowers 7.5–10 cm across; sepals less than 4.5 cm long, subcircular to broadly-ovate, spreading; staminode with bright yellow centre and central purple mark; leaves narrowly-oblong to oblong, 8–11 cm long, 3–3.5(4) cm wide, tessellated more or less contrastingly deep-green/whitish-green; plant of acidic substrates . . . . **1. *P. delenatii***

Flowers 10–12 cm across; sepals longer than 4.5 cm, broadly-oblong to broadly-ovate, down directed; staminode with a dull yellow or light olive-green centre; leaves ovate to oblong-elliptic, (8)11–20 cm long, (2.5)4–7 cm wide, tessellated glossy deep-green/silver-gray; plant of calcareous substrates . . . . **2. *P. vietnamense***

6. Flowers (8)9–12.2 cm across, pale yellowish-green to apple-green, dorsal sepal and petals with brown-purple stripes; dorsal sepal cuspidate-acuminate at apex, more than 4 cm long; petals ovate to narrowly-ovate, subacute at apex, longer than 4 cm; staminode more than 12.5 mm long, broadly ovate, convex at the base, apically with a deep fold along the midvein and involute side margins . . . . **3. *P. malipoense***

Flowers 6–8(8.7) cm across, pink to pink-purple dorsal sepal and petals with purple striped along veins; dorsal sepal obtuse or shortly acute at apex, less than 4 cm long; petals broadly obovate to subcircular, rounded at apex, less than 4 cm long; staminode less than 12.5 mm long, elliptic, longitudinally conduplicate . . . . **4. *P. micranthum***

7(3). Flowers commonly subcampanulate, sepals and petals light yellowish-green with purple-brown flushing at the base; lip yellowish-white; staminode broadly triangular, more than 17 mm wide, convex, broadly-rounded at apex, white or yellowish-white with purple anastomosing markings . . . . **5. *P. hangianum***

Flower commonly spreading, sepals and petals white, sometimes faintly flushed with purple at base; lip yellow to pink-orange, with a pink flush around the rim; staminode narrowly triangular, less than 12 mm wide, longitudinally sulcate, with a deep groove along midvein and a narrow apex, bright yellow with reddish markings and a white margin. . . . . **6. *P. emersonii***

8(1). Inflorescence one-flowered, very rarely with 2 flowers; petals narrowly-obovate to oblong-spathulate, more or less flat, undulate at margin but never spirally twisted, less than 8(8.5) cm long, distinctly less than six times longer than broad; medium-sized or small plants with more or less herbaceous leaves commonly less than 20 cm long, very rarely longer. **9**

Inflorescence multi-flowered, with (1)2–5 flowers; petals linear-tapering, strongly spirally twisted, much narrower than median sepal, 8–11 cm long, more than ten times longer than broad; large plants with coriaceous leaves 20–50 cm long, 2–5 cm wide **(*P.* sect. *Pardalopetalum*)** . . . . **18. *P. dianthum***

9. Leaves uniformly green, sometimes with deep green venation; often with purple-violet marks beneath near the base; staminode obovate, obcordate, subcircular or subquadrate with a central umbo or broad convex callus; petals without maroon-violet warts; lip with plain smooth surface of side-lobes **(*P.* sect. *Paphiopedilum*)** **10**

Leaves distinctly tessellated deep-green/light-green, without purple-violet marks beneath; staminode commonly lunate, with falcate side-lobes and a small central tooth, rarely broadly obcordate, 3- or 5-dentate at the apex, without a distinct central boss; petals with more or less prominent small maroon-violet warts along margin and near the base; lip with warted surface of side-lobes **(*P.* sect. *Barbata*)** **16**

10. Staminode subquadrate, smooth, glossy, with a broad central swelling, lacking a glossy umbo; petals purple in apical half. . . . . **8. *P. hirsutissimum***

Staminode obovate, obcordate or semicircular, rugulose, with a central glossy umbo; petals orange, yellow, yellow with a pink tint, yellowish-brown or light brown **11**

11. Flowers small or medium-sized, less than 6.5 cm across; dorsal sepal shorter than 4 cm; petals less than 4 cm long **12**

Flowers large, more than (7)7.5 cm across; dorsal sepal longer than 4.5 cm; petals more than 4.5 cm long **15**

12. Plants small with coriaceous, rigid, semi-succulent leaves 3–7(12) cm long; flowers bright yellow to dull yellowish-pink with an ochre-yellow to dull yellowish-pink lip; petals with straight margins, commonly ochre-yellow with orange veins . . . . **12. *P. helenae***

Plants medium-sized with more or less herbaceous leaves more than 7 cm long; flowers not coloured as above; petals with more or less undulate margins, light yellowish-brown or dull yellowish with a pink tint or brown veins **13**

13. Petals oblong, with strongly undulate margins, light yellowish-green with chocolate-brown veins and flushing; lip chocolate-brown with darker veins; dorsal sepal white with a greenish base and chocolate-brown veins towards the base . . . . **10. *P. tranlienianum***

Petals spathulate or narrowly obovate, with slightly undulate margins, light yellow-brown or dull yellowish with pink tint and large maroon spots; lip pink-purple or purple-brown; dorsal sepal white with a purple centre or dull yellowish with maroon spots **14**

14. Dorsal sepal subcircular to obovate, with reflexed margins at the base, white with a purple or brownish-purple centre; petals narrowly-spathulate, 0.8–1 cm wide, dull yellowish-brown with pink-purple tint; lip yellowish-brown to purple-brown . . . . **9. *P. barbigerum***

Dorsal sepal broadly-ovate to ovate, more or less flat, with flat margins at the base, light dull yellowish-green, heavily spotted with large maroon spots; petals broadly-oblong to narrowly-obovate 1.4–1.6 cm wide, dull yellowish, flushed with pink-purple, spotted with numerous large maroon spots; lip pink-purple . . . . **12. *P. henryanum***

15(11). Dorsal sepal 4.8–5.2 cm long, white, sometimes with light brownish or green at the base, more or less heavily spotted with brown-purple; petals spathulate, usually less than 5.5 cm long, 1.8–2.5 cm wide; peduncle commonly erect, shortly densely hairy with short purple-violet hairs; lithophytic plant . . . . . . **13. *P. gratrixianum***

Dorsal sepal (4.5)5–7 cm long, greenish with a white margin and glossy maroon central area, rare maroon spotted; petals obovate-spathulate, usually more than 5.5 cm long, 2.5–4.6 cm wide; peduncle commonly arching, densely villosely hairy with long whitish hairs; epiphytic plant, very rarely a lithophyte . . . . . **14. *P. villosum***

16(9). Dorsal sepal pale green with darker venation, forming a hood arching over the lip; staminode broadly cordate with 1–3 obtuse apical teeth or lunate with broad falcate side-lobes, 3- or 5-denticulate at the apex; petals spathulate, broadening toward apex; lip often with 2–6 obtuse teeth on front margin . . . . . **15. *P. appletonianum***

Dorsal sepal erect, more or less flat, with reflexed margins at the base, white with purple or green veins; staminode lunate with narrow falcate side-lobes and 1–3 small median teeth; petals oblong, sigmoid or elliptic; lip front margin without obtuse teeth **17**

17. Petals sigmoid, down-curved, green with a pink-purple apex and prominent deep maroon hairy warts along the upper margin; dorsal sepal subcircular to broadly-elliptic, white, green/purple-striped; lip pink-brown to dull purple; leaves commonly elliptic to obovate, 10–20 cm long . . . . **16. *P. callosum***

Petals elliptic, horizontally spreading, deep-purple, finely purple-violet spotted in lower part, without hairy warts along the upper margin; dorsal sepal broadly-ovate, white, deep purple-striped; lip brown-purple to deep purple-violet; leaves commonly oblong to elliptic, 7–14 cm long . . . . **17. *P. purpuratum***

Paphiopedilum subgen. Parvisepalum

Paphiopedilum subgen. **Parvisepalum** *Karasawa & Saito* in Bull. Hiroshima Bot. Gard. 5: 31 (1982). **Type:** *Paphiopedilum delenatii* Guillaum.

DESCRIPTION. *Plants* with or without long rhizomes. *Leaves* usually tessellated, rarely plain green. *Median sepal* and *synsepal* distinctly smaller than lateral petals. *Petals* broadly elliptic to subcircular. *Lip* inflated, more or less spherical, thin-textured, calceolate with involuted upper apical margin and lightly grooved along the veins. *Column* elongated, stigma and anthers long-stalked. *Staminode* large, varying in shape from cordate or trullate and deeply longitudinally grooved to circular, convex in most species but longitudinally conduplicate in *P. micranthum*. *Pollen masses* powdery or viscid.

Seven species distributed in southern China and Vietnam (Fig. 36).

Paphiopedilum sect. Parvisepalum

Paphiopedilum sect. **Parvisepalum:** *Cribb*, Gen. Paphiopedilum: 86 (1987).
Paphiopedilum subgen. *Parvisepalum* Karasawa & Saito in Bull. Hiroshima Bot. Gard. 5: 31, p.p. (1982).

DESCRIPTION. *Plants* often develop stolons. *Leaves* distinctly tessellated with dark green and pale grayish blotches above and heavily spotted or uniformly tinged with dirty-purple-violet beneath, usually 10–15 cm long (rarely longer), 1.5–4 cm wide. *Inflorescence* 1–2 flowered. *Flowers* pink, pink-purple, bright yellow or pale green. *Lip* larger or scarcely shorter than lateral petals.

5 species distributed in southern China and Vietnam (Fig. 36).

1. PAPHIOPEDILUM DELENATII Guillaum.

Paphiopedilum delenatii *Guillaum.* in Bull. Soc. Bot. France 71: 554, 558 (1924); id., Ann. Soc. Hort. Paris 1925: 127 (1925); id., Fl. Gen. Indo-Chine 6: 640 (1934); Cribb, Gen. Paphiopedilum: 89 (1987); id., Gen. Paphiopedilum, ed. 2: 77 (1998); Braem, Paphiopedilum: 31 (1988); Seidenfaden, Orch. Indochina: 16 (1992); Pham Hoang Ho, Ill. Fl. Viet. 3, 2: 946 (1993); Averyanov, Identif. Guide. Vietnam. Orch.: 16 (1994); Braem, C. Baker & M. Baker, Gen. Paphiopedilum 1: 70(1998). **Lectotype:** Tonkin, *Delenat*, cult. Mornay, Fleuriste de la ville de Paris (P!), selected here.
Cypripedium delenatii (Guillaum.) C.H.Curtis in Gard. Chron. ser. 3, 89: 208 (1931).

DESCRIPTION. *Lithophytic* or *terrestrial herb* with 5–7 distichous leaves. *Leaves* elliptic to oblong-elliptic, obtuse and minutely tridentate at apex, up to 11 cm long, 3–3.9 cm wide, margins ciliate near base, mottled dark and pale light green above, heavily purple-spotted below. *Inflorescence* 1–2, very rarely 3-flowered; peduncle up to 22 cm long, green, purple-spotted, white-hirsute; flower bracts elliptic to ovate, 1.2–1.5 cm long, 1 cm wide, green, spotted with purple, shortly pubescent; pedicel and ovary up to 5.5 cm long, green, spotted with purple, hirsute. *Flowers* 7.5–8 cm in diameter, pale pink with pink or purple-pink lip, red and yellow markings on staminode, pubescent both within and without. *Dorsal sepal* ovate, obtuse to subacute, 1.7–3.5 cm long, 1.8–2.5 cm wide. *Synsepal* similar, 1.9–3 cm long, cm wide. *Lip* ellipsoidal to subglobose, 2.5–3.8 cm long, 2.5–3 cm wide, margins incurved, minutely pubescent. *Staminode* somewhat convex, ovate, obtuse, 14–17 mm long, 13–16 mm wide, ciliate. 2n = 26. Fig. 40.

DISTRIBUTION. Southern Vietnam (Dac Lac and Khanh Hoa Provinces, E slopes of Hon Giao ridge and Bi Dup mountain system). Figs. 36, 41.

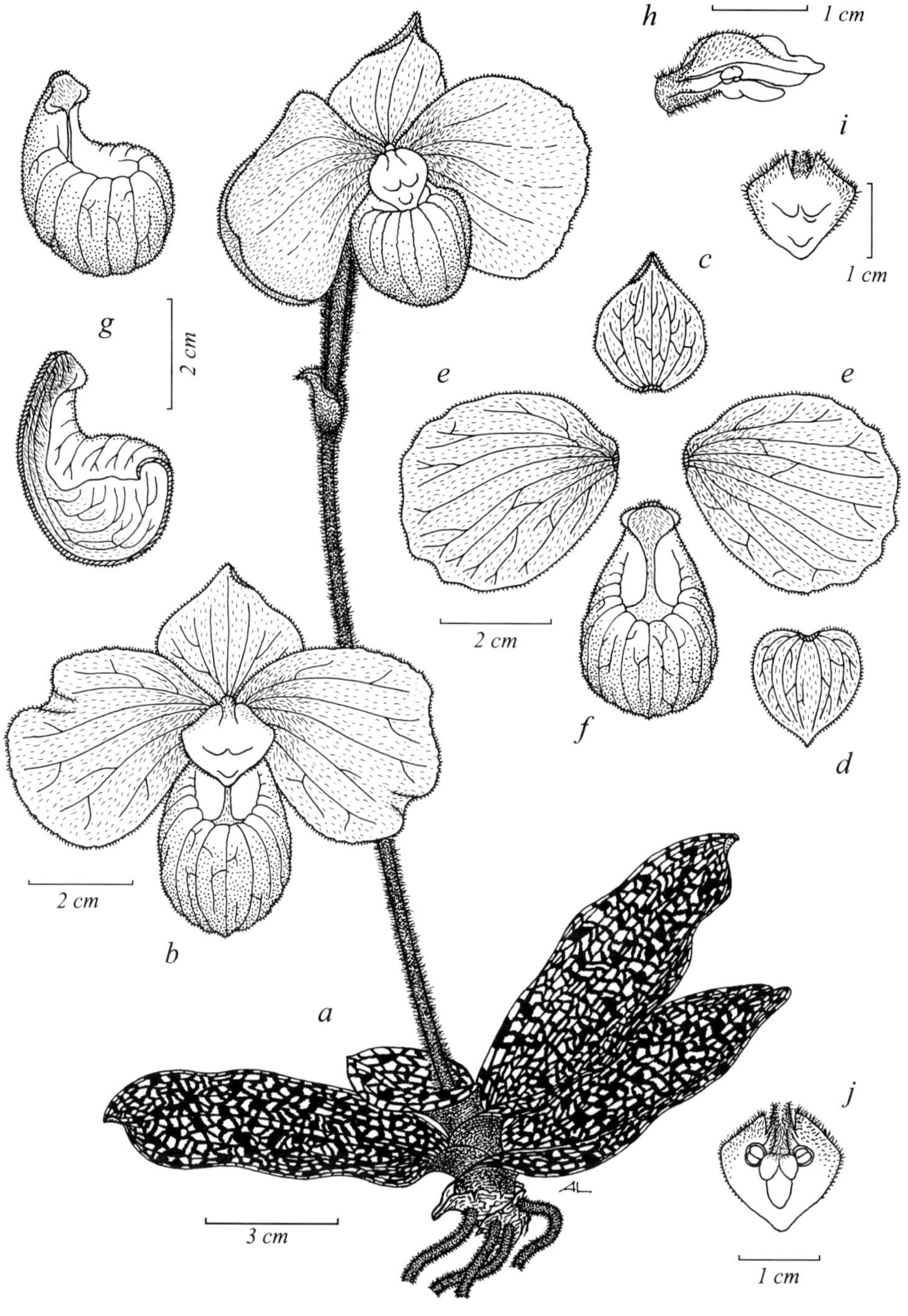

Fig. 40. *Paphiopedilum delenatii*. (Drawing: L. Averyanov).

a — flowering plant; **b** — flower; **c** — median sepal; **d** — synsepal; **e** — petals; **f, g** — lip, frontal, side and section views; **h–j** — column, ventral, side and dorsal views (Drawn from *Averyanov & Hiep* VH s.n., December 1995).

ECOLOGY. Primary, broad-leaved, evergreen forests on granite and gneiss along bluffs of montane river canyons at elevations of 750–1300 m. Fig. 42–44.

FLOWERING IN THE WILD. December.

IUCN STATUS. CR.

AFFINITIES. *Paphiopedilum delenatii* is one of the most geographically restricted Vietnamese endemics with a distribution area certainly of less than 100 square km. Its closest allies are to be found in southern China and northern Vietnam. In particular, *P. armeniacum, P. micranthum* and *P. vietnamense*, are closely related and vegetatively quite similar. *Paphiopedilum delenatii* is distinguished from them by its rose-pink flowers and distinctive ovate, convex, deep pink-purple staminode with a yolk-yellow central mark. The ecological preference of this species for silicate rocks stands in strong contrast to the substrate requirements of its closest relatives which occur exclusively on marble-like crystalline limestone.

HISTORY. The discovery of *Paphiopedilum delenatii* was attributed to a French army officer who bought to France in 1913 or 1914 pink-flowered *Paphiopedilum* specimens from unknown locality of northern Vietnam. It was reported that these plants were sent to France for cultivation in nurseries of M. Delenat and later by M. Mornay (Guillaumin, 1924). However, most probably they never flowered in France and died soon after importation (it is difficult to accept that these plants survived 10 or 11 years before flowering). No undoubted herbarium documentation has yet been found for these plants. It is possible that this first report of a pink-flowered slipper orchid from northern Vietnam was based on *Paphiopedilum vietnamense* or *P. micranthum*, both discovered here almost 85 years later. Anyway the report of *Paphiopedilum delenatii* by Guillaumin (1934) as a species found in northern Vietnam (Tonkin) is almost certainly mistaken.

Paphiopedilum delenatii was certainly discovered by the French explorer Poilane in 1922 near Nha Trang city of southern Vietnam. Without doubt some plants of this exciting discovery were exported to France shortly afterwards. Some of them were cultivated in France by M.Delenat and later by M.Mornay, the head gardener of the *"Fleuriste Municipal de la Ville de Paris"*. These plants flowered in 1924 and one was used as the basis of the description of the new species. The following year Mornay's plant received a *"Certificate de Merité avec felicitations"* from the National Society of France.

Until the mid 1990s all the plants in cultivation were derived from one of the few early introductions into France. Most of these proved difficult to grow and quickly died, but a plant in the collection of Messrs. Vacherot & Lecoufle, the eminent French nursery, survived, bore fruit and plants were successfully grown from the resulting seed. Many generations later, *P. delenatii* is relatively common in cultivation and now has a reputation of being fairly easy to grow. Koopowitz (pers. comm.) has suggested that successive inbreeding has eliminated lethal recessive genes to produce this apparent character change. What is certain, however, is that the story of *P. delenatii* in cultivation must rank as one of the finest conservation achievements by orchid growers and an example that should be emulated for other rarities.

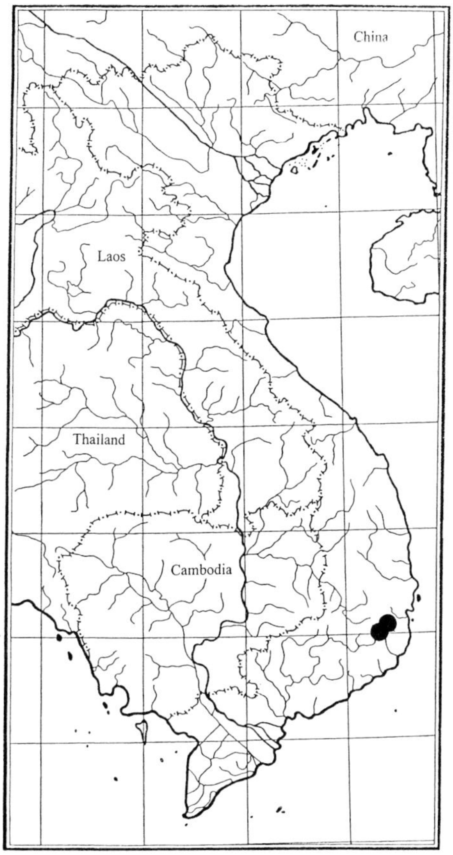

Fig. 41. Distribution of *Paphiopedilum delenatii*.
(Drawing: L. Averyanov).

Fig. 42 (above left). Primary broad-leaved evergreen forest on granite, east-facing cliffs at 1100–1200 m a.s.l. is the home of the remaining populations of *Paphiopedilum delenatii*. (Photo.: L. Averyanov).

Fig. 43 (above right). Typical habitat of *Paphiopedilum delenatii* —shelves on bluffs of canyons above streams. S. Vietnam, Khanh Hoa Prov. (Photo.: L. Averyanov).

Fig. 44. *Paphiopedilum delenatii* in its natural habitat.

a — Flowering size sample growing as a terrestrial herb among rocks on thin sandy soil; **b** — in intact habitats numerous seedlings of *P. delenatii* may be seen on shady vertical granite bluffs of mountain streams. Southern Vietnam, Khanh Hoa Prov., *Averyanov & Hiep* VH s.n., April 1995. (Photo.: L. Averyanov).

Many people searched for natural populations of *Paphiopedilum delenatii* after the war in Vietnam. However, nothing was found for a very long time. The landscapes around Nha Trang city were extensively deforested and as a result it was suggested that this attractive species was already extinct in the nature.

The rediscovery of *Paphiopedilum delenatii* in the early 1990s has been widely reported and documented in detail (Averyanov *et al.*, 1996b). The first inkling that it had been rediscovered came in 1992 when wild-collected plants began to appear in some quantity in trade in Taiwan and rapidly afterwards in Japan, Europe and the U.S.A. A few traders in Dalat city (southern Vietnam) had been in the habit of collecting *Cymbidium aloifolium* and *C. bicolor* for export to Taiwan where they were used for medicinal purposes. Knowing of the interest of their trading partners in unusual orchids they would include the occasional novelty in consignments. In 1990 or 1991 one of these happened to be *P. delenatii*. It had been rediscovered by local people in the Hon Giao Mountains in Khanh Hoa Province, 40–50 km west of the city of Nha Trang. It was suggested that as much as six tons of the orchid were collected for export by 1995 (Quang Dau, 1994), bought from local villagers in small lots for which they were paid US $1–3 per kg (Averyanov *et al.*,

Fig. 45. Flowering specimen of *Paphiopedilum delenatii.*

Paphiopedilum delenatii flowers in nature during the last weeks of rainy season and its flowers very are often damaged by heavy torrential rains. Southern Vietnam, Khanh Hoa Prov., *Averyanov & Hiep* VH s.n., December 1995. (Photo.: L. Averyanov).

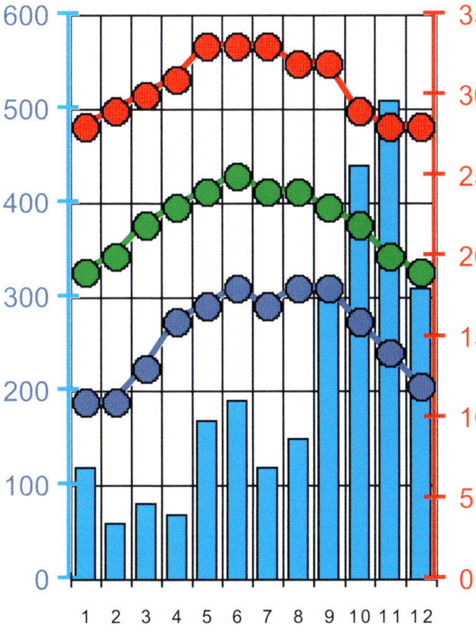

Fig. 46. Climatological data expected for the *Paphiopedilum delenatii* area based on extrapolation from data from the nearest weather stations.

Black figures at the chart bottom are months. Blue represents atmosphere precipitation in mm (columns designate mean monthly rainfalls); red, green and blue represent temperature in °C (red line — absolute maximuml monthly temperatures, green line — mean monthly temperatures and blue line means absolute minimum monthly temperatures).

Fig. 47. Colour variation of *Paphiopedilum delenatii*.

a, b — commonest form with light pink flowers, southern Vietnam, Khanh Hoa Prov., December 1995 (Photos.: L. Averyanov); **c** — flower with an unusually deep rose-purple lip, southern Vietnam, Khanh Hoa Prov., December 1995 (Photo.: L. Averyanov); **d** — albino form, cult. private collection, Germany (Photo.: P.J. Cribb).

1996b). Such quantities cannot but have made the orchid rare in the wild and prices had risen to US $10 per kg by 1995.

HABITAT. Why had this beautiful orchid not been rediscovered earlier? The answer undoubtedly lies in the habitat where it has now been found. Everyone assumed that, like its close allies, *P. delenatii* grew on limestone but it has now been shown to grow on acidic soils on granite and gneiss mountains, a habitat that has surprised slipper orchid experts. Rediscovery of wild populations of *P. delenatii* and a detailed description of its habitat was made possible through extensive field explorations supported by the U.S.A. National Geographic and the American Orchid Society (Averyanov *et al.*, 1996b). In 1995 Averyanov and N.T.Hiep found wild-growing specimens of *P. delenatii* at between 750 and 900 m elevation on very steep south-east facing slopes and cliffs of eroded granite mountains in the canyon system of the Lieng Ly

River along the western border of Khanh Hoa Province. Commonly plants grow in shade in mossy crevices on vertical cliffs or in pockets and depressions on shelves on steep granite slopes under montane evergreen broad-leaved forest above rivers and streams (Fig. 44, 45). The roots of the orchid are either in contact with the rock or in thin sandy soil with very little humus and leaf litter. The forest comprises *Barringtonia* sp., *Pandanus* sp., *Freycinetia* sp. and gymnosperms such as *Podocarpus neriifolius, Dacrycarpus imbricatus* and *Gnetum gnemon*, with an undergrowth of terrestrial ferns, including *Humata repens, Schizaea digitata* and *Gleichenia* sp. The habitat is rich in other orchids, such as *Acanthephippium striatum, Appendicula hexandra, Calanthe lyroglossa, Eria tomentosa, E. pannea, E. muscicola, Paphiopedilum appletonianum, P. villosum, Podochilus microphyllus, Tainia hookeriana, Collabium* sp., and several species of *Dendrobium* and *Bulbophyllum*. The climatic conditions in the area may be deduced by extrapolation of data from the two nearest available weather stations — Nha Trang and Hon Ba which are situated at elevations 5 m and 1484 m respectively (Tables. 6, 7).

The climate is warm tropical with two distinct seasons, a wet rainy period from June until December and a drier, sometimes very dry season from January until May (Fig. 46). During the dry season the monthly rainfall ranges varies from 50 to 175 mm, but in the wet period reaches 300–500 mm. Mean monthly temperatures in summer are about 23–24°C and in winter fall to 19–20°C. Maximum temperatures, which may reach 33°C, are observed during May-August. The coldest weather is between December-February with minimum expected temperatures about 11–12°C. These conditions are to be expected at 750–850 m but the temperatures are certainly lower and the rainfall higher where *P. delenatii* is found. The annual precipitation there may be about 2500–3000 mm.

In the wild *Paphiopedilum delenatii* commonly flowers in December during the rainy season. At this time, its flowers are very often damaged by heavy torrential rains that seriously decrease seed production (Fig. 45).

TABLE 6

CLIMATOLOGICAL DATA FOR NHA TRANG WEATHER STATION (12°15'N 109°12'E, ELEV. 5 m) ACCORDING TO BRUZON & CARTON (1930) AND NGUYEN KHANH VAN *ET AL.* (2000).

Month	01	02	03	04	05	06	07	08	09	10	11	12
1	31.9	32.6	34.2	35.9	38.5	39.5	39.0	38.1	38.3	34.8	33.1	32.8
2	23.9	24.5	25.7	27.3	28.3	28.4	28.3	28.2	27.5	26.4	25.0	23.9
3	15.4	14.6	16.4	19.4	19.7	19.8	20.6	21.5	21.5	19.7	17.4	16.1
4	45	18	30	36	64	49	39	53	163	312	356	160

TABLE 7

CLIMATOLOGICAL DATA FOR HON BA WEATHER STATION (12°05'N 108°45'E, ELEV. 1484 m). ACCORDING TO BRUZON & CARTON (1930) AND NGUYEN KHANH VAN *ET AL.* (2000).

Month	01	02	03	04	05	06	07	08	09	10	11	12
1	22.8	24.7	25.7	26.8	27.1	25.5	26.4	26.1	24.6	23.3	21.8	22.2
2	13.9	15.0	16.6	18.5	19.6	19.9	19.6	19.7	18.8	17.0	15.5	14.4
3	6.5	7.3	9.9	12.3	14.5	15.1	13.9	14.1	14.1	11.6	9.6	7.4
4	200	91	121	109	283	338	206	247	452	575	666	460

Key for Tables 6, 7:

1 — absolute maximum temperature (all temperatures in °C); 2 — mean temperature; 3 — absolute minimum temperatures; 4 — mean monthly atmosphere precipitation (in mm)

Usually, *Paphiopedilum delenatii* grows in country that is very difficult to explore. High vertical granite cliffs, very deep narrow canyons, numerous waterfalls and stream bluffs are typical elements of landscape here. This area also has an abundance of leeches and is a centre some of the most virulent forms of tropical malaria. Most accessible existing populations are nearly exhausted but some may still survive in less accessible places, despite the fact that collection from the wild continues. Due to its very small area of distribution and continuing commercial collecting, this species is certainly threatened with extinction in the wild. According to the latest IUCN criteria it can be regarded as a critically endangered (CR).

VARIATION. Although light pink quite uniform flowers (Fig. 47a) predominate in the wild populations of *Paphiopedilum delenatii*, some variation in depth of flower colour is nevertheless observed (Averyanov *et al.*, 1996b). Individual plants in some populations have a very dark pink-purple lip, usually slightly shorter than usual (Fig. 47b). Albino forms absolutely lacking anthocyanin pigments are very rarely found in the wild. These plants have pure white flowers with a light yellow-brown marking in the centre of the staminode (Fig. 47c) and light green leaves lacking purple markings. The new introductions into cultivation vary also in the shape and colour of their leaves. The morphological variation observed in recently imported samples is an important factor in stimulating the international demands for wild-collected plants of this species.

2. PAPHIOPEDILUM VIETNAMENSE Gruss & Perner

Paphiopedilum vietnamense Gruss & Perner in Die Orchidee Beihefte 5: 3 (11 Jan. 1999). **Type:** Vietnam, Cao Bang Province, 1000 m, November 1998, Dec. 1998, cult. O. Gruss s.n. (holotype K, photo. LE).

Paphiopedilum hilmari Senghas & Schettler in J. Orchideenfr. 6, 1: 4 (19 Jan. 1999). **Type:** N.Vietnam, Ha Tuyen Province, 1200 m, cult. Doll Orch–909 (holotype HEID).

Paphiopedilum mirabile Cavestro & Chiron in Orch. Cult. Protect., 32, 2: 32 (19 Feb.1999). **Type:** NE Vietnam, Tu Yen Quang Province, 700 m, cult. Bandung, Jan. 1999, *A. Parnata* s.n. (holotype P).

DESCRIPTION. *Lithophytic* and occasionally *terrestrial herb* with 3–5 distichous leaves. *Leaves* coriaceous-leathery, ovate to oblong-elliptic, obtuse at unequally bilobed apex, 8–17(20) cm long, 2.5–7 cm wide, glossy greyish green and dark green marbled, sometimes with a pronounced white margin, spotted or flushed with purple beneath, strongly keeled on lower surface, glabrous or sparsely white ciliated along margin at the base. *Inflorescence* erect, 1(2)-flowered; peduncle 15–25 cm long, 4 mm in diameter, purple-violet or purple-brown, densely white or light-brown hairy; bract broadly-ovate, acute, conduplicate, 1.8–3 cm long, 1.6–2 cm wide, dull pale green finely spotted with purple. *Flower* very large for size of plant, 10–12 cm across, 8–9 cm high; sepals white or pink with pink-purple flushing towards the tips; petals white, pink or purple strongly flushed purple in apical two-thirds to three-quarters; lip white, pink or purple with a deep-purple stain on front especially around the rim, spotted with deep purple within; staminode large, lemon-yellow, white on margin with an olive-green centre; pedicel and ovary 2.5–2.8 cm long, pale yellowish green with fine purple spotting, densely white-hairy. *Dorsal sepal* erect, slightly recurved on lateral margins, elliptic-obovate, acute and canaliculate towards apex, 4.2–5.1 cm long, 2.5–3 cm wide, white pubescent on outer surface, keeled on outer side. *Synsepal* elliptic-ovate, bidentate, 3–4.6 cm long, 3–3.3 cm wide, pubescent on outer surface, margin with white hairs, keeled on outer side. *Petals* incurved or spreading-deflexed, oblong-elliptic to obovate, obtuse to rounded at apex, 5–6 cm long, 3.3–4.5 cm wide, finely pubescent on both sides, white-villose at base, lateral lower margin sometimes strongly recurved. *Lip* subporrect, deeply urceolate, subglobose, 3.5–4 cm long, 2.5–4 cm wide, flared at base, with apical margin incurved and grooved along veins. *Column* short, 5–7 mm long; staminode convex, broadly rhombic or broadly sagittate, blunt, truncate, white-ciliate at margin, 17–25 mm long, 15–20 mm wide, curving forwards towards apex; stigma spathulate, very shortly papillose; anthers with dry pollen. Figs. 48, 55a,b.

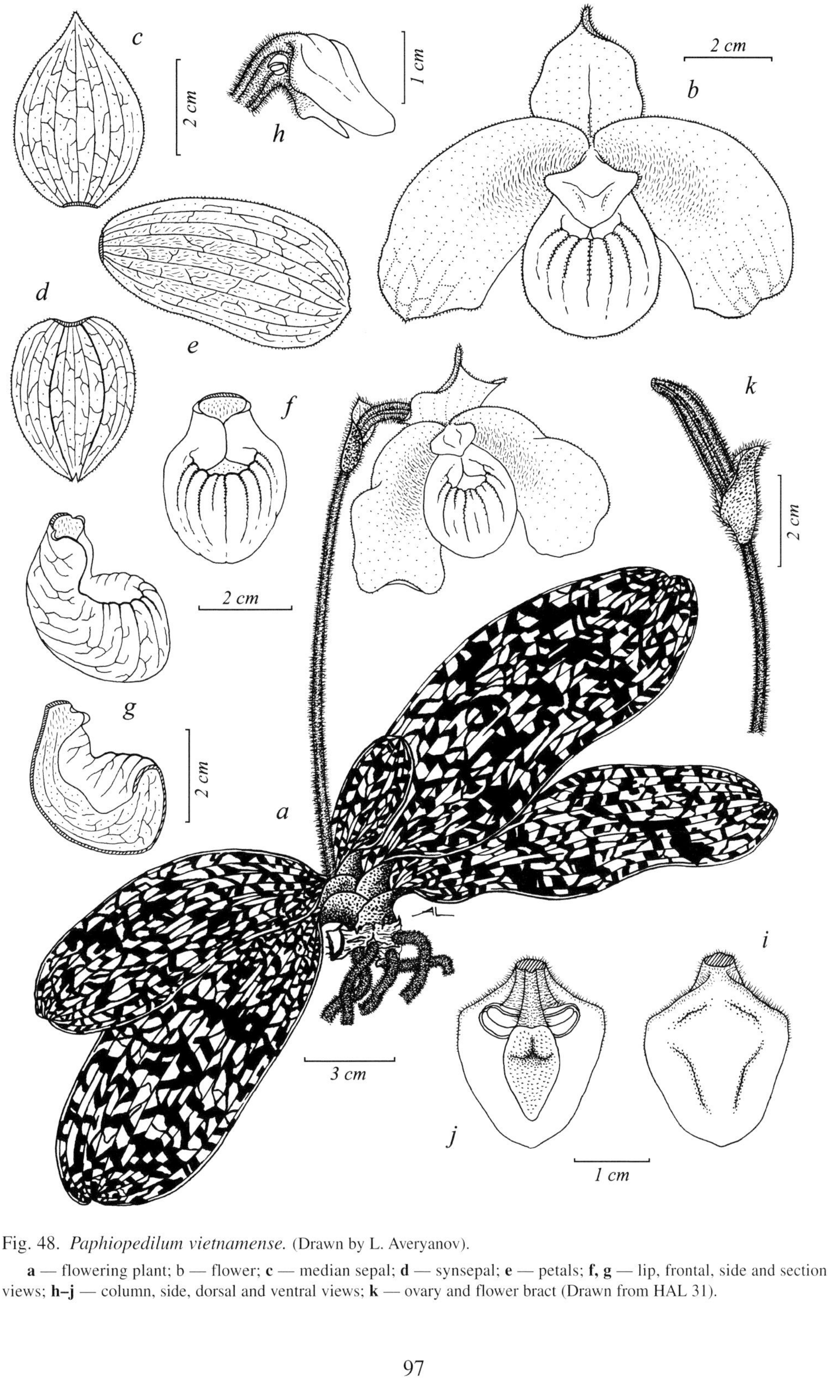

Fig. 48. *Paphiopedilum vietnamense.* (Drawn by L. Averyanov).

a — flowering plant; b — flower; **c** — median sepal; **d** — synsepal; **e** — petals; **f, g** — lip, frontal, side and section views; **h–j** — column, side, dorsal and ventral views; **k** — ovary and flower bract (Drawn from HAL 31).

DISTRIBUTION. Northern Vietnam (Thai Nguyen Province). Figs. 36, 49.

ECOLOGY. Primary, broad-leaved, evergreen, wet, closed forests on highly eroded crystalline limestone mountains, 350–550 m. Fig. 50.

FLOWERING IN THE WILD. March.

IUCN STATUS. EW.

INTRODUCTION. *Paphiopedilum vietnamense* is one of the most spectacular slipper orchid with tessellated leaves and a very large rose-pink or purple flower. It has obvious close affinities with *P. delenatii* which inhabits a very restricted mountain area in the southern Vietnam. Besides geographical isolation these species differ distinctly from each other in the form of their tepals and in their leaves. *Paphiopedilum vietnamense* probably had smallest range of all Vietnamese endemics. The history of the discovery and subsequent predation of this outstanding species in Vietnam is a sad example of the speed at which illegal international dealers can plunder natural plant resources.

HISTORY. The discovery of *Paphiopedilum vietnamense* was connected with the activity of a joint Japanese-Vietnamese trading company, based near Hanoi. In June 1997, following its establishment, this corporation organized the wide-scale collecting of Vietnamese *Paphiopedilum* species by engaging local villagers all over northern Vietnam. The first specimens of *P. vietnamense* were found among huge lots of wild-collected *Paphiopedilum* plants delivered to the company's nursery in May-June 1998 (Tanaka, 1998). A month later many more specimens appeared on the international black market and were illegally exported by orchid dealers to a number of countries. From observations of plants in the nursery and of related nurseries in Taiwan, Europe, Japan and America it seems likely that many thousands of plants of *P. vietnamense* have been collected since its discovery.

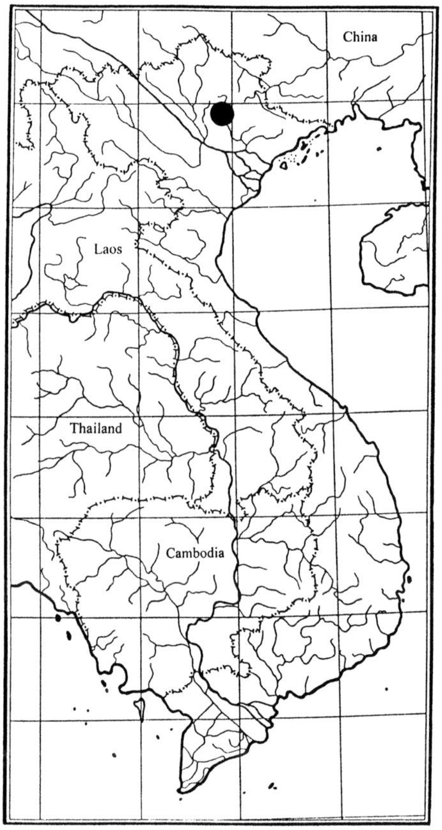

Fig. 49. Distribution of *Paphiopedilum vietnamense*. (Drawn by L. Averyanov).

The first flowering of *Paphiopedilum vietnamense* in private collections outside Vietnam brought the cascade of descriptions of this species. Almost simultaneously several names, *P. vietnamense* (Gruss & Perner, 1999a), *P. hilmari* (Senghas & Schettler, 1999a) and *P. mirabile* (Cavestro & Chiron, 1999), were published. The name *P. vietnamense* has priority over the others by a week and a month respectively. All these names were based on wild-collected plants that had been illegally exported from Vietnam. All specimens of *P. vietnamense* in international trade have undoubtedly originated from a single source (Averyanov *et al.*, 2000). However, the provenance information that accompanied exported plants is doubtful. Shrewd local collectors and dealers involved in the illegal orchid trade have good reasons for keeping secret any information on their sources (Senghas & Schettler, 1999b). According to the authors of the names cited above and others, their plants supposedly originated, respectively, from the rocky limestone areas of Cao Bang, Ha Giang or the northern part of Tuyen Quang province in Vietnam (Gruss & Perner, 1999a; Gruss, 2000; Tanaka, 1998, 1999; Senghas & Schettler, 1999a,b; Cavestro & Chiron, 1999).

Large exports of *Paphiopedilum vietnamense* are clear evidence that this rare Vietnamese endemic stands on the verge of extinction in nature because of commercial over-collecting. Funded by "Flora and Fauna International" and U.S.A. National Science Foundations, extensive exploration throughout northern Vietnam was undertaken in 2001 by Averyanov and Phan Ke Loc to discover wild populations of *P. vietnamense*. This species was found to be limited to a very restricted area of small low limestone ridges in the centre of Thai Nguyen Province. The primary range of the species does not exceeded 60–80 square km. This is a very good example of the restricted local endemism of some species in the Vietnamese flora.

The oldest large lead-mining enterprise in the country is situated close to where *Paphiopedilum vietnamense* grows. The mines were exploited by the French for a many years. This region was easily accessible and a striking species such as *P. vietnamense* would certainly have been known to the French. It is quite possible that the first mention of a pink-flowered slipper orchid from northern Vietnam was based on this species (Guillaumin, 1924).

Numerous discussions with local villagers in the area revealed a clear picture of the collection of *P. vietnamense* in recent years. Rapid uncontrolled collecting of *Paphiopedilum vietnamense* by local villagers engaged by dealers began a few days after its discovery. According to local collectors, *P. vietnamense* was locally very common and abundant in many places before collecting began. It formed large colonies with numerous flowering size specimens that were easily accessible from

Fig. 50. Primary wet broad-leaved evergreen forests on low highly eroded limestone ridges and mesae supported large populations of *Paphiopedilum vietnamense* only few years ago. One of the first places of mass commercial collecting of the species.
N. Vietnam, Thai Nguyen Prov. (Photo.: L. Averyanov).

Fig. 51. Seedling of *Paphiopedilum vietnamense* in its natural habitat.
N. Vietnam, Thai Nguyen Prov. (Photo.: L. Averyanov).

nearby local villages, roads and trails. During 1998, large quantities of *P. vietnamense* were collected and purchased by dealers at 1–2 US$ per kg. In the next two years most populations of *P. vietnamense* were exterminated by over-collecting in the small area of its primary distribution. Only a few plants were found after extensive searches by local people in 2001 and were offered for sale in local villages. During the course of the study in 2001, a few seedlings were observed in the area (Fig. 51) and they will certainly be collected for sale in the very near future due to the high prices on offer (Fig. 52a,b). In this way wealthy foreign "orchid lovers" continue destroy the last populations of this rare species. Presently, Vietnam lacks the means to prevent this process. Regrettably, without urgent and effective action for the protection of *P. vietnamense* it may well become extinct in the wild in the near future.

HABITAT. *Paphiopedilum vietnamense* grows in an area with a monsoon tropical climate with a cold winter and warm wet summer. Relatively dry winters with a drought period of 4–5 months, extending from November to February or March, are typical of the low mountains of this region. Summers there are wet, with the peak rainfall in June, July and August (Tabs. 8, 9; Fig. 53). The mean annual rainfall for the region is about 2050 mm but annual precipitation increases greatly on south- and south-east-facing slopes of mountain ridges and may reach 2200–2500 mm in home range of *P. vietnamense*. Very wet primary vegetation certainly indicates a higher than usual level of humidity in *P. vietnamense* habitats. Temperature regimes in the area have a strong seasonality. Winter conditions with rather cool temperatures last from November to April, dropping to lows of 14–16°C. Persistent drizzle is common from early February until the end of March. Summers are hot and humid, with maximum temperatures reaching to 35–40°C, but usual mean summer temperatures are about 26–27°C (Fig. 53).

TABLE 8

CLIMATOLOGICAL DATA FOR THAI NGUYEN WEATHER STATION (21°35'N 105°50'E, ELEV. 36 m) ACCORDING TO NGUYEN KHANH VAN *ET AL.* (2000).

Month	01	02	03	04	05	06	07	08	09	10	11	12
1	15.7	17.0	19.7	23.5	27.1	28.4	28.4	28.0	26.9	24.1	20.9	17.4
2	27	37	53	126	238	333	401	353	256	151	51	25

Key for Table: 1 — median monthly temperature (°C); 2 — mean monthly rainfall (in mm).

TABLE 9

CLIMATOLOGICAL DATA FOR LANG SON WEATHER STATION (21°50'N 106°46', ELEV. 259 m) ACCORDING TO BRUZON & CARTON (1930).

Month	01	02	03	04	05	06	07	08	09	10	11	12
1	31.6	36.4	36.7	38.2	39.8	36.1	37.0	36.4	36.2	34.3	32.6	31.1
2	14.2	14.8	18.3	22.2	26.1	27.4	27.4	27.3	25.9	22.8	18.8	15.4
3	0.9	0.4	5.4	6.6	11.1	16.0	19.6	19.2	13.6	7.6	2.3	0.4
4	26	53	51	88	170	202	283	272	165	71	38	20

The majority of *Paphiopedilum vietnamense* populations inhabit shady, mainly north-facing steep mountain slopes and cliffs at 350–450 m on low limestone ridges with the highest peaks 500–600 m (Fig. 50). Rocks are composed of highly eroded, hard, crystalline, marble-like, pure white and grey-black limestone that forms a spectacular karst landscape with numerous very deep narrow gorges, caves and holes (Fig. 54).

Fig. 52. A few seedlings of *Paphiopedilum vietnamense* still survive on inaccessible vertical cliffs.

N. Vietnam, Thai Nguyen Prov. (Photos: L. Averyanov).

a — the locality of these seedlings is known to local people who are only waiting until they mature before they are collected; **b** — the price of *P. vietnamense* is so high that orchid hunters usually plant the plants on large native rocks to keep them better before sale.

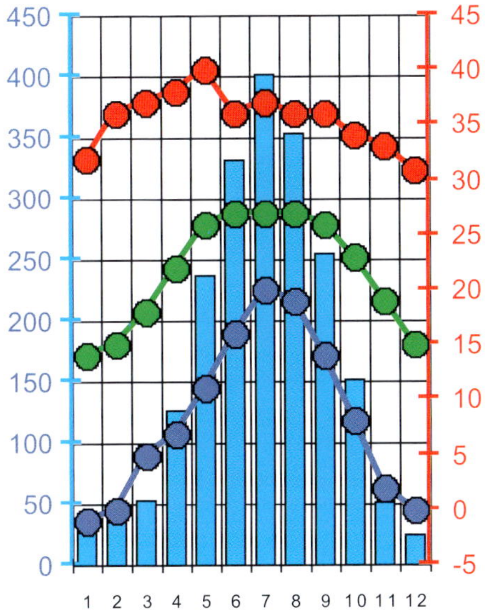

Fig. 53. Climatological data expected for the *Paphiopedilum vietnamense* area on the basis of extrapolation from the data of nearest weather stations.

Black figures at the chart bottom are months. Blue axis re presents atmosphere precipitation in mm (columns designate mean monthly rainfalls); red axis represents temperature in °C (red line — absolute maximum monthly temperatures, green line — mean monthly temperatures and blue line means absolute minimum monthly temperatures).

Luxuriant, very rich, humid, broad-leaved, evergreen, closed tropical forest supported populations of *P. vietnamense* before commercial exploration (Fig. 50). Large trees of *Burretiodendron hsienmu* (Tiliaceae), which reach 25–30 m tall often appear as a dominant in the canopy. In this stratum species of *Diospyros* (Ebenaceae), *Lithocarpus*, *Quercus* (Fagaceae), *Fernandoa* (Bignoniaceae), *Manglietia* (Magnoliaceae), *Cinnamomum*, *Persea* (Lauraceae), *Pterospermum truncatolobatum* (Sterculiaceae), *Chukrasia tabularis* (Meliaceae), *Choerospondias axillaris, Allospondias lakonensis* and *Dracontomelum duperreanum* (Anacardiaceae) are also commonly observed. The second and third forest layers contains a rich assemblage of warmth-loving tropical trees 5–20 m high characteristically with numerous species of Araliaceae (*Schefflera*, *Brassaiopsis glomerulata*), Elaeocarpaceae (*Elaeocarpus*), Euphorbiaceae (*Cleidion*), Lauraceae (*Cryptocarya*, *Litsea*), Moraceae (*Ficus*, *Streblus tonkinensis, S. ilicifolius*), Myrtaceae (*Syzygium*), Rubiaceae, Rutaceae (*Micromelum*, *Murraya)*, Sapindaceae (*Dimocarpus longan*), Staphyleaceae (*Turpinia*) and Sterculiaceae (*Sterculia*). Large palms, such as *Arenga pinnata* and *Livistona chinensis*, and large terrestrial and lithophytic plants, such as *Dracaena cambodiana* and *Pandanus* sp., are very typical of these forests. Among the numerous shrubs and shrublets are *Ilex* sp. (Aquifoliaceae), *Leea indica* (Leeaceae) and *Lysimachia insignis* (Primulaceae). Moisture-loving giant herbs such as *Aglaonema*, *Alocasia*, *Amorphophallus paeonifolius, Amorphophallus*, *Arisaema*, *Colocasia* (Araceae), *Alpinia*, *Amomum* (Zingiberaceae), *Costus speciosus* (Costaceae), *Curculigo* (Hypoxydaceae), *Musa* (Musaceae) and *Phrynium* (Maranthaceae) are prominent and abundant.

A very rich herbaceous flora is found in these places, including species of Acanthaceae, Balsaminaceae (*Impatiens*), Begoniaceae (*Begonia*), Convallariaceae (*Aspidistra*, *Ophiopogon*, *Pelyosanthes teta*), Primulaceae (*Lysimachia insignis*), Rubiaceae (*Ophiorrhiza*), Taccaceae (*Tacca plantaginea*) and Urticaceae (*Elatostema*, *Laportea*, *Pellionia*, *Procris*), and numerous moisture-lowing ferns, such as *Tectaria decurrens*, *T. deflexa* and *T. subpedata.* The terrestrial orchids *Cymbidium lancifolium, Malaxis acuminata, Phaius longicruris, Tropidia angulosa* and *T. curculigoides* are also very common. All these plants form a lush vegetation on steep rocky slopes and cliffs especially in unshaded places.

A rich and species specific composition is typical of lithophytic limestone floras of this region. Numerous rare species which grow in small narrow crevices of vertical cliffs may be seen here. Typical lithophytes include some species of Acanthaceae, Begoniaceae (*Begonia*), Gesneriaceae, Rubiaceae, Urticaceae, numerous ferns and fern allies (*Adiantum caudatum, A. gravesii, Antrophyum* sp., *Asplenium* spp., *Colysis* spp., *Nephrolepis cordifolia, Polypodium* sp., *Pyrrosia* spp., *Selaginella* spp.) and orchids (*Cheirostylis yunnanensis, Cleisostoma rostratum, Coelogyne fimbriata, Dendrobium chrysanthum, Hemipilia calophylla, Liparis mannii, Paphiopedilum hirsutissimum, P. tranlienianum, Pholidota rubra* and *Podochilus microphyllus*). Large lithophytic orchids, such as *Acampe rigida, Cleisostoma rostratum, Cymbidium aloifolium, Dendrobium chrysanthum* and *Vandopsis gigantea*, and large nest-like ferns, particularly *Asplenium anthrophoides* and *A. nidus* and creeping lithophytic vines, *Dischidia acuminata,*

Fig. 54. Limestone structure in habitats of *Paphiopedilum vietnamense.*

N. Vietnam, Thai Nguyen Prov. (Photo.: L. Averyanov).

Fig. 55. *Paphiopedilum vietnamense* (**a**) recently collected from the wild and offered for sale by local orchid hunters, March 2001, and (**b**) plant exhibited at an orchid show in Europe, 2001.

a — Private collection, N. Vietnam, Thai Nguyen Prov. (Photo: L. Averyanov); **b** — Photo P.J Cribb .

Dischidia sp., *Hoya* sp. (Asclepiadaceae), and *Pothos repens, Pothos* sp., and *Raphidophora* spp. (Araceae), commonly inhabit open mossy rocks. Lithophytic and epiphytic bryophytes are common especially in shady places, but they usually form a sparse or very thin cover. On the ground, mosses are usually absent.

Numerous large woody vines, including *Acacia*, *Bauhinia*, *Entada phaseoloides, Pterolobium microphyllum* (Fabaceae), *Calamus* (Arecaceae), *Embelia* (Myrsinaceae), *Jasminum* (Oleaceae), *Beaumontia grandiflora, Melodinus* (Apocynaceae), *Combretum* (Combretaceae) and *Tetrastigma* (Vitaceae) are typical of this forest, especially in shady humid conditions of north-facing mountain slopes. Epiphytes, mostly orchids, are common, including *Callostylis rigida, Cleisostoma fuerstenbergianum, C. striatum, Coelogyne fimbriata, Dendrobium lindleyi, Eria sutepensis, Luisia* sp., *Phalaenopsis mannii* and *Thelasis khasiana.* A creeping *Vanilla* sp., that reaches 10–12 m in length, is also commonly observed in the forest.

Paphiopedilum vietnamense commonly grows as a lithophyte in crevices and holes on very steep shady humid slopes and cliffs, mainly of north exposition on highly eroded limestone (Fig. 51). This species is usually found in the middle slopes of ridges between 350–450 m. The plants flower in nature in March (Fig. 55). In the past, the species formed large colonies often in the company of other warmth-growing *Paphiopedilum* species, such as *P. concolor, P. hirsutissimum* var. *esquirolei, P. tranlienianum* and other orchids, like *Cheirostylis yunnanensis, Cleisostoma rostratum, Coelogyne fimbriata, Dendrobium chrysanthum, Hemipilia calophylla, Liparis mannii, Pholidota rubra, Podochilus microphyllus* and *Vandopsis gigantea,* that are typical of the limestone orchid flora at low elevations.

Local collectors have noted the clear preferences of *P. vietnamense* for pure white limestone rocks rather than limestone stained with grey or grey-black. According to their information more than 80% of samples were collected on pure white limestone, and less than 20% were found on rocks stained with metal oxides. Very few specimens of *P. vietnamense* were found above 450 m, and then on rather open steep slopes near the ridge tops and on cliffs covered with light secondary forest communities with bamboo.

VARIATION. Numerous photographs of exported plants flowered in culture (Gruss & Perner, 1999a,b; Senghas & Schettler, 1999a,b) provide evidence of considerable variability in the flowers of *Paphiopedilum vietnamense*. This is mainly in the flower colouration and the form of the lateral petals and lip. Usually the petals are white or light pinkish, more or less intensive pink-purple toward the apex (Fig. 55), in other cases petals may be bright purple and white only in the basal portion. Their form ranges from oblong-elliptic to broadly obovate. Petals may spread horizontally, or may be more or less pendent with forward directed upper margins. The lip varies from obovate to almost spherical, and it may be white with crimson-purple marks at the front or entirely deep purple with a white basal part.

3. PAPHIOPEDILUM MALIPOENSE S.C. Chen & Z.H. Tsi

Paphiopedilum malipoense *S.C. Chen & Z.H. Tsi* in Acta Phytotax. Sinica, 22, 2: 119 (1984); Cribb, Gen. Paphiopedilum: 90 (1987); id., Gen. Paphiopedilum: 82 (1998); Braem, Paphiopedilum: 37 (1988); Aver., Identif. Guide. Vietnam. Orch.: 16 (1994); Braem, C. Baker & M. Baker, Gen. Paphiopedilum 1: 80 (1998). **Type:** Yunnan: Malipo, Hwang Jin In, 1300–1600 m, Nov. 11 1947, *K.M. Feng* 13161 (holotype PE!).

KEY TO THE VARIETIES

1. Petals broadly obovate to obovate; lip wider than petals, 3.8–5.4 cm wide; staminode with a deep brown-purple or maroon apex; leaves heavily spotted with purple-violet beneath . . . . **3a. *P. malipoense* var. *malipoense***

+. Petals obovate to narrowly obovate or lanceolate; lip of equal width or narrower than petals, usually less than 3.8 cm wide; staminode with a white, light yellowish or light greenish apex marked with finest brown-purple marks; leaves sparsely spotted with purple-violet beneath . . . . 2.

2. Petals obovate to narrowly obovate; lip subglobose, more than 1.8 cm wide, truncate at the apex . . . . **3b. *P. malipoense* var. *jackii***

+. Petals narrowly obovate to lanceolate; lip narrowly ovate, slightly bilaterally compressed, hooked at the apex . . . . **3c. *P. malipoense* var. *hiepii***

3A. PAPHIOPEDILUM MALIPOENSE VAR. MALIPOENSE

DESCRIPTION. *Lithophytic* or *terrestrial herb* with 4–6 distichous leaves and a more or less elongate creeping rhizome 2–3.5 mm in diameter. *Leaves* coriaceous, suberect, oblong or narrowly elliptic, acute at asymmetric apex, 10–16(20) cm long, 2.5–5(7) cm wide, dark green–light green tessellated above, heavily purple-marked and keeled on lower surface; petiole 2–4 cm long, ciliate along margin near the base. *Inflorescence* 30–50(65) cm long, 1(2)-flowered; peduncle green, purple-brown spotted or purple-brown, white hairy; bract narrowly ovate-lanceolate, about 1.5 cm long, white-ciliate on margins, white-pilose on outer surface. *Flower* 8–12.2 cm across with a light sweet fragrance; sepals and petals apple-green, striped and spotted garnet-purple; lip pale yellow-green greyish, spotted with garnet-purple on inner surface; staminode pure white, deep maroon-brown in apical half; pedicel and ovary about 4 cm long, white-hairy, rostrate at apex. *Dorsal sepal* broadly ovate to elliptic-lanceolate, cuspidate-acuminate, 4.4–7.1 cm long, 1.8–4.5 cm wide, sparsely white-pubescent on inner surface, white-villose and ciliolate without. *Synsepal* broadly-ovate to ovate-lanceolate, acuminate or sometimes

Fig. 56. Primary broad-leaved evergreen forests on steep slopes of highly eroded limestone ridges support habitat of *Paphiopedilum malipoense* in many regions of Vietnam.

N. Vietnam, Hoa Binh Prov. (Photo.: L. Averyanov).

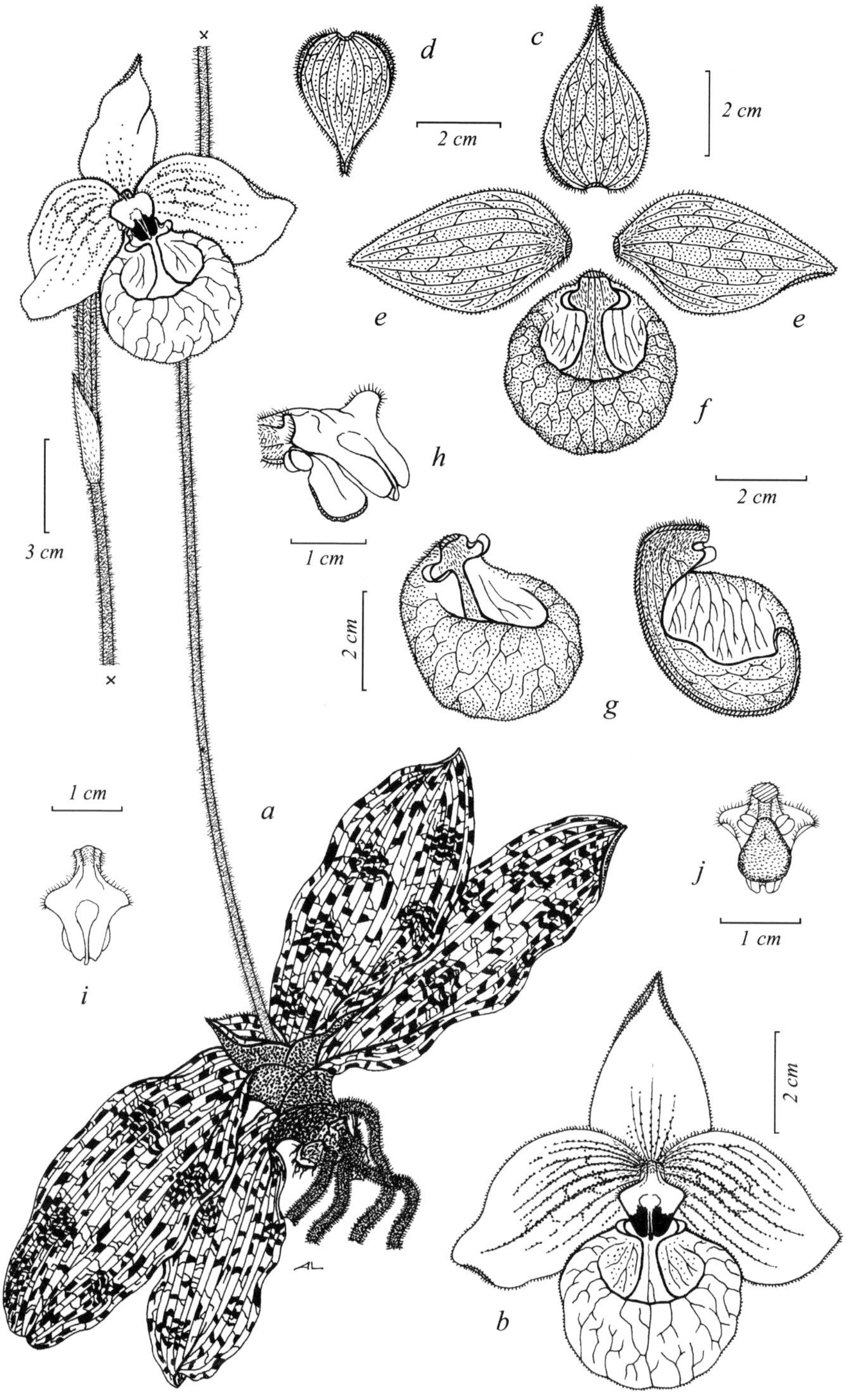

Fig. 57. *Paphiopedilum malipoense* var. *malipoense*. (Drawing: L. Averyanov).

a — flowering plant; **b** — flower; **c** — median sepal; **d** — synsepal; **e** — petals; **f, g** — lip, frontal, side and section views; **h–j** — column, side, dorsal and ventral views (Drawn from HG 52).

bidentate at apex, dorsally bicarinate, 3.8–5.3 cm long, 2.4–4.8 cm wide. *Petals* ovate, subacute, 4–7.1 cm long, 3.4–5.1 cm wide, white-villose and ciliolate at base, white-pubescent on inner surface. *Lip* horizontal, deeply saccate, 4.5–6.5 cm long, 3.8–5.4 cm wide, subglobose, inrolled margins erose, white-villose at base within, finely pubescent outside. *Staminode* convex, broadly ovate-oblong, 13–14 mm long, 11–13 mm wide, truncate at apex, subsessile at base, white-ciliolate and pubescent in basal half, dorsally carinate, rounded bosses on surface in apical half. Fig. 57.

DISTRIBUTION. Southern China (south-eastern Yunnan, western Guangxi, south-western Guizhou), northern Vietnam (Bac Can, Ha Giang, Hoa Binh, Lang Son, Quang Binh, Son La, Thanh Hoa and Tuyen Quang Provinces), north-eastern Laos ? Figs. 36, 58.

ECOLOGY. Primary, evergreen broad-leaved, mixed and coniferous wet, closed forests on highly eroded crystalline limestone mountains at elevations of (450) 600–1200 (1450) m. Fig. 56.

FLOWERING IN THE WILD. March–April.

IUCN STATUS. EN.

AFFINITIES. *Paphiopedilum malipoense* is a distinctive slipper orchid that has striking elegant flower with green sepals and petals, a pale green-greyish lip and white staminode with purple-brown apex. It is highly prized for its sweetish apple- or raspberry-scented flower. It has growths with large, dark and light green tessellated or chequered leaves, spotted purple-violet beneath, and a tall inflorescence usually bearing a single large flower. It is allied to *P. micranthum* and *P. armeniacum*.

HISTORY. It was first described by the Chinese botanists S.C. Chen and Z.H. Tsi in 1984 on the basis of an herbarium specimen which was collected by K.M.Feng in 1947 near Malipo in south-eastern Yunnan near the Vietnamese border. The original specimen is housed at the Herbarium of the Institute of Botany in Beijing. The species was previously thought to be endemic to south-west China. It remained little known until it attracted great attention around 1984, when it came into cultivation. At that time very large quantities of wild-collected living plants appeared in the Hong Kong orchid market.

Although large quantities were exported from China, its natural distribution and habitat remained unclear for a long time. It was originally collected in the mountains near the town of Malipo. Besides the type locality, natural populations of the species were observed along the southern border of the Guizhou limestone plateau between 800–1100 m in the west of Guangxi Province (Fowlie, 1990b, 1991a). The present known distribution of *P. malipoense* in China comprises karst limestone areas of western Guangxi, south-western Guizhou and south-eastern Yunnan (Cribb, 1998). This limestone country is a distinct bio-geographical region that is also found in large areas of northern Vietnam. As a result it was suspected that *P. malipoense* might also occur in Vietnam. Indeed, considerable quantities of plants exported from China actually originated from across the border in areas of northern Vietnam (Averyanov, 1994). A few years later, it was also discovered in Vietnam in numerous localities along the Chinese and Laotian borders (Averyanov *et al.*, 1997a). Recent field

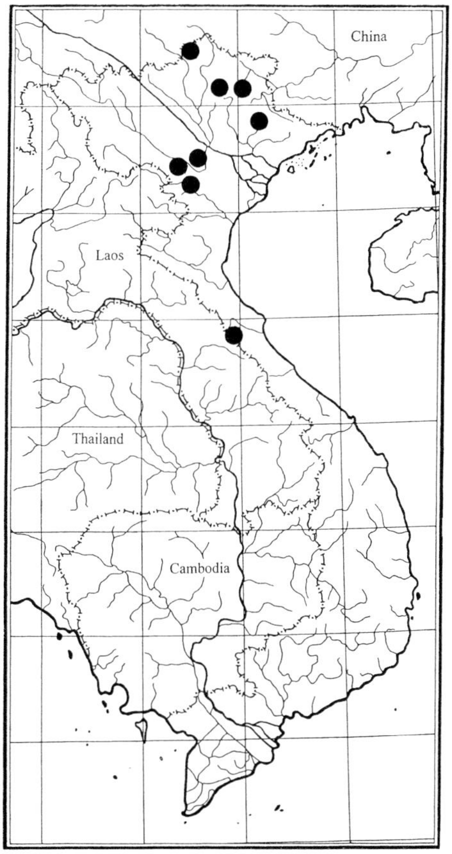

Fig. 58. Distribution of *Paphiopedilum malipoense* var. *malipoense* in Vietnam.
(Drawing: L. Averyanov).

exploration has provided clear evidence that this species is widely distributed in northern Vietnam (Fig. 58). The plant inhabits very steep inaccessible rocky limestone mountains, which explains why its wide distribution in Vietnam was not discovered earlier. Numerous populations of *P. malipoense* were observed in Vietnam on limestone ridges that cross the Laotian border. There is no doubt that this species also occurs in Laos, but no herbarium collection has been made from there.

The first descriptions of the natural habitat of *P. malipoense* were given by Fowlie (1990b, 1991a). He found it on karst limestone hills in south-western Guangxi province in China at between 760 and 1040 m elevation, growing at the base of limestone rocks in soil and leaf litter. Later, the species was observed in similar places in two sites in south-western Guizhou, three in western Guangxi and three in south-eastern Yunnan (Cribb, 1998). It was growing in deep shade in evergreen forest on the north-facing side of karst limestone hills at between 850 m and 1200 m. In Guizhou it grew at one locality with other orchids such as *Paphiopedilum concolor, Vanilla annamica, Phalaenopsis wilsonii, Vanda concolor* and *Cheirostylis sinensis*. At another, it grew with *P. micranthum* on the north-facing slope of a karst limestone hill at the slightly higher elevation of 1140 m, in soil and leaf litter just below the summit ridge and at the foot of a steep limestone boulder. The soil pH was 7.47. In Guangxi it was found it at 1300 m elevation in montane forest just below a ridge top, growing in deep shade between limestone rocks in more or less neutral, friable, brown soil but with little competition from other herbs. Elsewhere in the province, it was found in two localities at 1000–1050 m elevation where the forest cover had been removed. It just survived in deep shade around the roots of shrubs and dense bamboo on near vertical north-facing slopes. The soils were again more or less neutral but had lost much of their humic content (Cribb, 1999).

According to recent information based on thorough botanical exploration in Vietnam, *Paphiopedilum malipoense* is one of the most widespread slipper orchid species on limestone. It was observed on all the

Fig. 59. Very steep shady rocky slopes and narrow shelves of vertical limestone cliffs are typical habitats of *Paphiopedilum malipoense*. (Photos: L. Averyanov).

a — N. Vietnam, Prov. Son La; **b** — N. Vietnam, Prov. Ha Giang.

largest limestone rocky areas in several provinces of the northern Vietnam (Fig. 58). Environmental conditions are similar in Vietnam to those described for southern China (Fowlie, 1991a; Cribb, 1998). The species inhabits evergreen broad-leaved, mixed and coniferous forests that cover steep rocky slopes and cliffs of highly eroded limestone mountains (Fig. 56). In intact conditions it is common on mountain slopes of any exposition. However, in degraded habitats, most populations usually survive only on wetter cliffs facing the north or north-east. Commonly, *P. malipoense* occurs at elevations between 700–1200 m, however the altitudinal range of the species is surprisingly wide. Its populations in Vietnam were found in the mountain belt from 450 m up to 1450 m. Very steep shady rocky slopes, the shelves of vertical cliffs, deep narrow karst gorges and holes just below the limestone ridges are the most typical habitats of *P. malipoense* (Fig. 59). All populations of var. *malipoense* so far discovered have been on crystalline marble-like limestone with deep karst vertical erosion (Fig. 60).

The climatic conditions are typical for northern Vietnam. This is area of monsoon tropical climate with a cold winter and warm wet summer. Winter here is very dry with mean temperatures that lie between 11–15°C between 500–1000 m (Fig. 61). The coldest months are December and January when night temperatures

Fig. 60. *Paphiopedilum malipoense* grows exclusively in Vietnam on crystalline marble-like limestone with vertical erosion.

N. Vietnam, Prov. Thanh Hoa. (Photo.: L. Averyanov).

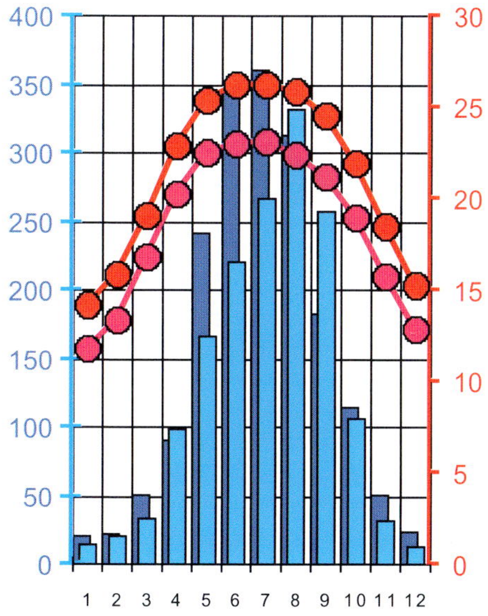

Fig. 61. Climatological data most typical for the northern part of the distribution in Vietnam of *Paphiopedilum malipoense* with monsoon tropical climate with cold winter and summer rains.

Black figures at the chart bottom are months. Blue axis represents atmosphere precipitation in mm (columns designate mean monthly rainfalls); red axis represents mean monthly temperature in °C. The deep blue and red are designated data from the Hoang Su Phi weather station, 22°45'N 104°40'E, elev. 553 m; light blue and purple are data from the Moc Chau weather station, 20°51'N 104°38'E, elev. 958 m, all according to Nguyen Khanh Van *et al.*, 2000.

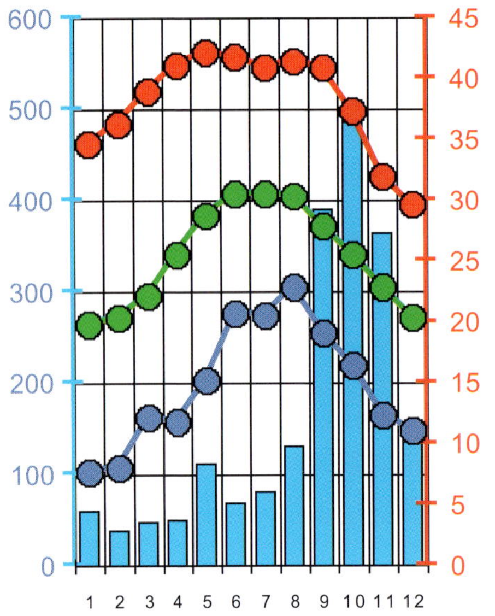

Fig. 62. Climatological data for the southern part of the distribution in Vietnam of *Paphiopedilum malipoense* with monsoon tropical climate with cold winter and summer-autumn rains. Dong Hoi weather station (17°29'N 106°36'E, elev. 7 m) according to Bruzon & Carton, 1930.

Black figures at the chart bottom are months. Blue axis represents atmosphere precipitation in mm (columns designate mean monthly rainfall); red axis represents temperature in °C (red line — absolute maximum monthly temperature, green line — mean monthly temperature and blue line means absolute minimum monthly temperature).

may fall to -2°C. Morning ground hoarfrost is not uncommon in winter in open places in this area. During March and first weeks of April, cold heavy fogs and persistent mists are very common. Most of the moisture at this time of year comes from this source. The summer lasts from mid April to September and October when it is hot with median temperatures of 23–27°C. Maximum temperatures may occasionally reach 38–40°C. The rainy season with heavy torrential rains is from June–August (Fig. 61). The humidity at this time is very high. Mean annual rainfall in the area ranges from 1600 to 2700 mm.

The southern portion of the range of *Paphiopedilum malipoense* in Vietnam is an area of monsoon tropical climate with a cold winter and warm summer-autumn rains. The climate in this region differs in having warmer winters and a rainy season shifted to the autumn (Fig. 62). The most intensive rains are observed here during September–November. Mean annual rainfall in the region varies between 2500–3000 mm. As a result of the high rainfall this species occurs in this area at lower elevations.

Primary, broad-leaved, evergreen, closed montane forests along limestone ridges are the usual home of *Paphiopedilum malipoense* in northern Vietnam (Fig. 56). This forest contains old, but not particularly tall, often gnarled specimens of *Acer* (Aceraceae), *Alangium* (Alangiaceae), *Burretiodendron hsienmu* (Tiliaceae), *Carpinus* sp. (Betulaceae), *Gleditchia* sp. (Fabaceae), *Lithocarpus*, *Quercus* (Fagaceae), *Pistacea weinmannifolia* (Anacardiaceae), *Platycarya strobilacea* (Juglandaceae), and *Ulmus* sp. (Ulmaceae). Many of them lose their leaves during the dry winter forming semi-deciduous and even open deciduous woods.

Broad-leaved forests with conifers support populations of *P. malipoense* at higher elevations, usually in the montane belt below 900 m. A number of rare woody conifers are often important subdominants in such forests. In the northern part of its range in Vietnam *Amentotaxus argotaenia, Cephalotaxus mannii, Nageia fleuryi, Pinus fenzeliana, Podocarpus neriifolius, P. pilgeri, Pseudotsuga sinensis* var. *brevifolia, Tsuga chinensis* and *Xanthocyparis vietnamensis* were all observed. In the southern part of its range the commonest conifers are *Calocedrus macrolepis, Nageia wallichiana, Pinus fenzeliana, Podocarpus brevifolius* and *Taxus chinensis.* The southern populations of *Paphiopedilum malipoense* are supported by rather open coniferous forests with *Dacrycarpus imbricatus* and *Dacrydium elatum.*

Numerous small trees and shrubs such as *Celastrus* sp., *Euonymus* sp. (Celastraceae), *Illicium jiadifengpi* (Illiciaceae), *Lespedeza* sp., *Sophora glauca* (Fabaceae), *Madhuca* sp. (Sapotaceae), *Photinia* sp., *Sorbus* sp. (Rosaceae), *Pittosporum* sp. (Pittosporaceae), *Rhododendron* spp., *Vaccinium* spp. (Ericaceae), *Schefflera pesavis* (Araliaceae), *Tirpitzia sinensis* (Linaceae), and *Wikstroemia* sp. (Thymelaeaceae) are very typical of the forest undergrowth in these habitats. A very rich species assemblage of lithophytic herbs and ferns is an integral component of these forests. Among them are common and numerous species of *Ainsleya* (Asteraceae), *Alpinia, Hedychium* (Zingiberaceae), *Arisaema* (Araceae), *Aspidistra, Disporum, Ophiopogon, Peliosanthes, Polygonatum, Tupistra* (Convallariaceae), *Begonia* (Begoniaceae), *Elatostema, Laportea, Pellea, Pellionia* (Urticaceae), *Impatiens* (Balsaminaceae), *Lysimachia* (Primulaceae), *Selaginella* (Selaginellaceae), and species of Acanthaceae, Gesneriaceae and Melastomataceae.

Orchids of outstanding diversity and abundance are found in these places. Regularly between 50–70 terrestrial, lithophytic and epiphytic orchid species may be seen growing with *P. malipoense.* Among them, the commonest are *Bulbophyllum ambrosia, B. macraei, B. purpureifolium, Calanthe argenteostriata, Callostylis rigida, Coelogyne fimbriata, C. malipoensis, Cymbidium floribundum, C. lancifolium, Dendrobium fimbriatum, Eria coronaria, Epigeneium amplum, Liparis averyanoviana, L. balansae, L. distans, L. mannii, Malaxis acuminata, Monomeria barbata, Pholidota leveilleana, P. roseans* and *P. yunnanensis.* Often they cover mossy rocks and the bases of large trees. Colonies of *P. malipoense* commonly occur with lithophytic orchid species such as *Eria rhombodalis* and *Oberonia cavaleriei* and with populations of other slipper orchids such as *Paphiopedilum dianthum, P. hirsutissimum* and *P. micranthum.*

In undisturbed conditions in primary forest, *Paphiopedilum malipoense* was observed in many places growing abundantly in limestone hollows, often along the bottom of small, narrow, deep canyons on deep soils rich in humus that are often covered by leaf litter. Often this species also grows as a lithophyte in crevices of vertical rocks. Populations of *P. malipoense* usually occupy slopes near the tops of ridges. On steep and vertical rocks, this species spreads easily by means of long creeping stolons (Fig. 63) so that chains of mature and young plants are formed along a common rhizome. It may form continuous cover and appear as a codominant of lithophytic herb communities in some places (Fig. 64).

Fig. 63 (above left). On steep slopes *Paphiopedilum malipoense* spreads easily by means of long creeping stolons. N. Vietnam, Hoa Binh. Prov. (Photo.: L. Averyanov).

Fig. 64 (above right). In primary forest *Paphiopedilum malipoense* can form a continuous ground cover and appear as a co-dominant of lithophytic herb communities. (Photo.: L. Averyanov).

Fig. 65. Morphological and colour variations of *Paphiopedilum malipoense* var. *malipoense* flowers in northern Vietnam.

a — Son La Prov.;
b, c — Tuyen Quang Prov., HAL 245;
d — Hoa Binh Prov;
e — Hoa Binh Prov, VH 2413

(Photos **a, d, e**: L. Averyanov; photos **b, c**: P.J. Cribb)

The flower buds of *Paphiopedilum malipoense* appear in nature during September and October (Fig. 64) but then low temperatures stop their development. In its habitats at 1000–1400 m it is not unusual for night temperatures during December, January and February to fall to within a degree or two of frost. The flowering period of *P. malipoense* in northern Vietnam coincides with the first spring warmth in March and April (Figs. 59, 60). Surprisingly, the inflorescence of this species can reach 60–65 cm long in the wild raising the flower above the windless conditions of the very deep shady limestone gorges and karst holes where it grows.

In Vietnam *Paphiopedilum malipoense* is still locally common in some localities. However, it is one of the most sensitive herbs of the primary forest. After logging of large trees the habitats dry out and this moisture-loving species very quickly disappears. Commercial collecting is extensive throughout its range. This plant is also regularly illegally exported to China from local markets along the border. An approximate estimation of the total export of *P. malipoense* specimens from Vietnam to China during 1990–1997 years may be as high as a few tons per year (Averyanov *et al.*, 1997a). In this connection the species surely warrants status as endangered species in Vietnamese flora (EN). In China, *P. malipoense* is extinct in most of its known historical localities due to degradation of primary vegetation and commercial collecting. Wild-collected plants of *P. malipoense* are still sold in the Wenshan and Kunming Flower Markets in Yunnan and similar flower markets in the region. In Kunming plants were fetching prices of between 30 and 50 yuen (compared with *P. micranthum* at 10 yuen a kilo). The price it now fetches in Kunming suggests that *P. malipoense* is now a very rare plant in the wild, possibly the rarest of the Chinese species (Cribb, 1999).

VARIATION. Clear morphological and colour variations exist between populations of var. *malipoense* in Vietnam (Fig. 65a–d). Flowers from different populations may differ from each other in size from 6 to 12 cm wide. Petals of northern populations are broader and shorter than those from the southern part of its distribution. Sepals usually have a uniform apple-green colouration, but sometimes they are finely purple-spotted at the base. The petals are light apple-green, later becoming pale yellowish-green, and are more or less heavily striped with purple-brown. Plants in nature usually produce a 1-flowered inflorescence but, from time to time, a smaller second flower may be produced (Fig. 65e). Albino forms, with unspotted pale yellow tepals and staminode, are very rare in populations but are highly prized by collectors. These also lack of any purple colouration on their leaves and inflorescence. A form with a white lip and brown-purple side lobes was recognized among plants recently imported into Germany from Vietnam or southern China as *P. malipoense* var. *tonnianum* J.Röth (in Die Orchidee 51, 5: 561 (2000)) *.

3B. PAPHIOPEDILUM MALIPOENSE VAR. JACKII (H.S. Hua) Aver.

Paphiopedilum malipoense var. **jackii** *(H.S. Hua) Aver.* in Orchids 66, 2: 153 (1997). **Type:** China, Yunnan, Wenshan, 12 March 1995, *S.S.Hu* 3268 A-S (holotype IBSC).
Paphiopedilum jackii H.S. Hua in Die Orchidee 46, 3: U4 (1996).

DESCRIPTION. *Lithophytic* or *terrestrial herb. Leaves* light green, lightly tessellated and veined dark green above, light green and sparsely spotted with purple-violet beneath. *Flowers* differ from the typical variety in having narrower petals, a smaller yellowish-green or green lip and a white staminode, usually with a yellow, yellow-green or green central mark with fine purple venation. Fig. 67.

DISTRIBUTION. China (Yunnan), northern Vietnam (Tuyen Quang Province). Fig. 68.

ECOLOGY. Primary evergreen broad-leaved dry, rather open forests on stratified soft limestone mountains and hills at elevations of 550–650 m. Fig. 66.

FLOWERING IN THE WILD. March–April.

IUCN STATUS. EN.

HISTORY. *Paphiopedilum jackii* Hua was described in 1996 on the basis of collection from southern

* **Type:** S.China/Vietnam, 9 Feb. 2000, cult. W. Tonn s.n., Neu-Eichenberg, Germany (holotype HAL 089456).

Fig. 66. Primary evergreen broad-leaved dry forest on stratified soft limestone is the typical habitat of *Paphiopedilum malipoense* var. *jackii*. (Photo.: L. Averyanov).

Yunnan made by S.S.Hu. Its provenance of Wenshan is, however, suspect because that city has a major orchid market that deals in orchids from the surrounding countryside as well as plants from adjacent Vietnam and Laos. It was named after the late Dr. Jack Fowlie, a well-known orchid enthusiast. The plant is similar to var. *malipoense* in habit and floral morphology, differing slightly in petal attitude and most obviously in its leaves and staminode colouration. Both varieties grow more or less sympatrically, but never form mixed populations in the same habitat.

HABITAT. Var. *jackii* has a distinct ecological preference, occurring in Vietnam on soft (presumably Mesozoic) stratified limestone (Fig. 66) and schist and shale with high content of lime. This taxon inhabits low mountain and hill areas between 550–650 m (Fig. 67). In Vietnam it has been found only in Tuyen Quang province in a restricted lowland limestone area delimited by the rivers Gam and Nang and on the high-silicate mountain ridges of Pia Bioc, Sam Sao and Lung Ni Ung (Fig. 68). This may explain its rarity in northern Vietnam. It has not yet found in any other areas of the country.

Paphiopedilum malipoense var. *jackii* prefers distinctly warmer habitats at lower elevations than the typical variety. Summer temperatures in the area from May to September usually lie between 25° to 30°C and in winter from 15° to 18°C. In winter, cold northeast winds are common during which temperatures may fall to 5–10°C. The rainy season lasts here from May to September, and the winter is very dry. Mean annual rainfall is about 1600–1700 mm. Available climate data for the area of its distribution are presented in Fig. 69.

Fig. 67. *Paphiopedilum malipoense* var. *jackii* in its natural habitat. (Photo.: L. Averyanov).

VARIATION. Flowers of *P. malipoense* var. *jackii* are remarkably variable in the form of its petals and colouration of its staminode. The petal shape varies from ovate to narrowly ovate from population to population (Fig. 70a–c), while the staminode may be pure white, or white with yellow, yellow-green or light green center, usually with fine purple with venation and marks (Fig. 71b,c). Leaves, as a rule, have a very light, often whitish, background above and very sparse purple spotting beneath (Fig. 72b,c).

3C. PAPHIOPEDILUM MALIPOENSE VAR. HIEPII (Aver.) P.J. Cribb

Paphiopedilum malipoense var. **hiepii** *(Aver.) P.J.Cribb*, Genus Paphiopedilum: 88 (1998). **Type:** Northern Vietnam, Prov. Tuyen Quang, Distr. Tuyen Quang, closed evergreen primary forest on limestone hills at 500–600 m, 14–15 June 1995, *N.T.Hiep* s.n. (holotype LE, photo. HN).

Paphiopedilum hiepii Aver. in Orchids 67, 3: 261 (1998).

Paphiopedilum angustatum Z.J.Liu & S.C.Chen, in Acta Phytotax. Sinica 38, 5: 464 (2000). **Type:** China, Yunnan, Malipo, 23 March 2000, *Z.H. Liu* 1505 (holotype Herb. Shenzhen City Wutongshan Nurseries).

Paphiopedilum jackii H.S.Hua var. *hiepii* (Aver.) Koop. in Orchid Digest 64, 4: 168 (2000).

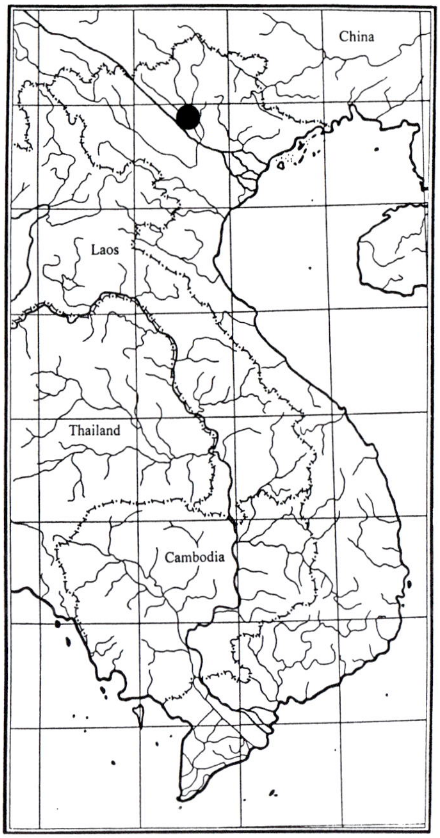

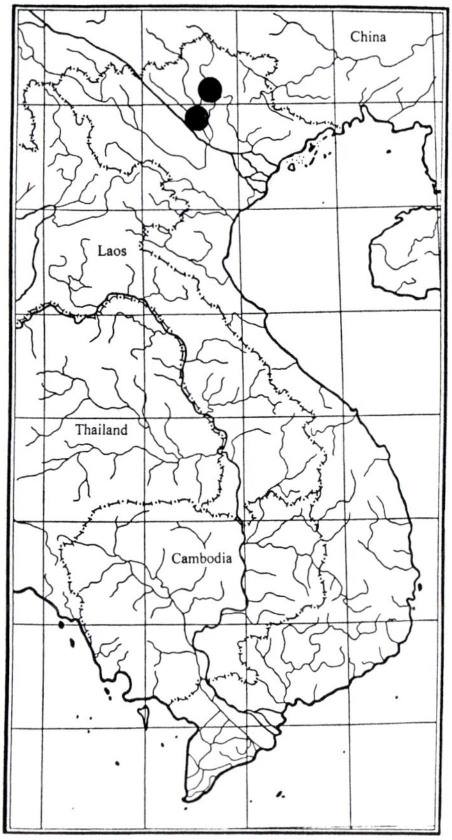

Fig. 68. Distribution of *Paphiopedilum malipoense* var. *hiepii* (above left) and var. *jackii* (above right) in Vietnam. (Drawing: L. Averyanov).

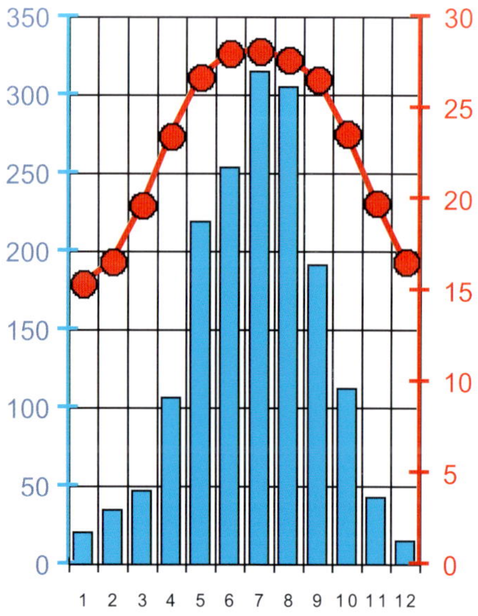

Fig. 69. Climatological data for habitats of *Paphiopedilum malipoense* var. *jackii* and var. *hiepii*. Dinh Hoa weather station (21°54'N 105°38'E, elev. 220 m) according to Nguyen Khanh Van *et al.*, 2000.

Black figures at the chart bottom are months. Blue axis represents atmosphere precipitation in mm (columns designate mean monthly rainfall); red axis represents temperature in °C (red line designates mean monthly temperature).

DESCRIPTION. *Terrestrial herb. Leaves* very pale green with light tessellation and dark green venation above and light green with sparse purple-violet spotting beneath. *Flower* differs from the typical variety in having narrow ovate or lanceolate petals hooked at the apex, a small narrow laterally compressed yellowish-green lip and a white staminode with few fine purple marks at the apex. Fig. 73.

DISTRIBUTION. China (south-eastern Yunnan), northern Vietnam (Tuyen Quang Province).

ECOLOGY. Primary, evergreen, broad-leaved, dry, closed hill forests soft limestone and schist rich in lime at 500–600 m.

FLOWERING IN THE WILD. March–April.

IUCN STATUS. EX.

HISTORY. *Paphiopedilum malipoense* var. *hiepii* is a curious taxon which differs distinctly from all other species of the subgenus *Parvisepalum* in its narrow sepals and petals, and narrow, laterally compressed hooked lip never seen in known slipper orchids (Fig. 73, 74). Nevertheless, it is certainly very closely related to var. *jackii.* Some specimens develop abortive pollen that gives reason to consider them as abnormal forms. It was discovered by Vietnamese botanist Dr. N.T.Hiep during field work on the American Orchid Society funded "Exploration of endangered Vietnamese Paphiopedilums" project in 1995.

HABITAT. Plants were found near a population of var. *jackii*, but as terrestrial plants in deep shade of closed dry primary forest. They grew on a hill composed of soft stratified limestone and schist rich in lime at 500–600 m. The natural habitat and ecological requirements of both varieties are very similar. Climatic data for the area of var. *hiepii* are presented in Fig. 69. Its flowering season in nature lasts from March to April.

Fig. 70 (right). Morphological and colour variations of flowers of *Paphiopedilum malipoense* var. *jackii* in northern Vietnam.

(Photo **a**: P.J. Cribb; photos **b**, **c**: L. Averyanov).

a — northern Vietnam, Tuyen Quang Prov.; **b, c** — plants collected in Tuyen Quang Province, in private nurseries, Hanoi.

Paphiopedilum angustatum is undoubtedly the same as *P. malipoense* var. *hiepii*. Its given provenance of Malipo in Yunnan is undoubtedly erroneous. Malipo is a Chinese border town where Vietnamese orchids are commonly sold.

During recent years the total destruction of the primary vegetation in the area where var. *hiepii* was discovered has resulted in the extinction of the only known population of this orchid. Despite extensive searches, including those of numerous commercial collectors, no further material of this unusual plant has been found. Regrettably, this variety is probably extinct in Vietnam.

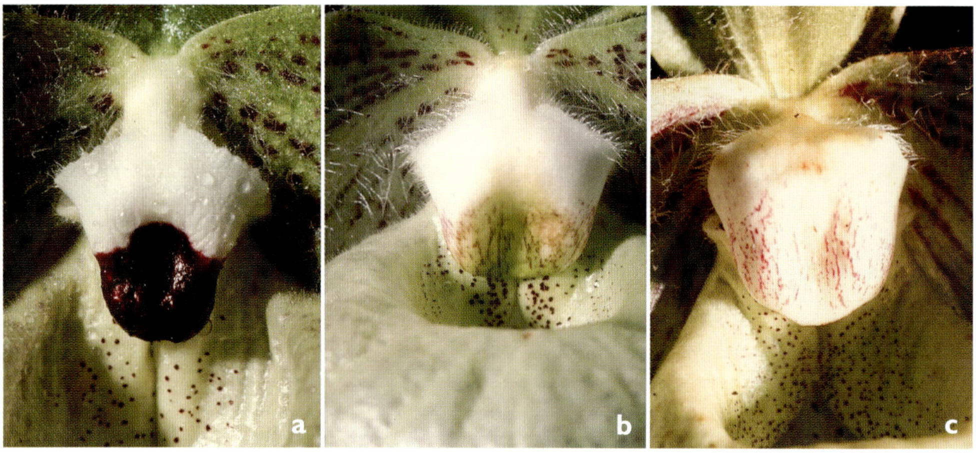

Fig. 71. Colour patterns of staminode in flowers of *Paphiopedilum malipoense* var. *malipoense* and var. *jackii* in northern Vietnam. (Photos: L. Averyanov).

a — uniform coloration of *P. malipoense* var. *malipoense*; **b, c** — most common variants of staminode observed in Vietnamese populations of *P. malipoense* var. *jackii.*

Fig. 72. Leaf colour patterns in *Paphiopedilum malipoense* var. *malipoense* and var. *jackii* in northern Vietnam. (Photos: L. Averyanov).

a — rather uniform coloration of the leaves of *P. malipoense* var. *malipoense*; **b, c** — usual coloration of leaves of *P. malipoense* var. *jackii.*

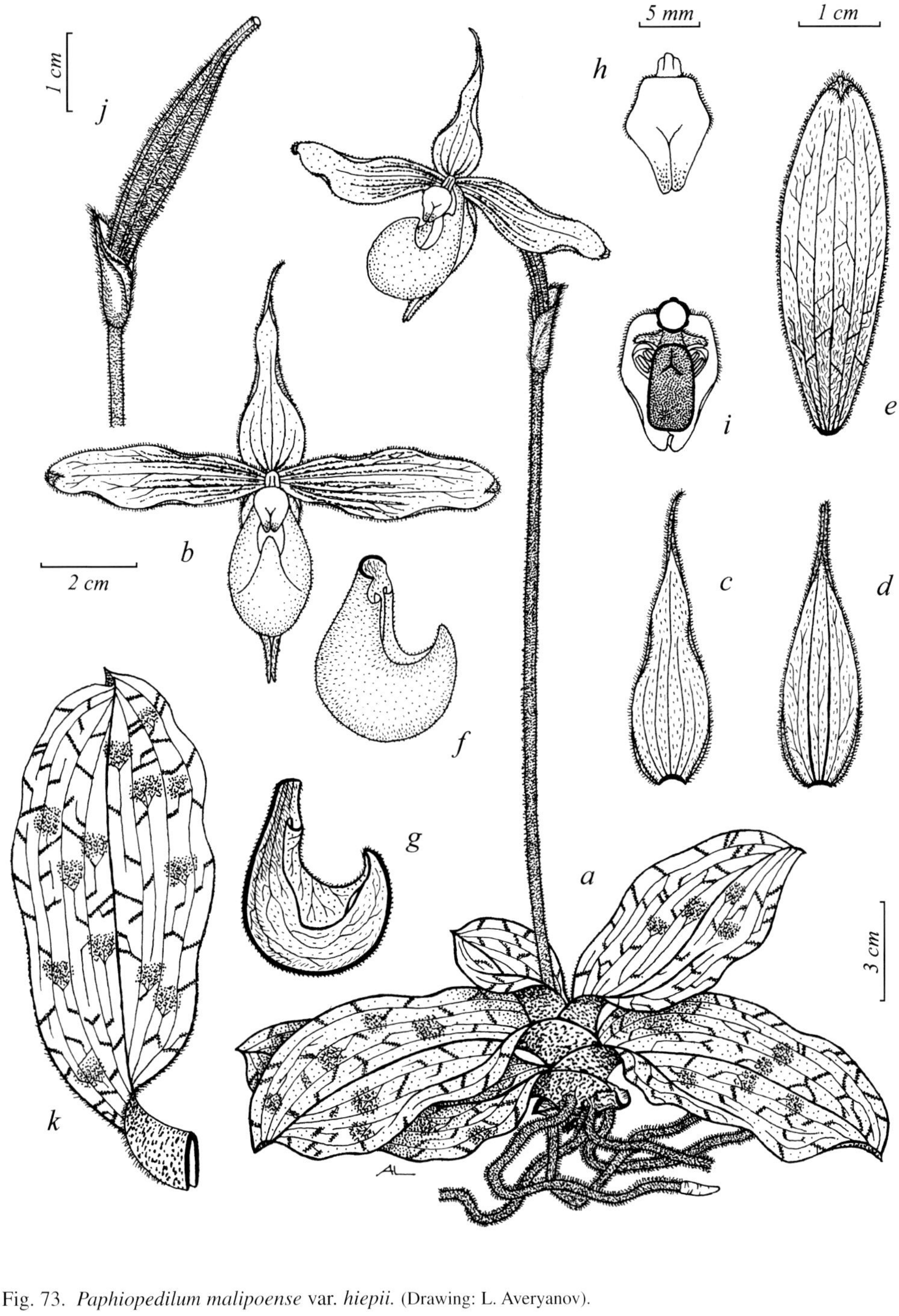

Fig. 73. *Paphiopedilum malipoense* var. *hiepii.* (Drawing: L. Averyanov).

a — flowering plant; **b** — flower; **c** — median sepal; **d** — synsepal; **e** — petal; **f, g** — lip side and section views; **h, i** — column, dorsal and ventral views; **j** — ovary and flower bract. (Drawn from the type).

Fig. 74. Type of *Paphiopedilum malipoense* var. *hiepii.* (Photos: L. Averyanov).
a — flower; **b** — flowering size plant in cultivation.

4. PAPHIOPEDILUM MICRANTHUM T. Tang & F.T. Wang

Paphiopedilum micranthum *T. Tang & F.T. Wang* in Acta Phytotax. Sinica 1, 1: 25, 56 (1951); Cribb, Gen. Paphiopedilum: 93(1987); id., Gen. Paphiopedilum: 91 (1998).; Braem, Paphiopedilum: 40 (1988); Aver., Identif. Guide. Vietnam. Orch.: 17 (1994); Braem, C.Baker & M.Baker, Gen. Paphiopedilum 1: 90 (1998). **Type:** Yunnan, *C.W. Wang* 86182 (holotype PE!).

P. globulosum Z.J. Liu & S.C. Chien in Acta Phytotax. Sinica 40(4): 365, fig. 1 (2002), **syn. nov.** Type: China, Yunnan, Wenshan, *Z.J. Liu* 21052 (holotype PE, isotype Shenzhen City Wutongshan Nurseries Herbarium).

P. micranthum var. *oblatum* Z.J. Liu & S.C. Chen, loc. cit. 366, **syn. nov.** Type: China, Yunnan, Wenshan, 2 May 2001, *Z.J. Liu* 21053 (holotype PE, isotype Shenzhen City Wutongshan Nurseries Herbarium).

DESCRIPTION. *Lithophytic* or *terrestrial herb* with 3–5 distichous leaves and a more or less elongate creeping rhizome 2–3 mm in diameter and up to 25 cm long. *Leaves* oblong-elliptic to elliptic, obtuse, 5–12(15) cm long, 1.5–2 cm wide, mottled dark and pale green above, purple-violet spotted below. *Inflorescence* erect, 1-flowered; peduncle 9–25 cm long, white-villose, dull greenish, purple-brown spotted. *Flower* odourless, large for size of plant, 5.8–9 cm tall, thin-textured; sepals and petals pale yellow, flushed pink above, veined with more or less deep red-purple; lip rose-pink rarely white, or pink-purple, paler towards base, purple-spotted within; staminode white, pink-flushed at the base, yellow above and below toward the apex, spotted with red; pedicel and ovary 3.5–4.6 cm long, brown-purple spotted, white-hairy. *Dorsal sepal* ovate to broadly ovate, obtuse or shortly acute, 1.5–3.6 cm long, 1.7–3 cm wide, white-hairy on outer surface. *Synsepal* broadly ovate to broadly elliptic, usually obtuse or shortly acutely bilobed at apex, 1.8–3.3 cm long, 1.1–2.5 cm wide, bicarinate. *Petals* broadly obovate or elliptic-subcircular, rounded at apex, 1.9–4.3 cm long, 2.3–4.4 cm wide, ciliate, covered with long white hairs on inner surface especially in basal half. *Lip* deeply inflated, broadly-ovate, ovate or elliptic-ovate, 5–10.4 cm long, 3.4–5.6 cm wide. *Staminode* convex, longitudinally conduplicate, broadly-oblong to elliptic, 10–12.5 mm long. 2n = 26. Fig. 76.

DISTRIBUTION. South-western China (south-eastern Yunnan, western and northern Guangxi and south and north-western Guizhou), northern Vietnam (Cao Bang, Ha Giang and Tuyen Quang Provinces). Fig. 36, 77.

ECOLOGY. Primary evergreen and semi-deciduous, mixed and coniferous, rather open, mossy, wet forest on highly eroded limestone mountains, (600)1000–1600 m. Fig. 75.

Fig. 75. Coniferous forests with *Pinus fenzeliana, Pseudotsuga brevifolia* and *Tsuga chinensis* on the top of narrow limestone ridges are the home of *Paphiopedilum micranthum* populations in Vietnam.

N. Vietnam, Cao Bang Prov. (Photo.: L. Averyanov).

FLOWERING IN THE WILD. March–May.

IUCN STATUS. EN.

AFFINITIES. *Paphiopedilum micranthum* is one of the most striking species in subgenus *Parvisepalum.* It has attractive flowers with very large inflated pink lip and incurved sepals and petals, the sepals are yellow, the petal pink, both being more or less veined with deep purple. The flowers look unusually large for the relatively small plant. *Paphiopedilum micranthum*, like its most closely related allies *P. malipoense* and *P. armeniacum*, has boldly chequered or tessellated leaves and long creeping stolons. Its leaves are usually rigid, but its roots are usually brittle, fleshy and succulent. The species has a limited distribution in the north of Vietnam but is more widespread in southern China in SE Yunnan, Guangxi and Guizhou.

HISTORY. *Paphiopedilum micranthum* was described in 1951, by the Chinese orchid specialists

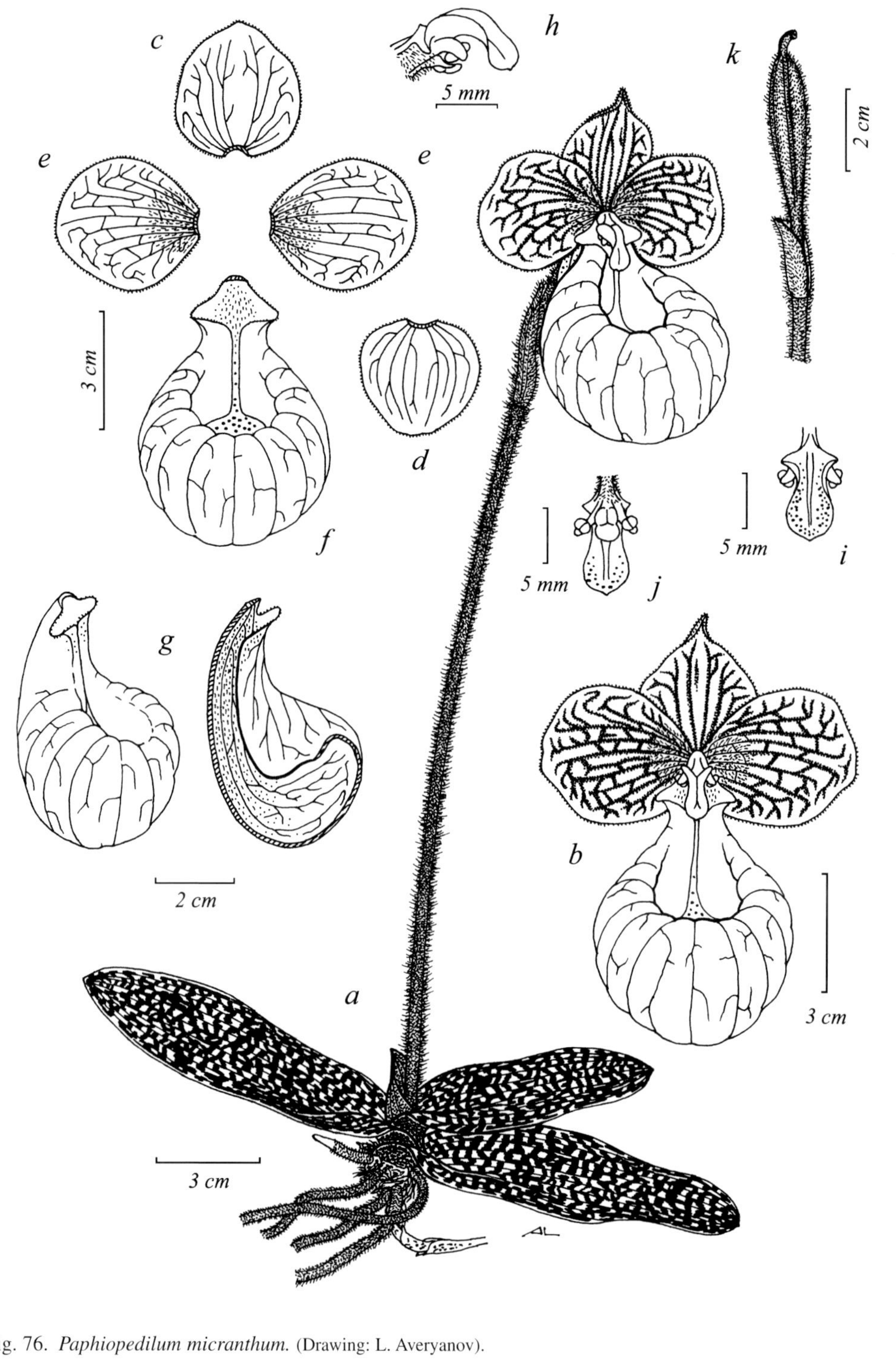

Fig. 76. *Paphiopedilum micranthum.* (Drawing: L. Averyanov).

a — flowering plant; **b** — flower; **c** — median sepal; **d** — synsepal; **e** — petals; **f, g** — lip, frontal, side and section views; **h–j** — column, side, ventral and dorsal views; **k** — ovary and flower bract. (Drawn from CBL 1577).

T.Tang and F.T.Wang, based on a collection by C.W.Wang from between Malipo and Shichou in southeastern Yunnan. Few plants have ever been given a less appropriate name than *P. micranthum*, which is remarkable for its very large flower with an inflated lip borne on a relatively small plant. Examination of the type suggests that the original description was based on a specimen with an aberrant small flower or from a specimen in bud (Cribb, 1998). Occasionally very small flowers in the closely related *P. armeniacum* have also been observed in natural habitats (Cribb, 1998).

S.C.Chen and F.Y.Liu (1982) recognized the anomalous description of *P. micranthum* and provided a new description of its flower in the same article in which they described *P. armeniacum*. It has a single large flower with purple-red-veined sepals and petals and a large inflated pink lip. The white and yellow staminode is longitudinally grooved and spotted with red. Each growth has chequered leaves that are purple-spotted beneath. It is obviously related to *P. malipoense* and *P. vietnamense* from which it differs in having small narrower acute leaves and a differently coloured flower with a distinctive conduplicate staminode. The latter feature also allows its ready distinction from the other pink-flowered species, *P. delenatii*, which also lacks the purple-veined petals and sepals of *P. micranthum*.

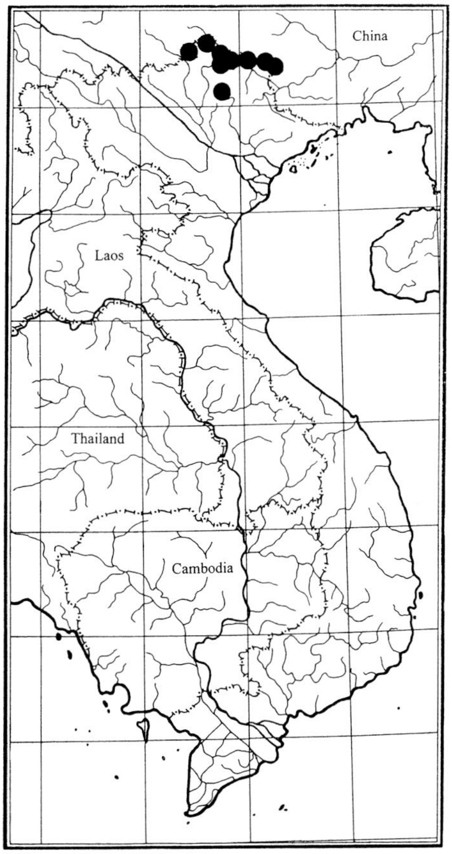

Fig. 77. Distribution of *Paphiopedilum micranthum* in Vietnam. (Drawing: L. Averyanov).

Although *Paphiopedilum micranthum* was described in 1951, it was not seen in cultivation until the early 1980s. Since then it has been exported from China in vast quantities. Plants, which in nature are rhizomatous and form sizeable clumps of many growths, are often cut into single growths for sale. Despite its remote habitat, *P. micranthum* has appeared in large quantities in cultivation since 1984 and it is declining in the wild. Such exploitation is particularly distressing when it has proved so easy to propagate it from seed and vegetatively. Many wild-collected plants are in poor condition by the time they reach nurseries and collections abroad and many, no doubt, fail to survive their first season in cultivation.

For many years *Paphiopedilum micranthum* was regarded as a local endemic distributed in a limited area of South China in south-eastern Yunnan, northern and western Guangxi and south-western Guizhou. It has recently been found in eastern and north-eastern Guizhou (Luo, pers. comm.). It was surprising and remarkable that such a spectacular and locally common orchid could have been overlooked in Vietnamese territory until recently. Numerous populations of this species were discovered in the north of Vietnam in 1998–1999 during field work sponsored by the National Geographic Society and the American Orchid Society (Averyanov & Averyanova, 2000; Averyanov *et al.*, 2000a,b; Nguyen Tien Hiep, Phan Ke Loc and Averyanov, 2000). *Paphiopedilum micranthum* was also recognized among large quantities of wild-collected slipper orchids delivered to a commercial trading company at the end of 1998 (Tanaka, 1998). These plants were presumably collected from Cao Bang Province.

HABITAT. In nature *Paphiopedilum micranthum* is found on very steep slopes and cliffs of mountains composed of crystalline, marble-like limestone exhibiting deep vertical erosion. In China its range extends from south-eastern Yunnan, through northern and western Guangxi to southern and north-eastern Guizhou. In Vietnam, it is distributed in Cao Bang, Ha Giang and the northern part of Tuyen Quang Provinces (Fig. 77).

Occasionally, *Paphiopedilum micranthum* may be found in northern Vietnam as low as 600–800 m in evergreen broad-leaved forests. However, it most commonly occurs between 900 to 1600 m. This is the highest elevation for slipper orchids that inhabit the limestone ridges and peaks in northern Vietnam. Conditions in this area are monsoon tropical climate associated with mountains. Winter here is cloudy but usually very dry with mean temperatures that lie between 8–12°C (Fig. 78). The coldest months are December and January when night temperatures may fall to -2°C. Morning ground hoarfrost and wet snow may be occasionally observed during the coldest winter nights in this area at elevations of 1200–1600 m. Cold north and north-east winds are typical here from late autumn to early spring. During March and the first weeks of April they bring persistent cold mists. Most of the moisture at this time of year comes from this source. The summer, lasting from April to September–October, is the rainy season with moderate fresh temperatures ranging around 18–20°C (Fig. 78). Maximum temperatures in summer occasionally reach 28–30°C. Maximum precipitation falls between May–September. This is a time of very high humidity. Mean annual rainfall ranges from 1800 to 2500 mm.

The primary evergreen and semi-deciduous mixed and coniferous forests that grow along narrow limestone ridges support most populations of *Paphiopedilum micranthum* in Vietnam (Fig. 75). Usual co-dominants in these forests are conifers such as *Pinus fenzeliana, Podocarpus brevifolius, P. neriifolius, P. pilgeri, Pseudotsuga sinensis* var. *brevifolia* and *Tsuga chinensis*. Other rare gymnosperms such as *Amentotaxus argotaenia, Cephalotaxus mannii, Fokienia hodginsii, Nageia fleuryi, N. wallichiana, Taxus chinensis* and *Xanthocyparis vietnamensis* are not uncommon in these forests. In China, *Calocedrus macrolepis* and *Keteleeria calcarea* are also often found in habitats where *P. micranthum* has been observed (Cribb, 1998). Some broad-leaved trees also sometimes appear as important co-dominants in these habitats. Most are deciduous or semi-deciduous trees from such genera as *Acer* (Aceraceae), *Carpinus* (Betulaceae), *Lithocarpus* and *Quercus* (Fagaceae), *Ulmus* (Ulmaceae), *Pistacia weinmannifolia* (Anacardiaceae) and *Platycarya strobilacea* (Juglandaceae). The tallest trees may reach 15–20 m in high.

In the second forest stratum numerous species of smaller trees were observed, including *Alnus nepalensis* (Betulaceae), *Celastrus* sp., *Euonymus* sp. (Celastraceae), *Cornus* sp. (Cornaceae), *Elaeocarpus* sp. (Elaeocarpaceae), *Eriobotrya* sp., *Photinia* sp., *Prunus* sp., *Sorbus* sp. (Rosaceae), *Illicium jiadifengpi* (Illiciaceae), *Garcinia* sp. (Clusiaceae), *Madhuca* sp. (Sapotaceae), *Myrsine kwangsiense* (Myrsinaceae), *Pittosporum* spp. (Pittosporaceae), *Rhododendron* spp., *Vaccinium* spp. (Ericaceae), *Schefflera pesavis* (Araliaceae), *Symplocos* (Symplocaceae), *Tirpitzia sinensis* (Linaceae) and *Wikstroemia* sp. (Thymeleaceae).

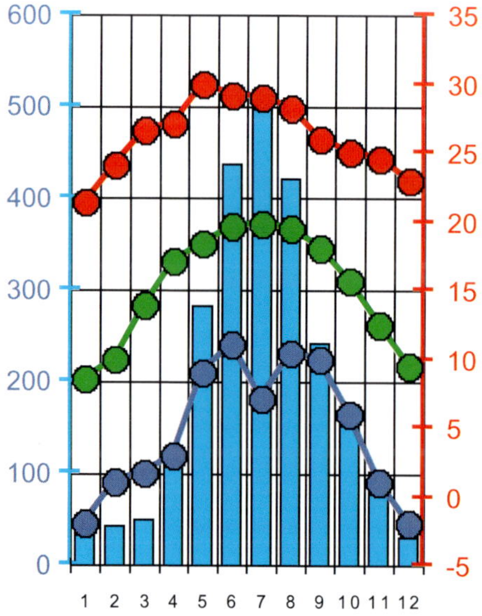

Fig. 78. Climatological data for the area of *Paphiopedilum micranthum* in Vietnam based on data from the nearest weather stations – Ha Giang (22°49'N 104°59'E, 118 m) and Sa Pa (22°20'N 103°50'E, 1570 m) according to Bruzon & Carton (1930) and Nguyen Khanh Van *et al.* (2000).

Black figures at the chart bottom are months. Blue axis represents atmosphere precipitation in mm (columns designate mean monthly rainfall); red axis represents temperature in °C (red line — absolute maximum monthly temperature, green line — mean monthly temperature and blue line means absolute minimum monthly temperature).

Numerous shrubby species are found in the undergrowth of coniferous forests in the area. Among these are lithophytic species of genera such as *Abelia, Lonicera* (Caprifoliaceae), *Agapetes, Rhododendron, Vaccinium* (Ericaceae), *Ardisia* (Myrsinaceae), *Berberis, Mahonia* (Berberidaceae). *Pyracantha, Rosa, Rubus, Spiraea, Stranvaesia* (Rosaceae), *Helwingia* (Helwingiaceae), *Ilex* (Aquifoliaceae), *Lespedeza*, *Sophora* (Fabaceae), *Pittosporum* (Pittosporaceae) and *Polygala* (Polygalaceae). Some dwarf palms, like *Guihaia* sp. and *Rhapis* sp., are also very typical of these rather open forests. Coniferous forests support a very rich flora of lithophytic and epiphytic bryophytes, ferns and herbs. Common herbs here are species of Acanthaceae, Araceae (*Arisaema*), Aristolochiaceae (*Asarum*), Asteraceae (*Ainsleya*), Balsaminaceae (*Impatiens*), Begoniaceae (*Begonia*), Convallariaceae (*Aspidistra, Disporum, Ophiopogon, Peliosanthes, Polygonatum, Tupistra*), Gesneriaceae, Liliaceae (*Lilium*), Melastomataceae (*Phyllogathis, Sonerila*), Podophyllaceae (*Podophyllum tonkinense*), Primulaceae (*Lysimachia, Primula*), Ranunculaceae (*Anemone*), Rubiaceae (*Ophiorrhiza, Hedyotis*), Urticaceae (*Elatostema, Pellea, Pellionia, Procris*) and Violaceae (*Viola*), and numerous sedges (*Carex, Scleria*) and selaginellas (*Selaginella*). The commonest ferns in these forests are species of *Asplenium, Certomium, Diplasium, Polypodium, Polystichum, Pyrrosia* and *Trichomanes*.

A great number of orchids may be seen in these coniferous forests, including numerous lithophytic species such as *Bulbophyllum ambrosia, B. macraei, B. purpureifolium, Coelogyne fimbriata, C. lockii, C. malipoense, Eria coronaria, E. rhombodalis, E. siamensis, Epigeneium amplum, Liparis distans, L. mannii, L. stricklandiana, Monomeria barbata, Pholidota roseans* and *P. yunnanensis*. These can totally cover the mossy rocks of the forest floor. In addition, lithophytic orchids, such as *Anoectochilus calcareus, Bulbophyllum maculosum, Cheirostylis bipunctata, C. yunnanensis, Cleisostoma paniculatum, Cryptochilus luteus, Cymbidium ensifolium, C. floribundum, C. lancifolium, Dendrobium*

Fig. 79. Vertical cliffs and very steep slopes of limestone mountains are typical habitats of *Paphiopedilum micranthum*.

S. China. (Photo.: P.J. Cribb).

Fig. 81. *Paphiopedilum micranthum* in natural habitats.

Specimens of *P. micranthum* usually grow as terrestrial herbs on steep rocky slopes (**a**) and narrow shelves of vertical cliffs (**b**) and, or as a typical lithophyte in karst crevices of limestone rocks (**c**). Northern Vietnam, Cao Bang Prov. (Photo.: L. Averyanov).

chrysanthum, D. nobile, D. thyrsiflorum, Eria globulifera, E. pusilla, E. sutepensis, Goodyera hispida, G. schlechtendaliana, G. viridiflora, Liparis averyanoviana, Malaxis acuminata, Oberonia cavaleriei, Paphiopedilum dianthum, P. henryanum, P. hirsutissimum, P. malipoense, Podochilus microphyllus and *Rhomboda petelotii*, can be locally very common. Epiphytes that are commonly observed here are *Bulbophyllum reptans, Ceratostylis himalaica, Cleisostoma striatum, Dendrobium acinaciforme, D. longicornu, Epigeneium amplum, E. chapaense, Eria pannea, Holcoglossum wangii, Luisia appressifolia, Monomeria barbata, Ornithochilus difformis, Pholidota missionariorum, P. yunnanensis, Sunipia scariosa* and *Thelasis khasiana.*

Paphiopedilum micranthum is found in Vietnam on the summit ridges and very steep slopes near the ridge-tops of rocky karst limestone mountains. In intact habitats under mossy primary old coniferous forest, colonies of this species are equally common on slopes of any exposition (Fig. 79, 80). In such conditions, numerous flowering specimens can commonly be seen on more open and lighter south-facing cliffs (Fig. 80). Plants usually grow on the mossy top edges of rocks, along narrow shelves and in crevices of vertical cliffs, or on the steeply sloping faces of the rocks in thin soils, mosses and leaf litter (Fig. 81,

Fig. 80 (opposite). At high elevation numerous flowering specimens of *Paphiopedilum micranthum* may be seen on open south-facing wet mossy rocky outcrops in primary communities of sedges.

China-Vietnam border. (Photo.: P.J. Cribb).

Fig. 82. In intact habitats *Paphiopedilum micranthum* easily spreads on steep slopes by sending out long stolons (**a**), often this species develops as continuous cover on the forest floor and appears as a dominant of lithophytic herb communities (**b**).

N. Vietnam, Cao Bang Prov. (Photos: L. Averyanov).

82, 83). In the wetter conditions of highland cloud forests it occasionally grows as an epiphyte on the mossy stems of large old trees near the ground. The soils in habitats of *P. micranthum* are brown and crumbly in texture and more or less neutral, ranging in pH from 6.68 to 7.8 (Cribb, 1999). Meanwhile plants sometimes grow directly on dry lime dust or travertine under the hanging rocks that protect colonies from the rain. In such dry conditions, *P. micranthum* develops long fleshy succulent roots and very rigid coriaceous leaves. Species can form large clumps in the wild by sending out long purple-spotted stolons (Fig. 82a) which appear as continuous rhizomes that connect numerous plants in one colony. Commonly, *P. micranthum* inhabits crystalline marble-like limestone with deep vertical karst erosion, in a few cases, however, it has also been seen on softer younger limestone presumably of Mesozoic age.

In some localities in Vietnam *Paphiopedilum micranthum* is still a common plant. In intact, primary, mixed and coniferous forests it often develops as a continuous cover on the forest floor and appears as a dominant of lithophytic herb communities (Fig. 82b). Sometimes, in Vietnam it grows with other slipper orchids, such as *P. dianthum, P. henryanum, P. hirsutissimum* var. *chiwuanum* or *P. malipoense*.

Paphiopedilum micranthum flowers in the wild from mid March to early May (Fig. 83). At this time, numerous flowering specimens may be seen on steep slopes of limestone mountains under light shade of intact primary forests (Fig. 79, 81a,b). Sometimes even more flowers appear in colonies in primary sedges communities that cover wet mossy open rocky outcrops (Fig. 80). Flowers distinctively continue their growth after anthesis and reach a very large size from relatively small buds.

The unscented flowers of *Paphiopedilum micranthum* have no nectar and are pollinated by means of food deception, probably as a result of mimicry of nectar-bearing *Rhododendron* flowers. Flowers of this

Fig. 83. The first flowers of *Paphiopedilum micranthum* in the wild appear in late March or early in April. (Photo: L. Averyanov).

N. Vietnam, Cao Bang Prov.

Fig. 84. Colour variation of *Paphiopedilum micranthum* flowers in Vietnam. (Photos: L. Averyanov).
a–d — N. Vietnam, Cao Bang Prov.

species certainly superficially resemble flowers of some *Rhododendron* species that are abundant with nectar that attracts numerous insects (Fig. 33), especially in their size and colour. Populations of *P. micranthum* are commonly seen growing under flowering *Rhododendron* species with large white or pink flowers. Most visitors to *Rhododendron* flowers also occasionally visit flowers of *P. micranthum.* Probably only a small number of these visitors acts as pollinators of *P. micranthum.* Insects, usually small flies and moths that are too small, are often drowned inside the lip where some rain water usually accumulates. Larger insects, such as some beetles (Cetoniinae), that cannot escape the lip trap by its narrow exits also die and are sometimes are found inside the lip. Without doubt other large insects trapped in the lip are able to gnaw through the lip wall to escape. Some flowers with nibbled holes in the lip have been observed in populations of *P. micranthum*. The water remaining inside the lip of *P. micranthum* after occasional rainfall may be an additional means of attracting some insects as the weather during the flowering period of this species is usually very dry. Some of these insects getting into the lip may act as unspecific pollinators. Between 1998–2000, very few seed capsules were observed in populations of the species.

Nowadays, *Paphiopedilum micranthum* is certainly one of the most endangered slipper orchids in Vietnam. Following logging of large trees in it habitats it vanishes very quickly. Sometimes small colonies and single plants survive in shady gorges and karst holes on the north-facing mountain slopes. But principally, populations in such degraded places are usually doomed to extinction due to irreversible desiccation of the habitat. Few accessible localities boast large populations nowadays. Felling of the forest trees for fuel wood exposes plants to the sun in most localities where dead and chlorotic plants are seen. Commercial collecting for sale is another important factor in the rapid extinction of this species in the wild. Thousands of wild-collected flowering examples were seen on sale in the streets of Hanoi and other Vietnamese cities of in the springs of 2000 and 2001. This wonderful species is very popular within the country, but the majority of specimens in trade die very soon after flowering. Successful cultivation of this highland plant in the hot climate of the Vietnamese lowlands is extremely difficult.

VARIATION. The size and colouration of *Paphiopedilum micranthum* flowers in Vietnamese populations are very variable (Fig. 84, 85). The flowers vary from about the same size as *P. delenatii*, with a dorsal sepal about 1.5–2 cm long and wide, and a lip about 5 cm long and 3.5 cm wide, to flowers that are twice the size with a 2.5–2.6 × 3 cm dorsal sepal, petals up to 3.5 cm wide and a very large lip, 7 cm long and 5 cm across. Lip colour and shape are also variable: from white to pink-purple (Fig. 84), and ovoid to spherical or slipper-shaped (Fig. 85). Leaves of *P. micranthum* in Vietnamese populations may be narrowly lanceolate, oblong and even broadly elliptic, their colour varying from silver-white with deep green venation to uniform very dark green with hardly pronounced tessellation.

Fig. 85. Variation of the lip shape in *Paphiopedilum micranthum* in Vietnam.
a–c — N. Vietnam, Prov. Cao Bang. (Photos: L. Averyanov).

Several forms of *Paphiopedilum micranthum*, originating from China, have apparently been introduced into cultivation in recent years. Fowlie (1989, 1993) attempted to recognize four infraspecific taxa in China: the typical large-flowered pink-lipped variant from south-east Yunnan; var. *marginatum* from south-west Guangxi with a small flower with a rounded lip with a darker lavender pink rim; var. *extendatum* from western Guangxi, with small flowers and a long stalked pink lip; and var. *eburneum* from north-west Guangxi with an ivory white lip and paler petals. Unfortunately, in the first article he omitted any Latin description or type designation, while in the second he contravened Article 33.1 of the International Code of Botanical Nomenclature. Thus all these names are invalid. From observations of large number of flowering plants in Chinese and Vietnamese populations we suspect that the taxonomical recognition of these taxa is unrealistic.

Albino forms of *Paphiopedilum micranthum* have also been reported from northern Vietnam. Usually albino plants have flowers with a white lip, white-yellowish petals, a pale yellowish median sepal and a yellow staminode. Their leaves have very few sparse purple-violet marks beneath. Similar plants exported from China have been described as var. *glanzeanum* O.Gruss & J.Roeth, (in Die Orchidee, Beiheft 2: 16–24, 1994) and as var. *albo-flavum* Braem (in Leafl. Schlecht. Inst. 1: 3, 1994). The occasional appearance of such plants in populations of *P. micranthum* represents a chance mutation and is unworthy of taxonomic differentiation. A spectacular albino cultivar 'Bob Weltz', with a white lip and yellow-marked dorsal sepal, petals and staminode and shown by Robert Weltz Jnr., received a First Class Certificate from the American Orchid Society in January 1994.

Paphiopedilum sect. Emersonianum Aver. & P.J. Cribb

Paphiopedilum sect. Emersonianum *Aver. et P.J. Cribb* **sect. nov.** affinis sect. *Parvisepali*, sed folia magna, 20–35 cm longa, homochroma viridis, inflorescentia uniflora, plantae exstolonosae, flori albo vel pallide viridi-flava, et labello distincto petalo minoris differt. **Typus:** *Paphiopedilum emersonii* Koop. & P.J. Cribb.

DESCRIPTION. *Plants* never develop stolons. *Leaves* leathery, usually 20–35 cm long, 4–6 cm wide, uniformly green on both sides, rarely with a few small purple-violet marks beneath near the base. *Inflorescence* 1 flowered. *Flowers* white or pale yellowish-green, rarely with a few small purple marks near the base of the tepals. *Lip* distinctly smaller than lateral petals.

From the allied section *Parvisepalum* this new section differs in always lacking stolons, having large uniformly green leaves, 20–35 long, a 1-flowered inflorescence, and has white or pale yellowish-green flowers and a lip which is distinctly smaller than the lateral petals. Two species distributed in southern China and northern Vietnam (Fig. 36).

5. PAPHIOPEDILUM HANGIANUM Perner & Gruss

Paphiopedilum hangianum *Perner & Gruss* in Die Orchidee, Beihefte 6: 3 (1999). **Type:** N. Vietnam, Bac Thai Province, *Tong Ngoc Hang* s.n., Nov. 1998, cult. Gruss (holotype HAL 072655, photo. K).

Paphiopedilum singchii Z.J.Liu & J.Y.Zhang in Acta Phytotax. Sinica 38, 5: 468 (2000), **syn. nov. Type**: China, Yunnan, without exact provenance, 12 July 2000, *Z.J. Liu* 2501 (holotype Herb. Shenzhen City Wutongshan Nurseries).

DESCRIPTION. *Lithophytic herb* with 4–7 distichous leaves. *Leaves* coriaceous, ligulate, obtuse or subacute, (12)15–30 cm long, 3–5 cm wide, glossy green above, pale green and distinctly keeled on lower surface. *Inflorescence* erect, 1-flowered (very rarely 2-flowered?); peduncle 6–12 cm long, 4–5 mm in diameter, bright green, shortly white-pilose; bract ovate, obtuse, conduplicate, 2.8–3 cm long, 1.2–2 cm

Fig. 86. Primary, broad-leaved, evergreen dry forests on highly eroded mountains support populations of *Paphiopedilum hangianum.*

N. Vietnam, Tuyen Quang Prov. (Photo.: L. Averyanov).

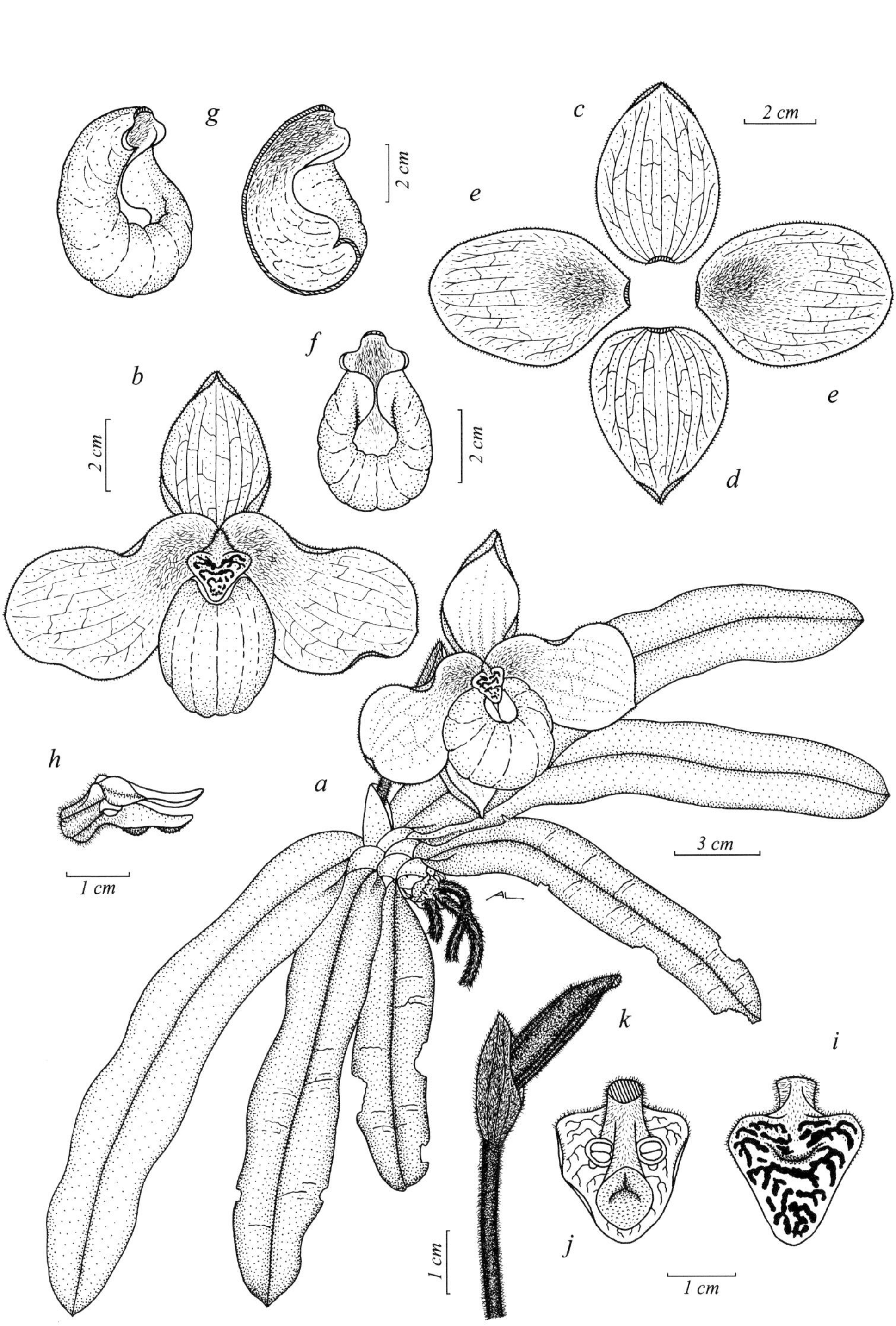

Fig. 87. *Paphiopedilum hangianum.* (Drawing: L. Averyanov).

a — flowering plant; **b** — flower; **c** — median sepal; **d** — synsepal; **e** — petals; **f, g** — lip, frontal, side and section views; **h–j** — column, side, dorsal and ventral views; **k** — ovary and flower bract. (Drawn from HAL 192).

wide, green, shortly white-pubescent, white-ciliate at the margin. *Flower* slightly sweetly fragrant, very large for size of plant, 9–12 cm wide, 6–8 cm tall, subcampanulate or spreading; sepals rather thick-textured, light yellowish or pale yellowish-green; petals light yellowish or pale yellowish-green with purple-flushing at the base, similar in texture to petals; lip yellowish-white, spotted with deep purple within; staminode white or yellowish-white with purple anastomosing markings; pedicel and ovary 3.5–4.2 cm long, bright green, shortly white-hairy. *Dorsal sepal* hooded at first, curving forward over lip, then suberect, recurved on lateral margins, elliptic-ovate, obtuse, 4–5(5.6) cm long, (2.2)3–3.9 cm wide, shortly pubescent on outer surface, keeled on outer side. *Synsepal* broadly ovate, obtuse, (3.5)4–5.5(6) cm long, (3)3.5–4(4.7) cm wide, shortly pubescent on both surfaces, keeled on outer side. *Petals* incurved or spreading, broadly elliptic to broadly ovate, obtuse to rounded at apex, (4)4.5–6(7.2) cm long, (2.5)3–3.5(4.5) cm wide, finely pubescent on both sides, white-villose toward the base, lateral margins sometimes recurved. *Lip* subporrect, deeply urceolate, (3.3)3.5–4.5(5) cm long, 2.3–3.4 cm wide, flared at base, with apical margin incurved and grooved along veins, surface velvety. *Column* short; staminode convex, broadly sagittate, rounded at apex, 11–17 mm long, 11–21 mm wide, with a transverse groove across the basal half; stigma spathulate, very shortly papillose; anthers with dry pollen. Fig. 87.

DISTRIBUTION. Southern China (southern Yunnan?), northern Vietnam (Tuyen Quang Province). Figs. 36, 88.

ECOLOGY. Primary broad-leaved, evergreen, closed forests on very steep slopes and shady cliffs of highly eroded mountains composed of stratified limestone, 450–750 m. Fig. 86.

FLOWERING IN THE WILD. April–May.

IUCN STATUS. EN.

AFFINITIES. *Paphiopedilum hangianum* is amazing recent discovery. This Vietnamese endemic has a very restricted area of distribution. It is closely related to the allopatric *P. emersonii*, and the relationship of these species needs further investigation. Both species obviously occupy a somewhat isolated taxonomic position and are regarded here as worthy of recognition as a separate section of the genus. *Paphiopedilum hangianum* differs from *P. emersonii* in its flower colour, relatively larger tepals and very distinctive white staminode with branching purple venation on it and which is as broad as long. The reported occurrence of this species in China (Liu Zhong-Jian & Zhang Jian-Yong, 2001) is still highly doubtful.

HISTORY. *Paphiopedilum hangianum* was described by Perner and Gruss on the basis of a plant exported from Vietnam in autumn 1998 and flowered in Germany the following spring. Following the increasing rise of interest in all Vietnamese *Paphiopedilum* species, very large quantities of wild-collected plants appeared in local Vietnamese orchid markets all over the country in 1998. At the same time commercial dealers intensively investigated the plants offered for sale for any novelties that showed promise in international trade.

The original samples of *Paphiopedilum hangianum* were exported by Mrs. Tong Ngoc Hang, a Vietnamese orchid grower, after whom it was named when it flowered in Germany. Plants certainly changed hands for high prices and consequently, the correct provenance of the species

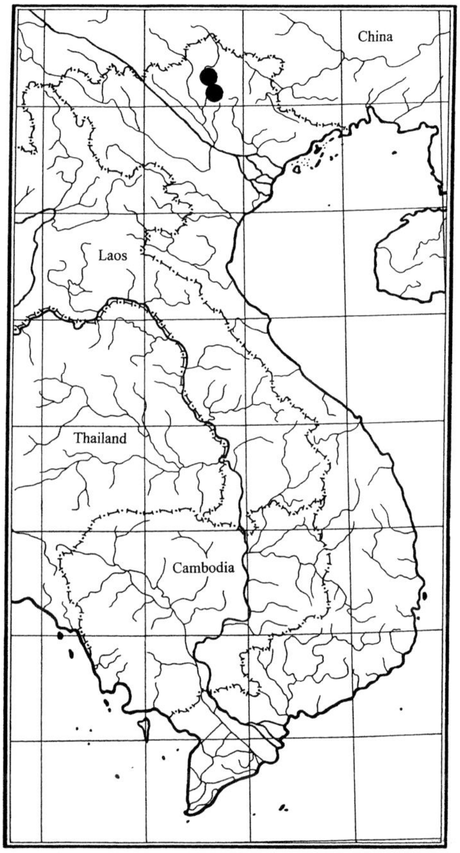

Fig. 88. Distribution of *Paphiopedilum hangianum*. (Drawing: L. Averyanov).

was deliberately hidden. The Vietnamese exporter reported that "the plant came from the province Bac Thai" (an old geographical name presently not presently used).

A little later, great quantities of *Paphiopedilum hangianum* appeared in Vietnamese flower markets and numerous specimens were exported to many countries. There is no doubt that *P. singchii* (Liu & Zhang, 2000), supposedly from southern Yunnan but without exact provenance, is based upon a plant exported from Vietnam. The authors stated that the plants were collected from a mountain bordering Vietnam.

The first populations of *Paphiopedilum hangianum* discovered in the wild by scientists were seen by Averyanov and Phan Ke Loc in the spring of 2001 during field work funded by Flora and Fauna International and the U.S.A. National Science Foundation.

HABITAT. *Paphiopedilum hangianum* inhabits primary, broad-leaved, evergreen, more or less closed forests on highly eroded relatively low limestone mountains between 450–750 m (Figs. 86, 89, 90). The rocks where it grows are composed of stratified, late Paleozoic or early Mesozoic limestone usually consolidated by numerous quartzite dykes of much younger age (Fig. 89).

Recent field exploration has determined the distribution of *Paphiopedilum hangianum* to be a very small area that lies entirely within Tuyen Quang Province (Fig. 88). This is region of monsoon tropical climate with a cold dry winter and warm summer rains. There is very little precipitation during November to early April and this is typical for the area of *P. hangianum*. The usual mean winter temperatures here range from 14 to 16°C with the minimum occasionally falling to 3–5°C in the coldest January and February nights. Summer is warm with maximum rainfall observed in June-August (Fig. 91). Total mean annual rainfall in the area is rather low for Indochina and ranges from 1200–1400 mm. This leads to the formation of rather dry open forests on the karst mountains in the area. The usual mean monthly summer temperatures are about 25–28°C and they rarely exceed 35°C. Heavy fogs and persistent drizzling rains are very common in the area of its distribution from early February until the end of March. Atmosphere humidity at this time is usually 100% but soil and leaf litter on the forest floor often stay absolutely dry.

Fig. 89. Vertical shady cliffs composed of stratified limestone are the typical habitat of *Paphiopedilum hangianum*.

N. Vietnam, Tuyen Quang Prov. (Photo.: L. Averyanov).

Primary, broad-leaved, evergreen, rather dry, more or less closed forests support habitats of *Paphiopedilum hangianum* in the area of its distribution (Fig. 86). The largest trees, reaching 20–25 m tall, commonly include warmth-loving tropical species such as *Burretiodendron hsienmu* (Tiliaceae), species of *Ficus* (Moraceae), *Lithocarpus*, *Quercus* (Fagaceae), *Cinnamomum*, *Persea* (Lauraceae), *Pterospermum truncatolobatum* (Sterculiaceae), *Chukrasia tabularis* (Meliaceae) and even *Dipterocarpus* sp. (Dipterocarpaceae), an indicator of very dry forests. Occasionally gymnosperms such as *Nageia* sp. (Podocarpaceae) are found in these woods. *Dacrydium elatum* (Podocarpaceae) sporadically occurs along ridge tops above habitats of *P. hangianum*. In the forest undergrowth small trees and shrubs are usually seen from families such as Araliaceae (*Schefflera*), Clusiaceae (*Calophyllum*), Euphorbiaceae (*Cleidion*, *Sauropus*), Lauraceae (*Litsea*), Moraceae (*Ficus*, *Streblus tonkinensis*, *S. ilicifolius*), Rubiaceae, Rutaceae (*Micromelum*, *Murraya*) and Sapindaceae (*Dimocarpus longan*). The large palms, *Arenga pinnata* and *Licuala* sp., and a *Pandanus* sp. may be commonly seen in shady canyons and deep rocky gorges on steep slopes near the habitat of *P. hangianum*. Small shrubs like *Ardisia* (Myrsinaceae) and *Biophytum* (Oxalidaceae) are also presents in lithophytic communities.

Representatives of families such as Acanthaceae, Araceae (*Aglaonema*, *Amorphophallus*, *Arisaema*), Aristolochiaceae (*Asarum*), Begoniaceae (*Begonia*), Convallariaceae (*Aspidistra*, *Colonia*, *Ophiopogon*, *Polygonatum*), Gesneriaceae (*Chirita*, *Paraboea*), Rubiaceae (*Ophiorrhiza*) and Urticaceae (*Elatostema*, *Pellionia*, *Procris*) are common terrestrial and lithophytic herbs in these forests. However, the main dominant of herb communities on steep dry slopes and cliffs are some sedges, such as *Carex indica* and *Scleria* sp. (Cyperaceae). Some rather xerophytic ferns and their allies, such as *Adianthum gravesii, Asplenium belangeri, A.*

Fig. 90. *Paphiopedilum hangianum* in its typical habitat.

N. Vietnam, Tuyen Quang Prov. (Photo.: L. Averyanov).

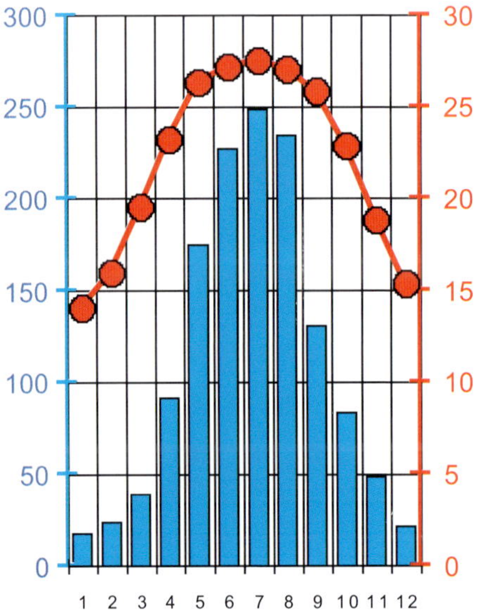

Fig. 91. Climatological data for the area of *Paphiopedilum hangianum*. Information from the nearest weather station — Cho Ra (22°27'N 105°43'E, 210 m) according to Nguyen Khanh Van *et al.* (2000).

Black figures at the chart bottom are months. Blue axis represents atmosphere precipitation in mm (columns designate mean monthly rainfall); red axis presents temperature in °C (red line designates mean monthly temperature).

Fig. 92. *Paphiopedilum hangianum* usually grows on vertical cliffs with its roots spreading along horizontal fissures of stratified limestone. Typical habitat.
N. Vietnam, Tuyen Quang Prov. (Photo.: L. Averyanov).

Fig. 93 (above left). Numerous seedlings and young plants of *Paphiopedilum hangianum* may sometimes be seen growing on soft powdery travertine accumulated on vertical cliffs.

N. Vietnam, Tuyen Quang Prov. (Photo.: L. Averyanov).

Fig. 94 (above right). Flower buds of *Paphiopedilum hangianum* appear in the wild in March but the peak flowering falls in May.

N. Vietnam, Tuyen Quang Prov. (Photo.: L. Averyanov).

griffithianum, A. tenuifolium, A. unilaterale, and species of *Cyclosurus*, *Pyrrosia*, *Tectaria* and *Selaginella*, are present but not abundant in such habitats.

Few orchid species are found in these habitats because of the rather dry natural conditions. Among terrestrial species found here are typical warmth loving, lowland, calcicolous species such as *Calanthe alismaefolia, C. triplicata, Corymborkis veratrifolia, Cymbidium lancifolium, Liparis nervosa, L. paradoxa, Neuwiedia* sp., *Tropidia angulosa* and *T. curculigoides*. Common lithophytes in this area are *Bulbophyllum purpureifolium, Flickingeria* sp., *Hemipilia calophylla, Liparis mannii, Liparis* sp., *Oberonia ensiformis, Pholidota rubra, Paphiopedilum henryanum, P. tranlienianum* and *Podochilus microphyllus*. Epiphytic orchids are rather rare but include *Cleisostoma rostratum, Dendrobium acinaciforme, D. lindleyi, Eria pubescens, Oberonia* sp. and *Phalaenopsis mannii*. *Didimoplexiopsis violacea* Aver., a remarkable species of mycotrophic orchid, was recently discovered growing with *P. hangianum*.

Paphiopedilum hangianum grows in its natural habitats exclusively as a typical lithophytic herb on high sheer shady cliffs with its roots spreading along more or less horizontal fissures of slatey stratified limestone (Fig. 92). Sometimes numerous seedlings and young plants may be seen growing on soft powdery travertine or tufas accumulated on vertical walls below hanging limestone rocky cliffs (Fig. 93). In these habitats it often grows in association with typical obligate lithophytes as *Hemipilia calophylla, Oberonia ensiformis, Paphiopedilum henryanum, P. tranlienianum, Podochilus microphyllus* (Orchidaceae), *Begonia* sp. (Begoniaceae), *Adianthum gravesii, Asplenium belangeri, A.*

griffithianum, A. tenuifolium, A. unilaterale (Polypodiaceae s.l.) and *Selaginella* sp. (Selaginellaceae).

Paphiopedilum hangianum inhabits the middle parts of ridge and mountain slopes between 450–750 m. It flowers in nature in May and early June during the warmest time in the area (Fig. 94, 95).

It is is a narrow endemic in the northern part of Tuyen Quang province in the region adjoining the southern borders of Ha Giang and Cao Bang provinces, as far as the western border of Bac Can province (Fig. 88). The range does not in total exceed 20 by 25 square km. Prior to 1999, it occurred in this area in great abundance, forming very large colonies of magnificent specimens in many places. However, during 2000–2001 the natural populations were almost obliterated by mass over-collection. Because of the high demand from numerous orchid dealers and traders, local people collected hundreds of thousands of specimens from throughout the area in this period. Most of the wild-collected plants were illegally exported abroad. Others were sold on the internal market. In May 2001, hundreds of flowering specimens of *P. hangianum* were seen on the streets of the larger cities of Vietnam. The relatively large natural populations of *P. hangianum* and the incredibly wide commercial collecting resulted in a relatively low market price for this orchid. In spring 2001, the cost in Hanoi flower markets for a flowering plant of *P. hangianum* was only 10,000 Vietnamese Dongs (about US$ 0.7). Regrettably, after two years of intense commercial exploration all known populations of this species in its primary distribution have been collected out. Surviving small colonies of *P. hangianum* may still be found very far from trails and roads in inaccessible mountains that have not yet been visited by commercial collectors. After two years of over-collection, *P. hangianum* is near extinction in the wild.

VARIATION. Observations of numerous flowering plants of *Paphiopedilum hangianum* from different populations give the impression of rather little variation in flower size, form and colouration. The size of flowers varies from 10 to 12(14) cm wide, with rather uniform, broadly ovate or obovate, usually dull yellow or pale greenish-yellow tepals. The petals are flushed with purple-brown at the base. The median sepal is pure yellow or bears sparse purple-brown marks at the base. The staminode is white or light yellowish, regularly with purple, irregular, transverse, anastomosing lines (Fig. 95).

Fig. 95. Typical flower of *Paphiopedilum hangianum* flowered in Hanoi.
N. Vietnam, Tuyen Quang Prov. (Photo.: L. Averyanov).

6. PAPHIOPEDILUM EMERSONII Koop. & P.J. Cribb

Paphiopedilum emersonii *Koop. & P.J. Cribb* in Orchid Advocate 12: 86 (1986). **Type:** China, cult. E.W. Charles (holotype K!).
Paphiopedilum huonglanae N.T.Tich in Hoa Canh 1998, 3: 10–11 (1998), ***nom. inval.***
Paphiopedilum emersonii Koop. & P.J. Cribb var. *huonglanae* (N.T.Tich) N.T.Tich in Hoa Canh 1998, 4: 12 (1998), ***comb. inval.***

DESCRIPTION. *Lithophytic herb* with 4–7 distichous leaves. *Leaves* coriaceous, ligulate, obtuse or subacute, 15–25 cm long, 3–5 cm wide, glossy green above, pale green and distinctly keeled on lower surface, sometimes with sparse purple-violet marks near the base. *Inflorescence* erect, 1-flowered, subtended by a white, papery, basal, sterile sheath; peduncle 10–15 cm long, 4–5 mm in diameter, yellowish-green, velvety white-pilose; bract broadly elliptic, acute, conduplicate, 2.8–3 cm long, 1.2–2 cm wide, white, pubescent, white-ciliate at the margin. *Flower* slightly sweetly fragrant, very large for size of plant, (7.5)8–9.5(10.8) cm wide, sub-campanulate or spreading; sepals rather thick-textured, white; petals white, sometimes faintly flushed pink at base, similar in texture to petals; lip pale yellow, pale pink-orange or creamy with a pink flush around the rim, spotted with deep purple within, turning yellow with age; staminode bright yellow with reddish markings and white margin; pedicel and ovary about 3 cm long, pale yellowish white, velvety white-pilose. *Dorsal sepal* shortly hooded, often curving forward over lip, recurved on lateral margins, elliptic to ovate, obtuse, (2.7)3–4.5(5.1) cm long, (1.5)2–3(3.5) cm wide, shortly pubescent on both surface, keeled on outer side. *Synsepal* broadly elliptic to subcircular, obtuse, (2)2.5–3.5(4.6) cm long, (1.8)2.4–3.5(4.2) cm wide, pubescent on both surfaces, keeled on outer side. *Petals* incurved or spreading, broadly obovate to subcircular, obtuse to rounded at apex, (3.5)4–4.5(5.4) cm long, (2)2.5–3.5(4) cm wide, finely pubescent on both sides, white villose at base, lateral margins sometimes recurved. *Lip* subporrect, deeply urceolate, (3)3.5–4.5(5) cm long, 2–3 cm wide, flared at base, with apical margin incurved and grooved along veins. *Column* short; staminode convex, trullate, truncate, about 1.6–2 cm long, 0.8–1.1 cm wide, deeply longitudinally sulcate with groove dilating towards apex; stigma spathulate, very shortly papillose; anthers with dry pollen. Fig. 96.

DISTRIBUTION. Southern China (southeastern Yunnan, Guangxi, southern Guizhou, and possibly Guangdong Provinces), northern Vietnam (Tuyen Quang Province). Figs. 36, 97.

ECOLOGY. Primary, broad-leaved, evergreen, closed forests on very steep slopes and shady cliffs of highly eroded mountains composed of crystalline limestone at 550–750 m.

FLOWERING IN THE WILD. May–June.

IUCN STATUS. EN.

AFFINITIES. *Paphiopedilum emersonii* is a showy species with very large white flowers and yellowish-orange lip. The flower has broad almost circular petals, a relatively small inflated lip and large conduplicate staminode marked with yellow and purple. It is closely related to *P. hangianum* from which it differ in having white flowers, distinct in the form and the colouration of its staminode. Vegetatively, both species look very similar. The leaves of *P. emersonii,* like those of *P. hangianum*, are uniformly green, entirely lacking tessellation on the upper surface. These two species form an isolated group recognised here as a separate section. For a long time *P. emersonii* was considered as a very isolated local endemic to southern China.

HISTORY. Orchid dealers from Guangxi brought about one hundred plants of this species to Maisie Orchids in Hong Kong in the spring of 1985. The plants were in very bad condition and the owner, Mr. Law Ka Kai, declined to purchase them. They were then taken to the Diamond Farm Company and subsequently exported to the collection of the well-known Californian orchid grower, Emerson Charles via Dr. Sandy Lee. Carefully cultivated, he was able to bring one plant into flower in April 1986 (Fowlie, 1990a). This plant was subsequently used for the description of this exciting novelty by Koopowitz and Cribb (1986). At the time, however, the exact provenance of the plant was unknown. It had been suggested that it had been originally collected at Hai Num, Guizhou Province in southeast China.

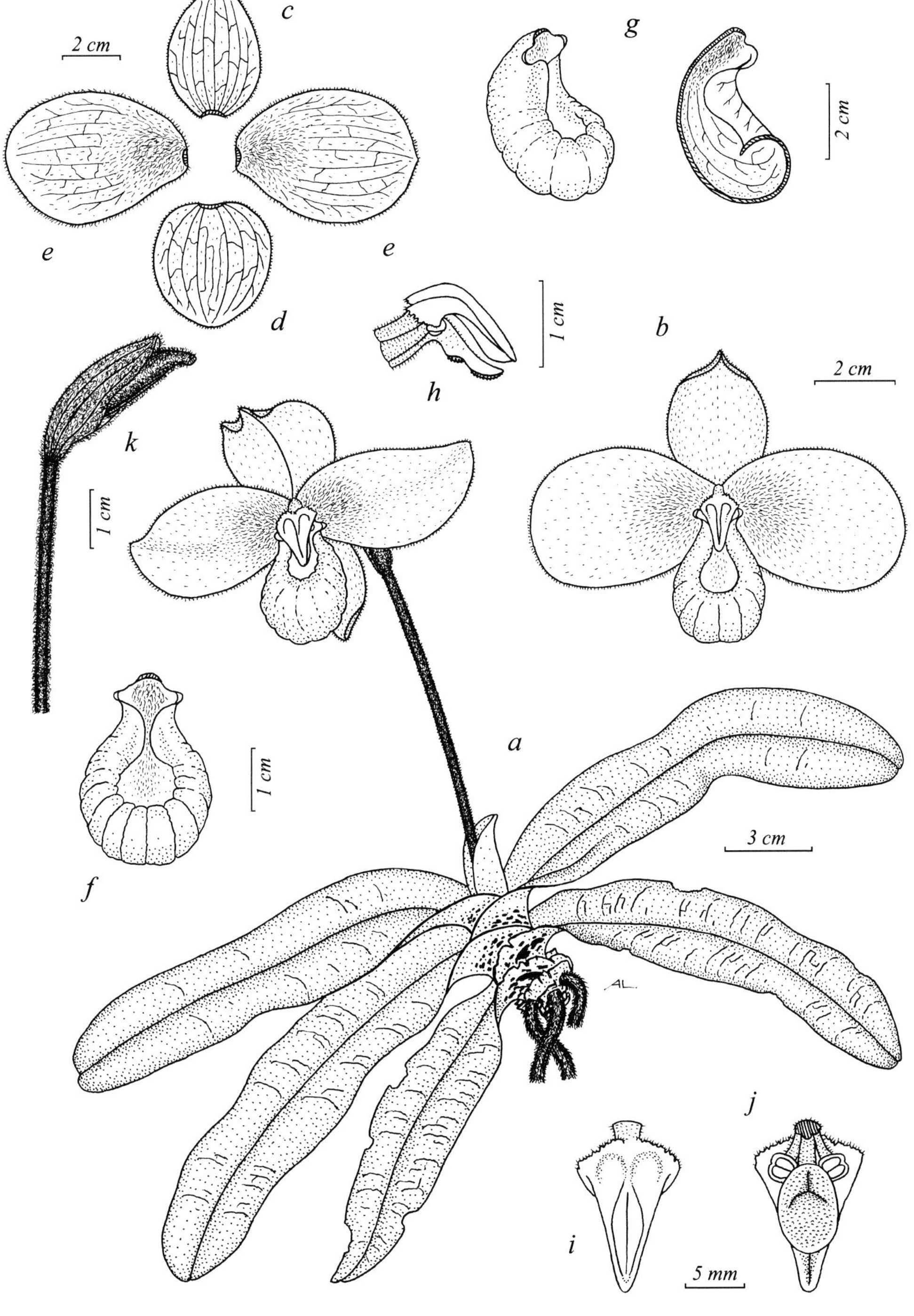

Fig. 96. *Paphiopedilum emersonii* (Drawing: L. Averyanov).

a — flowering plant; **b** — flower; **c** — median sepal; **d** — synsepal; **e** — petals; **f, g** — lip, frontal, side and section views; **h–j** — column, side, dorsal and ventral views; **k** — ovary and flower bract. (Drawn from HAL 31/1).

The well-known American orchid collector Jack Fowlie was the first to describe the habitat of *Paphiopedilum emersonii* in the wild on the border of Yunnan and Guangxi (Fowlie, 1990a). He saw plants on the middle part of mountain slopes at an elevation 600–700 m, growing on vertical limestone cliffs composed of eroded crystalline limestone alternating with remarkable concretions. The vegetation in the area was secondary spiny dense shrubs and open dry forest that replaced the primary forests of the area many years ago. Remnant colonies of *P. emersonii* were growing on shady north-facing cliffs and gorges associated with large mossy concretions with ground water seepages and protected from drought and occasional fires. The plants grew exclusively on rocks with their roots gaining purchase in the cracks and on the sills that abound in such habitats. It was suggested that the plants germinate in moss and send out roots into the clay, sand, calcareous soil and humus mixture in the cracks. Other orchids growing here were *Cymbidium kanran* and *C. forrestii* and a *Phaius* species (Fowlie, 1990a).

On the basis of information from Chinese dealers, Fowlie suggested that the distribution of *Paphiopedilum emersonii* extended from south-western Yunnan through north-western Guangxi to Guangdong Province. Recent field work by Tsi, Chen, Cribb, Luo and Siu has confirmed its occurrence in southern Guizhou (Chen, Tsi & Luo, 1999; Cribb, Luo & Siu, 2001).

The first evidence of the occurrence of *Paphiopedilum emersonii* in Vietnam came in June 1997 from a commercial trading company based near Hanoi who organized wide-scale collecting of Vietnamese *Paphiopedilum* species all over northern Vietnam. According to Tanaka (1998), who has detailed the activities of this company in his homepages, the first examples of *P. emersonii* were found "in December 1997 in Napa, the north side of Viet Nam". This locality is still unclear as it does not apparently exist in Vietnam. However, some indirect information indicates that plants may have been collected from the rocky limestone mountains around Na Phac town in the north of Bac Can Province near the southern border of Cao Bang Province.

The following year, Nguyen Thien Tich (1998a,b) published a description of *Paphiopedilum huonglanae* (a name given in honour of his daughter), in the magazine of the Vietnamese orchid society. It is certainly conspecific with *P. emersonii.* The plants which were used for the description were also obtained from the orchid dealers without any specific information about their origin. According to the author they were collected "in Cao Bang Province in a limestone area near (the) border of Vietnam and China".

Until recently, *Paphiopedilum emersonii* was known in Vietnam only from wild-collected specimens offered for sale, presumably collected in the north of the country (Averyanov *et al.*, 2000). Large quantities have been also illegally exported from Vietnam for international trade. Populations of this rare species were found in the wild in Vietnam for the first time by Averyanov and Phan Ke Loc in the spring of 2001 during field work funded by Flora and Fauna International and the U.S.A. National Science Foundation.

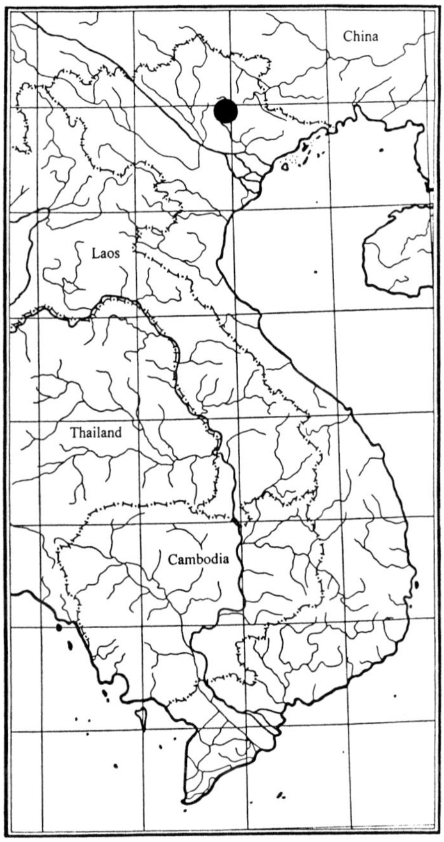

Fig. 97. Distribution of *Paphiopedilum emersonii* in Vietnam. (Drawing: L. Averyanov).

HABITAT. *Paphiopedilum emersonii* has ecological requirements that are very similar to those that we observed for *P. hangianum.* This species inhabits primary, broad-leaved, evergreen, more or less open, dry forests on highly eroded, relatively low, limestone mountains at between

550–750 m. The rocks in these habitats are composed of hard limestone usually consolidated by numerous quartzite dikes and alternating with concretions mainly of chert or schist.

Preliminary field studies suggest that *Paphiopedilum emersonii* currently occupies a small area of limestone in Bac Can and the northern part of Thai Nguyen provinces. In the past it might also have been found in Lang Son Province where its Vietnamese distribution abuts the Chinese part of its range. Unfortunately, primary vegetation there is nowadays nearly totally destroyed and most native species are extinct.

Climatic conditions in the area are very similar to climatic patterns observed in habitats of *P. hangianum*. As usual in the area of monsoon tropical climate with a cold winter and warm wet summer, the winter is dry and cool. Usual mean winter temperatures from November to early April range here from 13 to 15°C. Minimum temperatures are observed during January and February, when night temperatures may occasionally fall to 3–5°C. Summer is relatively warm with maximum rains during June-August (Fig. 98). Total mean annual rainfall in the area is rather low for Indochina at around 1500 mm. This leads to the formation of rather dry, more or less open forests in the area. Usual mean monthly summer temperatures are 24–27°C. They rarely exceed 30–35°C. Cold heavy fogs and persistent drizzling rains are very common in the area from early February until the end of March.

Habitats that support *Paphiopedilum emersonii* in Vietnam are similar to those supporting *P. hangianum*. Main dominants here are trees such as *Burretiodendron hsienmu* (Tiliaceae), *Ficus* spp. (Moraceae), *Lithocarpus* sp., *Quercus* sp. (Fagaceae), *Cinnamomum* sp., *Persea* sp. (Lauraceae), *Pterospermum truncatolobatum* (Sterculiaceae), and *Chukrasia tabularis* (Meliaceae). They reach 20–25 m and form dry, more or less open, primary woods on very steep slopes of highly eroded limestone ridges. The forest structure and basic species composition are very similar to those described for habitats of *P. hangianum*. As with that species, *P. emersonii* grows exclusively as a lithophyte on vertical shady or open mossy cliffs with its roots spreading into crevices of the crystalline limestone. It is usually found on the middle slopes below ridges at elevations of 550–750 m. Flowering in nature falls on May and the first part of June. Despite their similar habitats, *P. emersonii* and *P. hangianum* are found in Vietnam as allopatric species and probably never grow together. Anyway, further investigation of their relationship is necessary.

Paphiopedilum emersonii is a very rare species in Vietnam, all known populations having been predated heavily by commercial collecting. Demand for this attractive species within Vietnam remains high. This has led to the collection of large quantities of plants in recent years. Available information suggests that *P. emersonii* is rarer in Vietnam than *P. hangianum* and is certainly on the verge of extinction in the country.

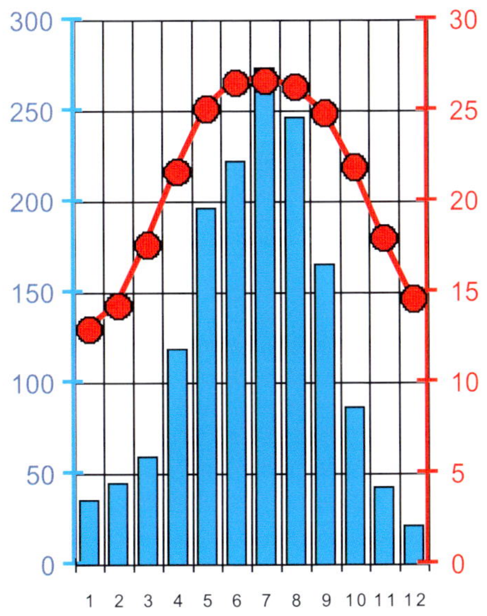

Fig. 98. Climatological data for the area of *Paphiopedilum emersonii* in Vietnam. Information from the nearest weather station — Cao Bac Son (21°54'N 106°19'E, 400 m) according to Nguyen Khanh Van *et al.* (2000).

Black figures at the chart bottom are months. Blue axis represents atmosphere precipitation in mm (columns designate mean monthly rainfall); red axis represents temperature in °C (red line designates mean monthly temperature).

VARIATION. *Paphiopedilum emersonii* is characterized by its green, untessellated leaves and its large, whitish flower, with a pale yellow lip and yellow and red staminode, borne on a long scape and subtended by a white papery bract. It is a variable species, particularly in petal shape and attitude. Fowlie (1990a) listed three variants: var. *kwangnanense* (as *kwangnanensis*) which closely resembles the type specimen; var. *guangxiense* (as *guangxiensis*), from north-western Guangxi, which has narrower, apically somewhat reflexed petals and a paler lip; and var. *angustipetalum*, from Guangdong Province, with a floppy flower with petals reflexed on the lateral margins. None of these names is validly published. More or less similar plants are observed in Vietnamese populations of this species. Most probably they reflect natural variation of the species and do not warrant taxonomic differentiation.

Flowers of *Paphiopedilum emersonii* in Vietnamese populations have usually pure white petals. In some cases, however, petals bear more or less numerous purple marks, especially near the base. The lip may be yellow, yellow-pink or pink-orange. Leaves are often sparsely spotted with purple-violet or are uniformly pale green at the base (Fig. 99). Flowers also vary greatly in size and form which may be subcampanulate of nearly flat (Fig. 100).

Fig. 99. Underside of leaves of *Paphiopedilum emersonii*.

N. Vietnam, Thai Nguyen Prov. (Photo: L. Averyanov).

Fig. 100. Morphological and color variation of the flowers of *Paphiopedilum emersonii* from Vietnamese populations.

Private nurseries, Hanoi. (Photos: L. Averyanov).

Paphiopedilum subgen. Brachypetalum (Hallier) Pfitzer

Paphiopedilum subgen. **Brachypetalum** *(Hallier) Pfitzer* in Engler, Pflanzenr. Orch. Pleon. 12: 55 (1903). **Type:** *Paphiopedilum concolor* (Bateman) Pfitzer.
Paphiopedilum Emerantha Tessellata Pfitzer in Engl. Bot. Jahrb. Syst. 19: 40, p.p. (1894).
Cypripedium sect. *Concoloria* Kraenzl., Orchid. Gen. Sp. 1: 13, 73 (1897).
Paphiopedilum Aphanoneura Brachypetalum Hallier f. in Ann. Jard. Bot. Buitenzorg 14: 34 (1897).
Paphiopedilum sect. *Concoloria* (Kraenzl.) V.A.Albert & Börge Pett. in Lindleyana 9: 137 (1994); Aver., Identif. Guide Vietnam. Orch.: 17 (1994).
Paphiopedilum sect. *Brachypetalum* (Hallier) P.J.Cribb, Gen. Paphiopedilum: 75 (1987).

DESCRIPTION. *Plants* of calcareous substrates, without stolons. *Leaves* are markedly tessellated above and usually purple-marked beneath. *Flowers* pale yellow-cream or white, usually variously spotted with purple-brown or brown. *Median sepal* and *synsepal* more or less similar to the petals in colour, size and shape. Petals broadly elliptic, broadly obovate or ovate to subcircular. *Lip* is ovoid, plain on surface and rather thick in texture, with an involute apical margin. *Column* rather short, stigma and anthers shortly stalked. Staminode is flat and distinctly tridentate at the apex. *Pollen masses* viscous.

Four species distributed in mainland SE Asia, but only one in Vietnam (Fig. 36).

7. PAPHIOPEDILUM CONCOLOR (Bateman) Pfitzer

Paphiopedilum concolor *(Bateman) Pfitzer* in Engler & Prantl, Naturl. Pflanzenf. 2, 6: 84 (1889); Guillamin in Bull. Soc. Bot. France 71: 552 (1924); id., Fl. Gen. Indo-Chine 6: 643 (1934); Karasawa & Saito in Bull. Hiroshima Bot. Gard. 5: 33 (1982); Cribb, Gen. Paphiopedilum: 78 (1987); id., Gen. Paphiopedilum, ed. 2: 110 (1998); Braem, Paphiopedilum: 48 (1988); Seidenfaden, Orch. Indochina: 16 (1992); Aver., Identif. Guide. Vietnam. Orch.: 17 (1994); Braem, C. Baker & M. Baker, Gen. Paphiopedilum 1–2: 98 (1998–1999); Pham-Hoang Ho, Ill. Fl. Viet. 3: 763 (2000). **Type:** Burma, cult. Rucker, *Parish* s.n. (holotype K!; isotype W!).
Cypripedium concolor Bateman in Curtis's Bot. Mag. 91: t. 5513 (June 1865).
Cypripedium concolor Lindl. in Gard. Chron. 1865: 318, (April 1865), *nom. nud.*
Cordula concolor (Bateman) Rolfe in Orchid Rev. 20: 2 (1912).

DESCRIPTION. *Lithophytic herb* with 4–6 distichous leaves and clustered growths. *Leaves* oblong to elliptic-oblong, rounded and minutely emarginate at apex, up to 10–16 cm long, 2.5–4 cm wide, upper surface contrastingly tessellated dark and pale green, usually finely spotted with purple-violet or uniformly purple-violet on lower surface. *Inflorescence* 1–2(3) flowered; peduncle up to 8 cm long, finely white-pubescent, purple or green, heavily spotted purple; bracts ovate, 1–1.5 cm long, 1–1.6 cm wide, white pubescent, green, spotted purple. *Flowers* with a pleasant sweet apple fragrance, relatively large, white-pubescent, 5–7 cm in diameter, dull yellow, yellowish-white, rarely ivory-white, finely spotted all over with brown-purple; ovary 3–5 cm long, white-pubescent, dull purple or green, spotted purple. *Dorsal sepal* broadly obovate, obtuse to emarginate, 2.5–3.7 cm long, 2.5–3.5 cm wide. *Synsepal* concave, elliptic to ovate, obtuse to retuse, 2.1–3.5 cm long, 2.1–3.1 cm wide. *Petals* elliptic, rounded at apex, 3.5–4.5 cm long, 2.3–2.7 cm wide. *Lip* ellipsoidal or narrowly ovoid, very fleshy, with incurved margins, 3–3.8 cm long, 1.3–1.8 cm wide. *Staminode* shortly trullate to subtriangular, toothed to subacute at apex, minutely ciliate, 10–13 mm long, 10–12 mm wide. 2n = 26. Fig. 103.

DISTRIBUTION. Southeast Burma, Southwest China (Yunnan, Guizhou and Guangxi Provinces), northern Vietnam (Hanoi and Hai Phong city districts, Bac Can, Bac Giang, Ha Nam, Lang Son, Nam Dinh, Ninh Binh, Quang Ninh, Son La, Thai Nguyen, Vinh Phuc Provinces), Thailand, Laos and Cambodia? (Figs. 36, 104).

Fig. 101. Vertical high cliffs composed of highly eroded limestone covered by secondary shrubs are the typical habitat of *Paphiopedilum concolor* in Vietnam.

N. Vietnam, Quang Ninh Prov. (Photo.: L. Averyanov).

Fig. 102. *Paphiopedilum concolor* on vertical cliffs composed with soft stratified limestone in primary dry broad-leaved evergreen forest.

China-Vietnam border. (Photo.: P.J. Cribb).

Ecology. Primary and secondary, broad-leaved, evergreen forests and scrub on very steep slopes and cliffs composed of highly eroded limestone of elevations 10–500(600) m. (Figs. 101, 102).

Flowering in the wild. May–June.

IUCN status. VU.

Affinities. *Paphiopedilum concolor* is a small slipper orchid with relatively large pale yellowish fragrant flowers finely spotted with purple. This attractive plant is widely cultivated in Vietnam as it is quite tolerant of the warm climate of the lowlands. It is still common in Vietnam. In the past it occurred in great abundance on numerous limestone hills in the Song Hong River delta near Hanoi.

Paphiopedilum concolor has a wide distribution in limestone areas of mainland South-east Asia. It is allied to *P. godefroyae* from Peninsular Thailand and *P. bellatulum* from Burma, northern Thailand and southwestern China, but differs in the yellow ground colour of its flowers, which are usually minutely purple-spotted, and in its relatively narrower petals, longer lip and rather narrow, deltoid staminode.

HISTORY. *Paphiopedilum concolor* was one of the earliest of the slipper orchids to be found in Vietnam. Balansa and Bon collected it near Hai Phong, while Regnier reportedly found it near Saigon. The last report is doubtful because it is hardly likely that this obligate calcicole was found in the alluvial Mecong delta that lacks any limestone hills. Most probably this record was based on a cultivated plant that came from northern Vietnam.

Paphiopedilum concolor was originally discovered in Burmese Tenasserim by the Rev. Charles Parish, that most meticulous of early collectors. His delightful watercolour of his original collection is preserved amongst his other illustrations in the collections of the Royal Botanic Gardens, Kew. James Bateman first described this species in *Curtis's Botanical Magazine.* However, John Lindley named it, as *Cypripedium concolor*, a few weeks earlier, his account appearing on 8 April 1865, although he failed to publish a valid description. Both accounts were based on a plant from Parish's original collection that was flowered and exhibited in London by Sigismund Rucker.

For many years *Paphiopedilum concolor* was the commonest slipper orchid sold in flower markets all over Vietnam. It was certainly a very common plant on rocky limestone mountains that are distributed

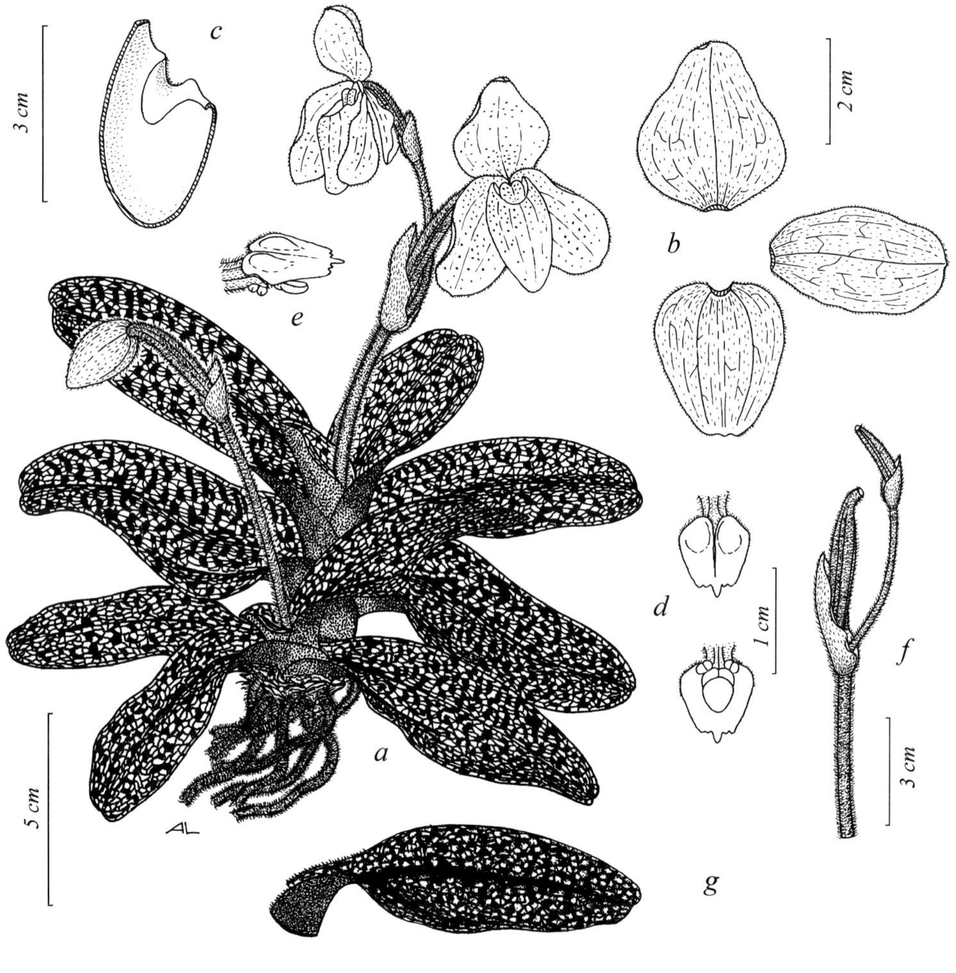

Fig. 103. *Paphiopedilum concolor.* (Drawing: L. Averyanov).

a — flowering plant; **b** — median sepal, synsepal and petal; **c** — lip section; **d, e** — column, dorsal, ventral and side views; **f** — inflorescence with ovaries and flower bracts; **g** — leaf. (Drawn from NTH 2922).

throughout the lowlands of northern Vietnam. However, old herbarium collections of this species from Vietnam are few. Only botanical explorations of recent years have revealed how widespread it really is. Regrettably, this species is now extinct in most of the localities mentioned in old literature.

HABITAT. *Paphiopedilum concolor* is the most widespread species in subgenus *Brachypetalum*. It is also widely distributed in lowland limestone areas of the northern Vietnam.

Fowlie (1978) reported *Paphiopedilum concolor* growing on limestone bluffs from 90–150 m above sea level in peninsular Thailand. He suggested that the seeds germinate on the roots and rootlets of trees growing on the limestone and then send their roots down into the cracks where humus accumulates. In China, this species grows on limestone cliffs and hills, often in karst areas (Cribb, 1998). It can cling to vertical rocks with its roots in crevices in the rock or in deeper soil on steep slopes at the base of rocks where the water runs off. In southern China, it is usually found on north- or north-east-facing slopes, between 300 and 950 m elevation. The soil pH varies from 7.2 to 8 and it usually favours shady places, but mature plants can survive full sun for at least part of the day. In south-east Thailand and Cambodia, Fowlie (1977) reported that *P. concolor* grows in low nutrient, sandy washes near sea level while further north, at Khao Yai, it grows on limestone hills. It is possible that the sandy washes are flushed in the monsoon season by calcareous water.

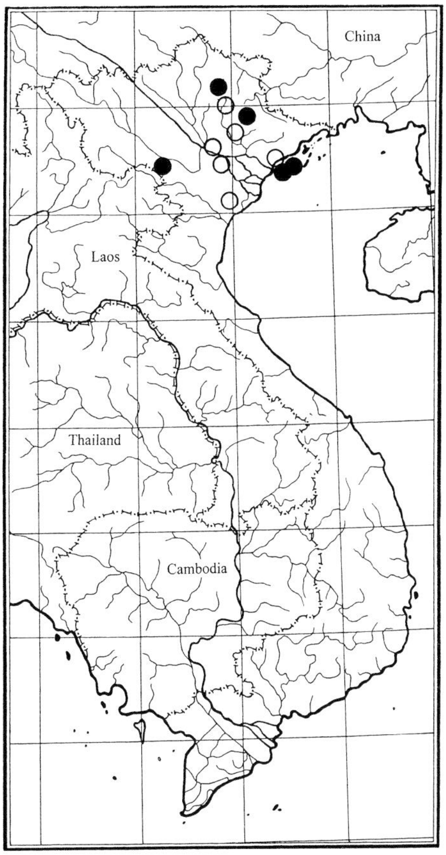

Fig. 104. Distribution of *Paphiopedilum concolor* in Vietnam. (Drawing: L. Averyanov).
Studied populations are marked with dots, rings designate localities from literature .

In Vietnam, *Paphiopedilum concolor* is associated with rocky, low, karst limestone mountains and hills, growing near the summit or on ridges, commonly on north- or north-east-facing rocks. Usually, it grows as a lithophyte on shelves of vertical cliffs or very steep slopes at elevations from sea level to 500(600) m (Fig. 105). However, in undisturbed places this species may be seen on vertical cliffs just above the sea (Fig. 101). It is not rare in rocky limestone regions all over northern Vietnam especially on lowlands facing the sea.

Paphiopedilum concolor is a typical plant of highly eroded limestone mesas and pinnacles widely scattered on the maritime plains and shallow continental shelf of Vietnam. It inhabits an area of typical monsoon tropical climate with a cold, dry winter and warm wet summer where the influence of the summer and winter monsoon is very strong.

The winter in this area is sunny, cool and rainless. The dry winter season lasts from November to early April. Rain at this time is very rare, however humidity is constantly high due to the influence of the sea. Usual median winter temperatures are about 15–18°C with absolute minima usually not lower than 5–6°C (Fig. 106). The coldest months are December and January. The summer is moderately warm to hot with mean temperatures of 26–30°C. Maximum observed temperatures may reach 41–43°C (Nguyen Khanh Van *et al.*, 2000). Hot weather coincides with the rainy season. Maximum rainfall occurs here from June-September. Total mean annual rainfall in the area ranges from 1700–2000 mm.

Paphiopedilum concolor is usually found in Vietnam from the sea level up to 450–500 m. Relatively rarely it occurs up to 550–600 m. It always grows as an obligate lithophyte on rocky, highly eroded, hard or soft, stratified limestone. Primary, broad-leaved, evergreen, dry rather open forests support *P. concolor*.

Fig. 105. Shelves on vertical limestone cliffs are a typical habitat of *Paphiopedilum concolor* in Vietnam. N. Vietnam, Bac Can Prov. (Photo.: L. Averyanov).

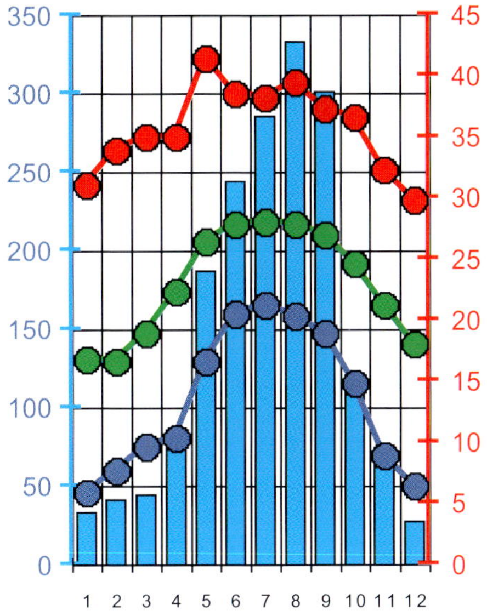

Fig. 106. Climatological data for the area of *Paphiopedilum concolor* distribution in Halong Bay and on maritime plains of Vietnam. Phu Lien weather station (20°48'N 106°38'E, 5 m) according to Bruzon & Carton, 1930.

Black figures at the chart bottom are months. Blue axis represents atmospheric precipitation in mm (columns designate monthly rainfall); red axis represents temperature in °C (red line — absolute maximum monthly temperature, green line — mean monthly temperature and blue line — absolute minimum monthly temperature).

Usual dominants in these forests in Vietnam are *Burretiodendron hsienmu* (Tiliaceae), *Calophyllum* sp. (Clusiaceae), *Choerospondias* sp., *Dracontomelum* sp. (Anacardiaceae), *Chukrasia tabularis* (Meliaceae), *Cinnamomum* sp. (Lauraceae), *Diospyros* sp. (Ebenaceae), *Erythrophleum fordii* (Fabaceae), *Ficus* spp., *Streblus tonkinensis, S. ilicifolius* (Moraceae), *Madhuca* sp. (Sapotaceae), *Pterospermum truncatolobatum* (Sterculiaceae), and *Terminalia* sp. (Combretaceae). These lowland, tropical forests have a very rich assemblage of smaller trees and shrubs, belonging to the numerous families such as Anacardiaceae, Annonaceae, Apocynaceae, Celastraceae, Euphorbiaceae, Flacourtiaceae, Rubiaceae, Rutaceae, Sapotaceae etc. The palms *Livistona halongensis*, *Livistona* sp., *Rhapis* sp., *Arenga pinnata*, and *Licuala* sp. and large *Pandanus* spp. are very common in these forests. On steep open slopes some cycads, such as *Cycas brachycantha* and other *Cycas* spp., are not uncommon. Xerophytic herbaceous vines, including species of *Dischidia* and *Hoya*, and creeping lithophytic ferns, such as *Pyrrosia* spp., are common in *P. concolor* habitats.

The commonest orchids observed in these habitats are lithophytic species such as *Acampe rigida, Calanthe argenteostriata, C. triplicata, Cheirostylis chinensis, C. yunnanensis, Cleisostoma rostratum, Cymbidium aloifolium, Liparis viridiflora, Malaxis acuminata* and *Oberonia ensiformis*. Other typical lithophytes here are succulent gesneriads, such as *Chirita drakei* and *C. hiepii*. Epiphytic orchids are not numerous due to the rather dry conditions, but *Dendrobium acinaciforme, D. lindleyi, D. parishii, Luisia morsei, Ornithochilus difformis, Thelasis khasiana* and *Vanda concolor* are not uncommon. The terrestrial orchids are represented by warm-loving species such as *Calanthe alismaefolia, Corymborkis veratrifolia, Habenaria* spp., *Nervilia* spp., *Tropidia angulosa* and *T. curculigoides.*

Paphiopedilum concolor is the only Vietnamese slipper orchid capable of surviving in highly degraded vegetation. Sometimes remnant populations may be found in highly degraded, dry, open secondary forests and even xerophytic secondary scrub and thickets (Fig. 107). In such conditions plants may survive on open eroded limestone rocks exposed to full sun (Fig. 109). However, its optimum habitat is on shelves of vertical limestone cliffs under light shade of tree canopies in open dry primary forest (Fig. 108, 110). Due to its high tolerance of drought, *P. concolor* may sometimes be seen on high inaccessible vertical limestone cliffs among secondary dry shrubs and grasses close to roads and villages.

Paphiopedilum concolor is very popular in Vietnam as an ornamental plant. Fortunately, collecting of this plant in the wild is limited to the internal market and no large scale collecting has been observed. It is the easiest native species in cultivation and, under appropriate conditions, will grow and flower in the hot climate of the lowlands where the most Vietnamese live.

Fig. 107 (above left). *Paphiopedilum concolor* on open vertical limestone cliffs covered by secondary shrubs.
N. Vietnam, Prov. Quang Ninh. (Photo.: L. Averyanov).

Fig. 108 (above right). In its optimal natural conditions on vertical limestone cliffs under light shade of dry primary forest, *Paphiopedilum concolor* produces large clumps with numerous flowers.
N. Vietnam, Prov. Son La. (Photo.: L. Averyanov).

Paphiopedilum concolor flowers in the wild from last weeks of April to the end of June. The flowering period of wild-collected plants in cultivation in Hanoi begins in March and continues until the first weeks of August.

VARIATION. *Paphiopedilum concolor* is a very variable species in its leaf markings, flower colouring, the shape and length of the petals and the number of flowers produced on each spike. Some of these variations have received formal taxonomic recognition: for example, var. *hennisianum* (*album*) lacks any purple spotting on the flowers or leaves; var. *chlorophyllum* lacks the characteristic purple markings on the underside of the leaf; var. *longipetalum* has narrowly spatulate petals 5 cm long and 1.5 cm wide; var. *striatum* has petals streaked rather than spotted with purple; and var. *sulphurinum* lacks purple-spotting on both flowers and leaves.

Guillaumin (1924, 1934) recognized two varieties of *Paphiopedilum concolor* in Vietnam: var. *regnieri* Rchb.f. from near Saigon with inflorescences of four to five deep yellow flowers with large petals; and var. *tonkinense* Guillaum. with a larger dorsal sepal and petals that are densely spotted towards the base. The former is deeply suspicious. Nobody has seen *P. concolor* with 4–5 flowers in nature. If such a plant really ever existed in the Ho Chi Minh city (Saigon) area it would certainly have been known to commercial collectors years ago. We have not heard about any such orchid. The latter falls within the normal range of natural variability of *P. concolor* and scarcely warrants formal recognition. Albinistic variants are also occasionally brought into cultivation, e.g. 'Albino Jamboree' was awarded by the American Orchid Society in 1993. A similar plant was named as *P. concolor* var.

Fig. 109. *Paphiopedilum concolor* may survive in open eroded limestone rocks exposed to full sun.
N. Vietnam, Quang Ninh Prov. (Photo: L. Averyanov).

Fig. 110. Leaves of *Paphiopedilum concolor* in optimal conditions.
N. Vietnam, Lang Son Prov. (Photo: L. Averyanov).

Fig. 111. Commonest variants of *Paphiopedilum concolor* in Vietnam. (Photos: L. Averyanov).
a — northern Vietnam,. Son La Prov;
b — northern Vietnam, Quang Ninh Prov.;
c, d — northern Vietnam, Hai Phong Prov.

album Braem. The staminode also is very variable in the flower of *P. concolor*. It is usually longer than broad and tapers to an obscurely tri-dentate apex that is well illustrated by Seidenfaden (1972).

In Vietnam, flowers of *Paphiopedilum concolor* commonly vary from pure white to dull yellow (Fig. 111). In some populations the tips of the tepals in flowers are light greenish. Usually the petals and lip have fine purple-brown spotting, but occasionally tepals are unspotted. In shady wet conditions the leaves of *P. concolor* may be contrastingly marked above with bright silver and deep green tessellations (Fig. 110) but, in sunny places, the leaves are usually dull green with indistinct whitish tessellation (Fig. 109). On their lower side, the leaves in Vietnamese plants are usually pale dull green with more or less dense purple-violet spotting. Occasionally, they may be pure dull green, or uniform dark violet.

Paphiopedilum subgen. Paphiopedilum

Paphiopedilum subgen. **Paphiopedilum:** Karasawa & Saito in Bull. Hiroshima Bot. Gard. 5: 39 (1982). **Type:** *Paphiopedilum insigne* (Wall. ex Lindl.) Pfitz. (*Cypripedium insigne* Wall. ex Lindl.).

Paphiopedilum subgenus *Polyantha* (Pfitzer) Brieger in Schlechter, Die Orchideen, ed. 3: 175 (1971).

Description. *Plants* lacking stolons. *Leaves* glossy green or tessellated green and pale green above, with sparse to dense purple-violet marks at the base beneath. *Flowers* odourless or with a slight smell of urine, have different colouration, the median sepal and petals often spotted or striped. *Petals* tapering, oblong or spathulate, more than twice as long as broad. *Lip* goblet-like, plain on surface and rather thick in texture, with incurved side-lobes, apical margin of the pouch is either straight or slightly recurved. *Column* rather short; stigma and anthers shortly stalked; staminode ovate, broadly obovate, subquadrate, obcordate or rarely more or less lunate, usually with three apical teeth, flat on dorsal surface, sometimes with a central raised glossy boss (umbo), glabrous or pustulate; the pollen masses viscous.

About 62 species distributed in mainland tropical Asia, the Malay Archipelago, the Philippines, New Guinea and the Solomon Islands (Fig. 37).

Paphiopedilum sect. Paphiopedilum

Paphiopedilum sect. **Paphiopedilum**: Karasawa & Saito in Bull. Hiroshima Bot. Gard. 5: 39 (1982). **Type:** *Paphiopedilum insigne* (Wall. ex Lindl.) Pfitz. (*Cypripedium insigne* Wall. ex Lindl.).

Cypripedium sect. *Insignia* Kraenzl., Orch. Gen. Spec. 1: 78 (1897).

Paphiopedilum sect. *Stictopetalum* Hallier f. ex Pfitz. in Engler, Pflanzenr. Orch. Pleon.: 69 (1903); Karasawa & Saito in Bull. Hiroshima Bot. Gard. 5: 41 (1982).

Paphiopedilum sect. *Neuropetalum* Hallier f. ex Pfitz. in Engler, Pflanzenr. Orch. Pleon.: 71 (1903), *nom. illeg.*

Paphiopedilum sect. *Thiopetalum* Hallier f. ex Pfitz. in Engler, Pflanzenr. Orch. Pleon.: 75 (1903); Karasawa & Saito in Bull. Hiroshima Bot. Gard. 5: 41 (1982).

Paphiopedilum sect. *Cymatopetalum* Hallier f. ex Pfitz. in Engler, Pflanzenr. Orch. Pleon.: 76 (1903).

Paphiopedilum sect. *Ceratopetalum* Hallier f. ex Pfitz. in Engler, Pflanzenr. Orch. Pleon.: 77 (1903); Karasawa & Saito in Bull. Hiroshima Bot. Gard. 5: 42 (1982).

Characteristic features. *Leaves* uniformly green, sometimes with deep green venation, usually with sparse purple-violet marks at the base beneath. *Inflorescence* erect or arching, single-flowered (very rare 2-flowered). *Median sepal* and synsepal distinctly different from petals in colour, size and shape. *Petals* oblong to narrowly obovate, much narrower than median sepal. *Lip* with large, well developed incurved side-lobes, its apical margin straight (not involuted). *Staminode* usually obcordate with rugose surface and central shiny boss (umbo).

14 species distributed from south and north-east India to south-west China, Indochina and southern Thailand (Fig. 37).

8. PAPHIOPEDILUM HIRSUTISSIMUM (Lindl. ex Hook.) Stein

Paphiopedilum hirsutissimum *(Lindl. ex Hook.) Stein*, Orchideenbuch: 470 (1892); Pfitzer in Engler, Bot. Jahrb. 19: 41 (1894); id., in Engler, Pflanzenr. Orch. Pleon.: 69 (1903); Karasawa & Saito, Bull. Hiroshima Bot. Gard. 5: 40 (1982); Cribb, Gen. Paphiopedilum: 138 (1987); id., Gen. Paphiopedilum, ed. 2: 220 (1998); Braem, Paphiopedilum: 125 (1988); Seidenfaden, Orch. Indochina: 19 (1992); Aver., Identif. Guide Vietnam. Orch.: 19 (1994); Aver. et al. in Orchids 66, 4: 378, 380 (1997); Braem, C.Baker & M.Baker, Gen. Paphiopedilum 1–2: 258 (1998–1999); Pham-Hoang Ho, Ill. Fl. Viet. 3: 764 (2000). **Type:** Without provenance, cult. Parker (holotype K!).
Cypripedium hirsutissimum Lindl. ex Hook. in Curtis's Bot. Mag. 83: t. 4990 (1857).
Cordula hirsutissima (Lindl. ex Hook.) Rolfe in Orchid Rev. 20: 2 (1912).

DESCRIPTION. *Lithophytic* or *terrestrial herb* with 5–7 distichous leaves and clustered growths. *Leaves* linear-ligulate, obtuse at obliquely bilobed apex, up to 45 cm long, 1.5–2 cm wide, green, spotted with purple on lower surface near the base. *Inflorescence* 1-flowered (very rarely 2-flowered); peduncle 17–25 cm long, densely covered in long hairs, subtended at base by green leaf-like sheath up to 11 cm long; bract elliptic, subacute to obtuse, 1.5–2.8 cm long, pubescent. *Flower* (6)10–14(16) cm across; sepals pale yellow to pale green, heavily suffused with glossy dark brown almost to margins; petals pale yellow, spotted purple-brown in lower half, flushed rose-purple in apical half; lip pale yellow to pale olive-green, spotted with rose-purple; staminode pale yellow, purple-spotted towards base, glossy dark brown towards middle; pedicel and ovary 5–7.5 cm long, densely covered by long black hairs. *Dorsal sepal* broadly ovate or ovate-elliptic, obtuse to emarginate, (1.8)3–4.5(5.2) cm long, (1.8)2.3–4(4.5) cm wide, with undulate margins, ciliate. *Synsepal* similar, (1.8)3-(4)4.5 cm long, (1.1)1.5–3(4) cm wide. *Petals* horizontal to deflexed, spathulate, rounded at apex, (2.6)5–7(8) cm long, (1)1.3–2.5(3) cm wide, half twisted towards apex, usually strongly undulate on basal margins, pubescent, ciliate. *Lip* (1.8)3–4(4.5) cm long, (1.2)1.5–2.2(2.8) cm wide. *Staminode* subquadrate, obtuse, convex, about 10 mm long, 8 mm wide. 2n = 26.

DISTRIBUTION. Northeast India (Meghalaya, Manipur, Mizoram and Nagaland), Myanmar, southwest China (Yunnan, Guizhou, Guangxi), northern Thailand, northern Laos, and Vietnam (Cao Bang, Ha Giang, Hoa Binh, Lao Cai, Son La, Thai Nguyen and Thanh Hoa Provinces). Fig. 37.

The species is represented by three varieties, namely, var. *hirsutissimum*, var. *esquirolei* and var. *chiwuanum*. The typical variety is distributed in north-eastern India and Myanmar. In Vietnam, both var. *esquirolei* and var. *chiwuanum* occur.

KEY TO THE VARIETIES

1. Flowers large, (10)12–16 cm wide; petals (5)7–8 cm long, in basal half strongly undulate on margins . . . . **8a. *Paphiopedilum hirsutissimum* var. *esquirolei***
+. Flowers small, 6–8(10) cm wide; petals (2.6)3–4(5) cm long, straight or scarcely undulate on margins . . . . **8b. *Paphiopedilum hirsutissimum* var. *chiwuanum***

Fig. 112. Primary evergreen broad-leaved forest on steep slopes of highly eroded crystalline limestone mountains still support large populations of *Paphiopedilum hirsutissimum* var. *esquirolei* in Vietnam.

N. Vietnam, Cao Bang Prov. (Photo.: L. Averyanov).

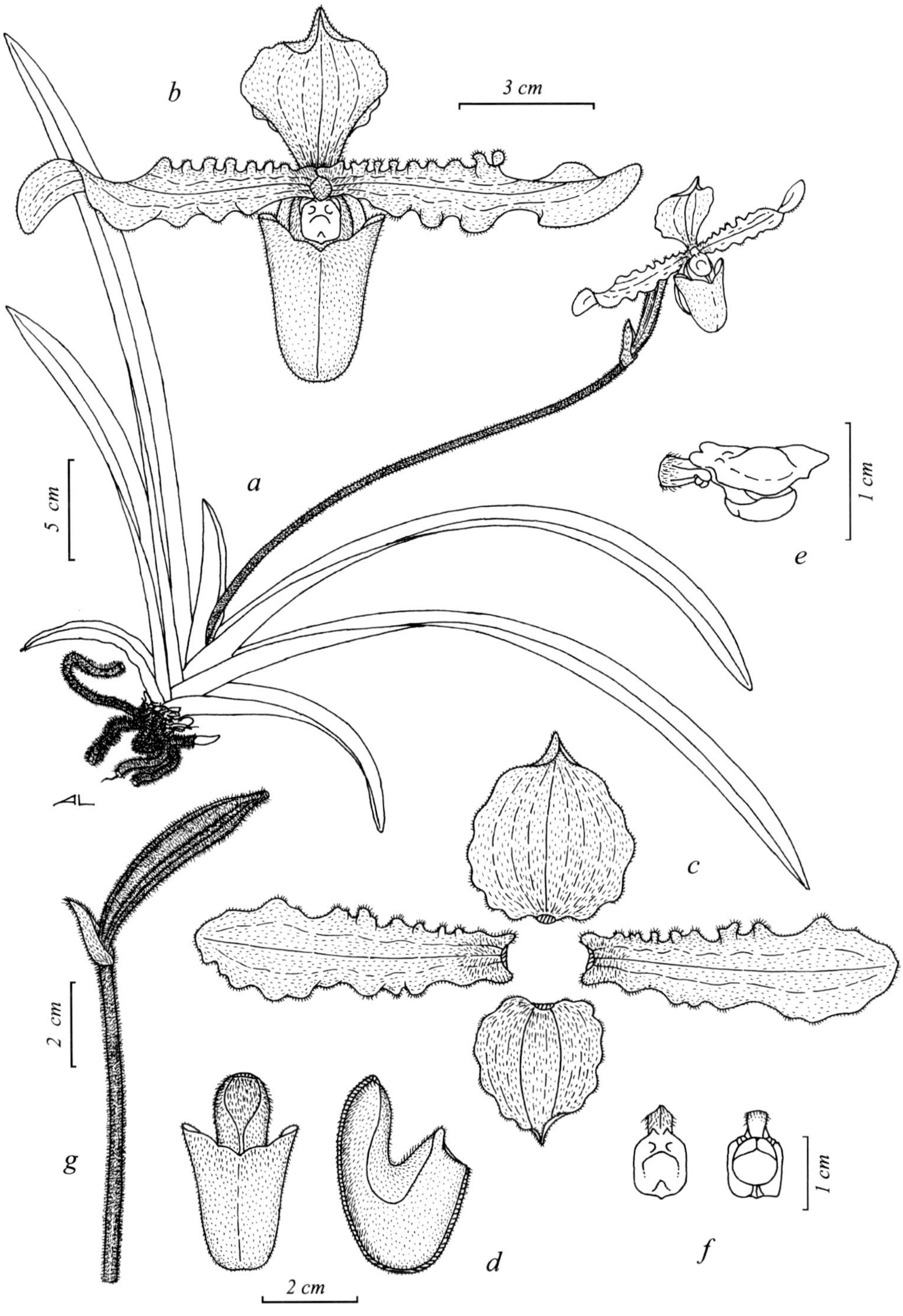

Fig. 113. *Paphiopedilum hirsutissimum* var. *esquirolei*. (Drawing: L. Averyanov).

a — flowering plant; **b** — flower; **c** — flattened sepals and petals; **d** — lip frontal and side section views; **e, f** — column, side, dorsal and ventral views; **g** — ovary and flower bract. (Drawn from Averyanov *et al.* s.n., 30 April 1995).

8A. PAPHIOPEDILUM HIRSUTISSIMUM VAR. ESQUIROLEI (Schltr.) Karasawa & Saito

Paphiopedilum hirsutissimum var. **esquirolei** *(Schltr.) Karasawa & Saito* in Bull. Hiroshima Bot. Gard. 5: 40 (1982); Cribb, Gen. Paphiopedilum: 140 (1987); id., Gen. Paphiopedilum, ed. 2: 225, p.p. (1998); Aver. et al. in Orchids 66, 4: 378, 380 (1997). **Type:** China, Guizhou (Kweichow), *Esquirol* 3277 (holotype B†; isotype P!).

Paphiopedilum esquirolei Schltr. in Repert. Spec. Nov. Regni Veg. Beih. 4: 39 (1919); Braem, C. Baker & M. Baker, Gen. Paphiopedilum 1–2: 254(1998–1999).

Cordula esquirolei (Schltr.) Hu in Rhodora, 27: 105 (1925).

Paphiopedilum saccopetalum Hua in Die Orchidee 49, 1: 38 (1998), *forma aberr.*

DESCRIPTION. Differs from the typical variety in its large *flowers*, (10)12–16 cm wide; peduncle and ovary shortly purple-black pubescent. *Dorsal sepal* 4.2–5.2 cm long, 3.4–4.5 cm wide. *Synsepal* 4.2–4.5 cm long, 2.9–4 cm wide. *Petals* (5)7–8 cm long, 2.1–3 cm wide. *Lip* 3.8–5.2 cm long, 1.5–2.8 cm wide. Fig. 113.

DISTRIBUTION. Southwest China (south-eastern Yunnan, Guizhou, southern Guangxi), northern Thailand, northern Laos, Vietnam (Cao Bang, Ha Giang, Hoa Binh, Lao Cai, Son La, Thai Nguyen and Thanh Hoa Provinces). Fig. 114.

ECOLOGY. Primary and secondary, evergreen, broad-leaved, mixed and coniferous forests on highly eroded crystalline limestone mountains at elevations of (350) 500–1100 (1250) m. (Fig. 112).

FLOWERING IN THE WILD. March–May.

IUCN STATUS. VU.

AFFINITIES. *Paphiopedilum hirsutissimum* is one of the most distinctive species in the genus, readily recognized by its very hairy peduncle, long spathulate petals, undulate on the upper margin, relatively small, blunt dorsal sepal and convex, subquadrate staminode.

The typical variety comes from north-eastern India (Assam) and Burma. Pradhan (1976) gives its distribution in India as Jowai (Meghalaya), Mizo (Mizoram) and Naga Hills (Nagaland and Manipur) where it grows as an epiphyte or lithophyte at elevations between 1200 m and 1800 m. A record from 200 m elevation in Moreh, Nagaland, is reported by Ghatak and Devi (1986). Its range in north-eastern India is subject to monsoon rains from June to September and the temperature between December and February can drop to 1°C (G. Pradhan, 1972).

In the east and south-east part of its range, the typical variety is replaced by var. *esquirolei*, another vicarious variety that is widely distributed in northern Thailand and southern China, from both of which large importations have originated in recent years. They differ from typical *P. hirsutissimum* mainly in having shorter hairs on the

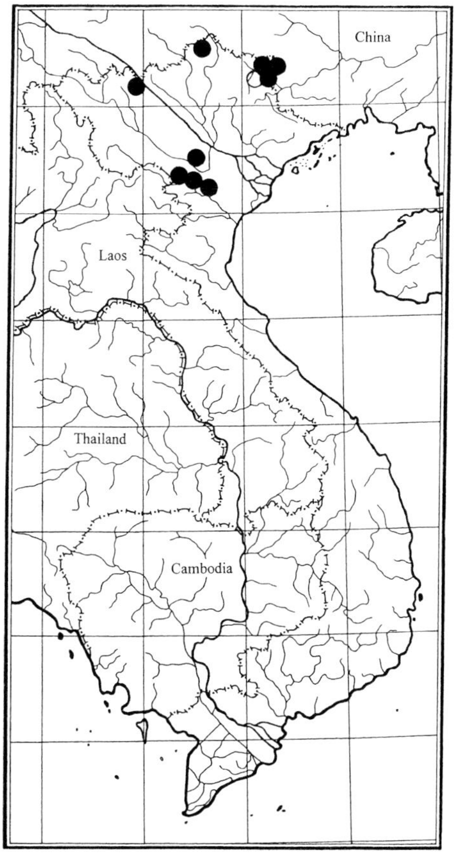

Fig. 114. Distribution of *Paphiopedilum hirsutissimum* var. *esquirolei* in Vietnam. (Drawing: L. Averyanov).

Studied populations are marked with dots, open circle designates locality from literature.

ovary and peduncle and less hairy flowers. The type of var. *esquirolei* was collected by Esquirol in Guizhou Province, southern China, and it differs in only minor ways from typical *P. hirsutissimum*. The minor nature of the differences between it and *P. hirsutissimum* seem insufficient to recognize it at specific rank and we agree with Karasawa & Saito (1982) in considering it a variety of *P. hirsutissimum*.

Small flowering forms of *P. hirsutissimum* discovered in south-eastern Yunnan were described as *Paphiopedilum chiwuanum* T.Tang & F.T.Wang. These plants, distributed in southern China and northern Vietnam, are very closely related to var. *esquirolei*. They are the result of altitudinal differentiation of the species, but they scarcely warrant recognition at higher than varietal rank.

HISTORY. *Paphiopedilum hirsutissimum* was described in 1857 by John Lindley on the basis of a plant that originated from north-eastern India and was cultivated at Kew. The distribution of this species in the broad sense encompasses the extensive mountainous regions of northeastern India, Myanmar, northern Thailand, southern China, northern Laos and northern Vietnam (Cribb, 1998). However, only one wild-collected specimen from Vietnam had been reported prior to recent field studies of this species (Averyanov, 1998b, 2000b; Averyanov & Averyanova, 2000; Averyanov *et al.*, 1997b; Averyanov *et al.*, 2000a,b). It had been recorded from of Cao Bang province in Northern Vietnam but without exact indication of the locality (Averyanov, 1990d, 1994). Samples of *P. hirsutissimum* of reputed Vietnamese origin have been cultivated for many years in a number of private Vietnamese collections, mostly in Dalat in central Vietnam (Pham Hoang Ho, 1993). At the same time, numerous samples of wild-collected *P. hirsutissimum* are occasionally seen in local ornamental plant markets near Hanoi and by the roadside in limestone regions in the north of the country. Unfortunately, all this indirect evidence of *P. hirsutissimum* in Vietnam lacks documentation and precise locality data.

Fig. 115. Vertical shady limestone cliffs in primary and secondary evergreen broad-leaved humid forests are the typical habitat of *Paphiopedilum hirsutissimum* var. *esquirolei* in northern Vietnam.

a, b — N. Vietnam, Cao Bang Prov. (Photos: L. Averyanov).

Numerous large populations of *Paphiopedilum hirsutissimum* have been found in many localities throughout the limestone regions of northern Vietnam during recent field work (Averyanov, 1998b, 2000b; Averyanov & Averyanova, 2000; Averyanov *et al.*, 1997b). It is now apparent that *P. hirsutissimum* is the commonest and most widespread slipper orchid species in northern Vietnam. It is a mystery why this striking orchid remained unknown to French botanists who had extensively explored the region.

In most of its important morphological features, including flower colouration, undulation of petals, and vegetative and floral dimensions, the Vietnamese plants generally seem to be slightly different from typical *P. hirsutissimum*. They have dense, but rather short, dark brown hairs on the flowering stalk, ovary and dorsal surface of the median sepal and synsepal. Moreover, the petals of the plants usually appear to be narrower on the average.

The description of *P. esquirolei* (Schlechter, 1919), from the south of China and from northern Thailand, is in full agreement with the morphology of the Vietnamese plants. It is clear, however, that *P. esquirolei* differs little from typical *P. hirsutissimum*. It represents an eastern vicariant race of the widely distributed *P. hirsutissimum*, and is reasonably treated at the rank of variety (Karasawa & Saito, 1982; Cribb, 1987, 1998; Averyanov *et al.*, 1997).

HABITAT. *Paphiopedilum hirsutissimum* var. *esquirolei* is quite common in southern China and northern Vietnam. Without doubt, it is also very common in limestone areas of northern Laos, but documentary evidence is still unavailable.

In Vietnam it is found in areas of deeply eroded limestone plateaus, partially degenerated into karst and more or less interrupted by river valleys. These zones of karst topography have numerous limestone mountains and mesas with steep cliffs, large caves, craters and lakes (Fig. 112). Forests near the top of limestone mountains are usually rather dry, as rainwater disappears quickly into numerous holes and crevices in the rock. At the elevations where *P. hirsutissimum* var. *esquirolei* is found, however, the natural conditions are much wetter and the primary vegetation produces rather dense shade.

Primary, evergreen, broad-leaved, mixed and coniferous, usually humid forests on highly eroded, crystalline, limestone mountains support populations of var. *esquirolei* in Vietnam. In a number of places, until recently, intact primary vegetation still harboured very large numbers of these slipper orchid (Fig. 115). The species is most frequent between about 500–1100 m, but occasionally may be found as low as 350–400 m and as high as 1200–1250 m.

The climate is typical for the regions of northern Vietnam with a cold, dry winter and with warm summer rains. The summer, which lasts here from May to September or October, is moderately warm to hot and wet with temperatures about 22–27°C (Fig. 116). Winter is the dry period, with the coldest months

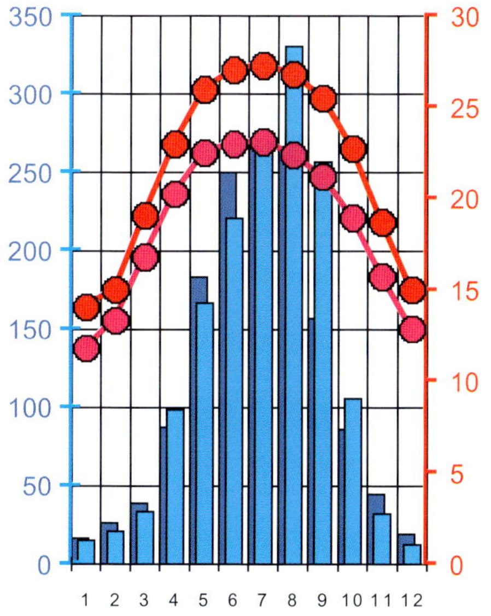

Fig. 116. Climatological data most typical for the area of *Paphiopedilum hirsutissimum* var. *esquirolei* in Vietnam.

Black figures at the chart bottom are months. Blue axis represents atmosphere precipitation in mm (columns designate mean monthly rainfall); red axis represents mean monthly temperature in °C. The deep blue and red are designated data from the Cao Bang weather station, 22°39'N 106°14'E, elev. 258 m; light blue and purple are data from the Moc Chau weather station, 20°51'N 104°38'E, 958 m, all according to Nguyen Khanh Van *et al.*, 2000.

Fig. 118. Typical flower of *Paphiopedilum hirsutissimum* var. *esquirolei*.
N. Vietnam, Cao Bang Prov. (Photo.: L. Averyanov).

in December and January when night temperatures may plunge to 5°C. During March and the first weeks of April, cold heavy fogs and mists are common. Mean annual rainfall usually varies in this area between 1400–1600 mm. Typical climatic data for the area in Vietnam are presented in Fig. 116.

Paphiopedilum hirsutissimum var. *esquirolei* usually grows in Vietnam as a lithophyte in large colonies on vertical or near vertical, rocky slopes. In primary vegetation it occurs in light shade on slopes of any exposition (Fig. 115). The orchid is commonly found in the rather damp crevices and hollows of moss-covered limestone rocks where leaf litter accumulates and the soil is rich in humus. Its roots find a precarious purchase in narrow cracks or sills, eventually forming a basket that catches leaf litter from the shading forest to fertilize the orchid as it grows. Plants can develop to a considerable size with 50 or more growths being not uncommon. In densely inhabited regions plants usually survive only on inaccessible high cliffs where collectors cannot reach them (Fig. 117). The leaves spread or are pendent and the flowers are borne almost horizontally from the rock-face (Fig. 115). The pH of the friable soil around the roots ranges from 7.5–7.86 (Cribb, 1998). In nature, its flowering period lasts from the middle of March to May.

Its usual habitats are very rich in orchids and it is often found in association with other rarer slipper orchids such as *Paphiopedilum dianthum, P. helenae, P. henryanum, P. malipoense* and *P. tranlienianum.* Other epiphytic, lithophytic and terrestrial orchids are also often abundant here, particularly *Aerides odorata, Bulbophyllum purpureifolium, B. ambrosia, Bulbophyllum* spp., *Calanthe argenteostriata, C.*

Fig. 117 (opposite). In remote mountain regions of Vietnam large colonies of *Paphiopedilum hirsutissimum* var. *esquirolei* may still be seen on high shady cliffs inaccessible to commercial collectors.
N. Vietnam — China border area. (Photo.: P.J. Cribb).

triplicata, Cheirostylis eglandulosa, Cleisostoma striatum, Cleisostoma spp., *Coelogyne fimbriata, Coelogyne* spp., *Cymbidium ensifolium, C. lancifolium, C. suavissimum, Dendrobium aduncum, Dendrobium* spp., *Eria coronaria, E. globulifera, E. pannea, E. pusilla, E. sutepensis, E. thao, Flickingeria* spp., *Gastrochilus minutiflorus, Liparis conopea, L. distans, L. mannii, L. nervosa, L. viridiflora, Luisia morsei, Malaxis acuminata, Oberonia ensiformis, Oberonia* spp., *Pholidota chinensis, P. convallariae, P. rubra, P. roseans, Podochilus cultratus, P. microphyllus, Thrixspermum* spp. and many others. The structure and species composition of woody plants in these forests in Vietnam are similar to those described for habitats of *P. malipoense* var. *malipoense.* The main forest dominants here are commonly broad-leaved trees such as *Burretiodendron hsienmu* (Tiliaceae), *Carpinus* sp. (Betulaceae), *Chukrasia tabularis* (Meliaceae), *Cinnamomum* sp., *Persea* sp. (Lauraceae), *Ficus* spp. (Moraceae), *Garcinia* sp. (Clusiaceae), *Lithocarpus* spp., *Quercus* spp. (Fagaceae), *Madhuca* sp. (Sapotaceae), *Pistacia weinmannifolia* (Anacardiaceae), *Platycarya strobilacea* (Juglandaceae) and some others. Conifers such as *Amentotaxus argotaenia, Calocedrus macrolepis, Cephalotaxus mannii, Nageia fleuryi, Podocarpus neriifolius* and *P. pilgeri* occasionally appear in these forests but only at high elevations, commonly above 900–1000 m. Many lithophytic herbs also occur here especially species from the Acanthaceae, Araceae (*Aglaonema, Amorphophallus, Arisaema*), Begoniaceae (*Begonia*), Convallariaceae (*Aspidistra, Disporum, Ophiopogon, Polygonatum*), Gesneriaceae (*Chirita*), Rubiaceae (*Ophiorrhiza*) and Urticaceae (*Elatostema, Pellionia, Procris*). Very common here are also numerous sedges (*Carex* spp., *Scleria* spp.) and lithophytic ferns (*Adiantum* sp., *Asplenium* spp., *Cyclosorus* spp., *Pyrrosia* spp., etc.).

VARIATION. The form and size of flowers vary somewhat. Most flowers are 14–16 cm across with densely pubescent petals 7–8 cm long, strongly undulate on the upper margins (Fig. 118). It is worth remarking that undulation of the petal margins in *P. hirsutissimum* var. *esquirolei* varies greatly during the lifespan of the long-lasting flowers. Flowers that are just opening usually have nearly flat straight petals, but soon their margins became characteristically undulate. Very rarely, 2-flowered inflorescence can be found in populations of this normally single flowered orchid (Fig. 119). Like other slipper orchids, albino forms are also known for this species. The first albino variant lacking any purple in the flower, reported by van Delden (in Paphiopedilum World 1: 36 (1971)), received an Award of Merit as var. *alba* Hort. from the Royal Horticultural Society in June 1986 when shown by Paul Phillips of Ratcliffes, Chilton, Oxfordshire. A similar plant was later named as *P. hirsutissimum* var. *esquirolei* f. *viride* by O.Gruss and J.Roeth (in Caesiana 12: 57, 1999).*

Paphiopedilum saccopetalum was described by Hua in 1998 based upon a cultivated plant

Fig. 119 (above). Specimens of *Paphiopedilum hirsutissimum* var. *esquirolei* that develop a 2-flowered inflorescence may be found very rarely in the wild.
N. Vietnam, Cao Bang Prov. (Photo.: L. Averyanov).

* **Type:** Without provenance, cult. Paphanatics, Los Angeles, USA, 10 April 1998, leg. Olaf Gruss & Juergen Röth (holotype HAL 078846).

Fig. 120. Highland primary coniferous and mixed forest on highly eroded crystalline limestone mountains is the home of *Paphiopedilum hirsutissimum* var. *chiwuanum* in Vietnam.

N. Vietnam, Cao Bang Prov. (Photo.: L. Averyanov).

collected in Guangxi Province in China and flowered by M.H.Chao. This plant obviously represents a teratological form of *P. hirsutissimum* var. *esquirolei* which is very common in this area. It has a flower with sack-like petals that resemble the lip. Similar teratological flowers are not uncommon in the genus (Cribb, 1998).

The widespread deforestation of limestone mountains is the main factor leading to the disappearance of *P. hirsutissimum* var. *esquirolei* everywhere in northern Vietnam. Small isolated colonies of slipper orchids may remain for a while on inaccessible vertical limestone cliffs, particularly above large karst holes and caves and especially on the wetter and more shaded slopes with a northern exposure similar to those observed in southern China (Fowlie, 1990c). In most cases, however, the subsequent desiccation of the habitats associated with irreversible degradation of primary vegetation by man leads to the eventual extinction of orchids in these places.

In limestone areas of northern Vietnam large quantities of plants of this variety are still collecting for sale for the internal flower market and for export (Tanaka, 1998). This is additional important factor in the rapid disappearance of this slipper orchid in Vietnam. As a result, this rather common species has become rather rare nowadays.

8B. PAPHIOPEDILUM HIRSUTISSIMUM VAR. CHIWUANUM (T. Tang & F.T .Wang) P.J. Cribb

Paphiopedilum hirsutissimum var. **chiwuanum** *(T. Tang & F.T. Wang) P.J. Cribb*, Gen. Paphiopedilum: 140 (1987); Aver. et al., Orchids 66, 4: 378, 380 (1997). **Type:** China, southeast Yunnan, Foo-ning Hsien, Ban-loun, 700 m., 15 April 1940, *Wang* 88252 (holotype PE!). *Paphiopedilum chiwuanum* T. Tang & F.T. Wang in Acta Phytotax. Sin. 1: 56 (1951).

DESCRIPTION. Differs from the typical variety in having smaller *flowers*, 6–8(10) cm wide and a shortly purple-black pubescent peduncle and ovary. *Dorsal sepal* 1.8–2.6 cm long, 1.8–2.5 cm wide. *Synsepal* 1.8–2.4 cm long, 1.1–1.5 cm wide. *Petals* shorter, straight or hardly undulate on margins, 2.6–4(5) cm long, 1–1.3 cm wide. Lip 1.8–2.6 cm long, 1.2–1.5 cm wide. Fig. 121.

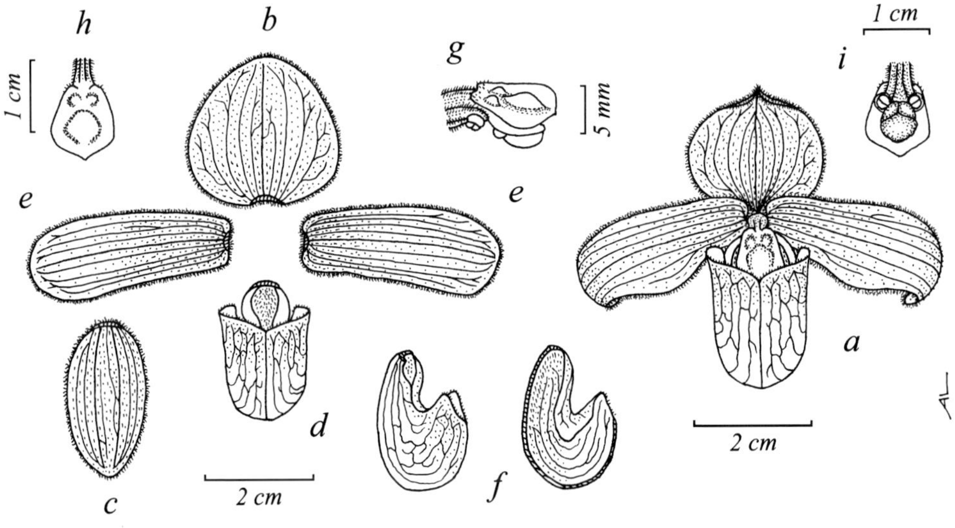

Fig. 121. *Paphiopedilum hirsutissimum* var. *chiwuanum*. (Drawing: L. Averyanov).

a — flower; **b** — median sepal; **c** — synsepal; **d** — lip, frontal view; **e** — petals; **f** — lip side and section views; **g–i** — column, side, dorsal and ventral views. (Drawn from CBL 1588).

DISTRIBUTION. Southwest China (southeastern Yunnan), northern Vietnam (Cao Bang, Hoa Binh and Son La Provinces). Fig. 122.

ECOLOGY. Primary and secondary evergreen broad-leaved, mixed and coniferous forests on highly eroded crystalline limestone mountains at elevations of (750)900–1450 m. Fig. 120.

FLOWERING IN THE WILD. March–May.

IUCN STATUS. VU.

AFFINITIES AND HISTORY. Small-flowered samples of *Paphiopedilum hirsutissimum* occurring in southern China have been called *Paphiopedilum chiwuanum*. The flowers of such samples rarely exceed 6–8 cm in width with the petals usually plane or slightly undulate on the margins. Additionally these orchids are obviously smaller in all other parts of the plant.

Paphiopedilum chiwuanum was described in 1951 by the Chinese botanists T.Tang and F.T.Wang based on a C.W. Wang collection from Foo-ning Hsien, Ban-Loun in southeast Yunnan where it was growing on rocks at 700 m altitude. This plant appears to differ from the typical variety and from var. *esquirolei* in its much smaller flowers with petals only 2.6 cm long that lack undulate margins. On the basis of examination of the type collection, Cribb (1987) recognized it as a distinct variety of *P. hirsutissimum* but it is apparent that occasional plants with flowers that have short petals lacking distinct undulations on their upper margins occur from time to time in populations of var. *esquirolei*. Therefore, Chen & Liu (1982) relegated both *P. chiwuanum* and *P. esquirolei* to the synonymy of *P. hirsutissimum*. However, we are able to recognize var. *esquirolei* and var. *chiwuanum* without any problem.

HABITAT. The known distribution and habitat preference of *P. hirsutissimum* var. *chiwuanum* in Vietnam suggests that this taxon may have arisen as an ecological variant of var. *esquirolei*. Without doubt both plants are very closely related and often connected by intermediate forms in sympatric populations. However, *P. hirsutissimum* var. *chiwuanum* is usually found at higher elevations in Vietnam between about 900–1450 m and most frequently around 1100–1400 m (Fig. 123). At these elevations, typical *P. hirsutissimum* var. *esquirolei* rarely occurs.

Climatic conditions in typical habitats of *P. hirsutissimum* var. *chiwuanum* are much cooler and wetter than in habitats of var. *esquirolei*. Annual mean rain precipitation here is usually higher than 1700 mm. The summer at elevations where this plant occurs is wet and moderately warm, but winter is dry and cold (Fig. 124). The temperature during coldest winter nights may fall to -2–4°C.

Conifers, such as *Calocedrus macrolepis*, *Nageia* sp., *Pinus fenzeliana*, *Podocarpus neriifolius*, *P. pilgeri*, *Pseudotsuga brevifolia* and *Tsuga chinensis* often comprise the first stratum of the primary forests that support populations of *P. hirsutissimum* var. *chiwuanum* in Vietnam (Fig. 120). Numerous semi-deciduous and deciduous broad-leaved tree species are also found here, including *Acer* (Aceraceae), *Carpinus* (Betulaceae), *Lithocarpus*, *Quercus* (Fagaceae), *Ulmus* (Ulmaceae), *Pistacea weinmannifolia* (Anacardiaceae) and *Platycarya strobilacea* (Juglandaceae). The orchid flora of these forests is very rich in highland species, such as *Anoectochilus*

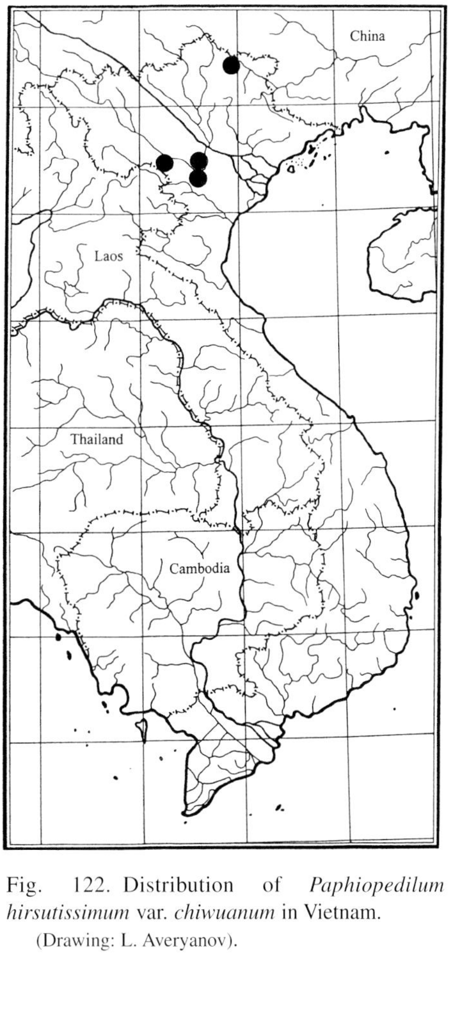

Fig. 122. Distribution of *Paphiopedilum hirsutissimum* var. *chiwuanum* in Vietnam. (Drawing: L. Averyanov).

Fig. 123. Shady limestone cliffs are the typical habitat of *Paphiopedilum hirsutissimum* var. *chiwuanum* in Vietnam.

a, b — N. Vietnam, Cao Bang Prov. (Photos. **a**: L. Averyanov; **b**: P.J. Cribb).

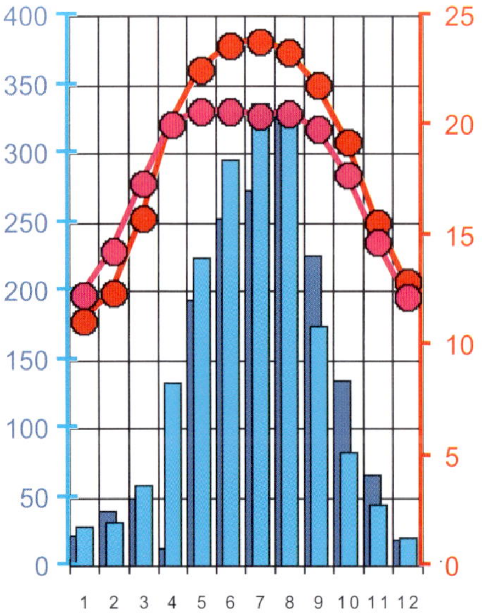

Fig. 124. Climatological data most typical for the area of *Paphiopedilum hirsutissimum* var. *chiwuanum* in Vietnam.

Black figures at the chart bottom are months. Blue axis represents atmospheric precipitation in mm (columns designate mean monthly rainfall); red axis represents mean monthly temperature in °C. The deep blue and purple are designated data from the Bac Ha weather station, 22°32'N 104°17'E, elev. 957 m; light blue and red are data from the Pha Dinh weather station, 21°34'N 103°30'E, 958 m, all according to Nguyen Khanh Van *et al.*, 2000.

Fig. 125. Typical flower of *Paphiopedilum hirsutissimum* var. *chiwuanum*.
N. Vietnam, Cao Bang Prov. (Photo.: L. Averyanov).

Fig. 126. Intermediate form between *Paphiopedilum hirsutissimum* var. *esquirolei* and var. *chiwuanum* in its natural habitat. (Photos: L. Averyanov).

a, b — N. Vietnam, Cao Bang Prov.

calcareus, Bulbophyllum ambrosia, B. reptans, Calanthe argenteostriata, Ceratostylis himalaica, Cheirostylis bipunctata, Coelogyne fimbriata, Cymbidium floribundum, C. suavissimum, Epigeneium chapaense, Eria coronaria, Dendrobium chrysanthum, D. fimbriatum, D. longicornu, D. thyrsiflorum, Goodyera schlechtendaliana, Holcoglossum wangii, Nephelaphyllum tenuiflorum, Pholidota missionariorum, P. yunnanensis, Podochilus microphyllus and *Rhomboda petelotii*. Sometimes *P. hirsutissimum* var. *chiwuanum* occurs in association with other highland Vietnamese slipper orchids such as *P. micranthum* and *P. malipoense.*

VARIATION. *Paphiopedilum hirsutissimum* var. *chiwuanum* shows wide variation in the size of its flowers, leaves and plant in Vietnamese populations (Fig. 125). These plants certainly are closely allied and possibly derived from *P. hirsutissimum* var. *esquirolei.* In some localities, both varieties and more or less numerous intermediate forms co-exist (Fig. 126). According to some observations the varietal diagnostic features have a clinal character and their variation is continuous. It is quite possible that further field studies will reduce this taxon to the synonymy of var. *esquirolei* as suggested by Cribb (1998).

9. PAPHIOPEDILUM BARBIGERUM T. Tang & F.T. Wang

Paphiopedilum barbigerum *T. Tang & F.T. Wang* in Bull. Fan. Mem. Inst. Bot. Ser. 10: 23 (1940); Cribb, Gen. Paphiopedilum: 140 (1987); id., Gen. Paphiopedilum, ed. 2: 234 (1998); Braem, C. Baker & M. Baker, Gen. Paphiopedilum 2: 216 (1998–1999). **Type:** China, Kweichow (Guizhou), *Cavalerie & Fortunat* 1794 (holotype PE!, isotype P!).

Paphiopedilum insigne (Wall. ex Lindl.) Pfitzer var. *barbigerum* (T.Tang & F.T.Wang) Braem, Paphiopedilum: 113 (1988).

DESCRIPTION. Small *lithophytic herb* with 4–6 distichous leaves and clustered growths. *Leaves* suberect or horizontally spreading, curved, narrowly lanceolate to linear, obtuse, often minutely tridenticulate at apex, 8–14 cm long, 0.7–1.2 cm wide, uniformly green. *Inflorescence* 1-flowered, erect, 12–16 cm long; peduncle 9–11 cm long, slender, shortly brown to dull yellowish pubescent; bract elliptic, subacute to obtuse, 1.8–3.5(4) cm long, shortly pubescent near the base. *Flower* about 6.5 cm across; dorsal sepal white with green base and less commonly a central pinkish vein; petals honey-coloured or ochre with a cream or green margin; lip tan-brown; staminode yellow with orange umbo; pedicel and ovary 2.7–3 cm long, shortly brown to dull yellowish pubescent. *Dorsal sepal* subcircular, obtuse, 3–3.2 cm long, 2.5–3.3 cm wide. *Synsepal* elliptic, obtuse, 2.5–3.5 cm long, 1.3–1.7 cm wide. *Petals* oblong-ligulate to ligulate-spathulate, obtuse, 3.3–3.4 cm long, 0.8–1 cm wide, hairy at base, margin undulate and sparsely ciliate. *Lip* glabrous on outer surface 2.7–3 cm long, 1.5–2.8 cm wide. *Staminode* transversely elliptic to obovate-obcordate, 6–9 mm long, 8–11 mm wide, umbonate in center, finely papillose.

DISTRIBUTION. The typical variety is found in southwest China (southern Guangdong, Guangxi, southern Guizhou (Kweichow), south-eastern Yunnan).

In Vietnam this species is represented only by the variety *lockianum.*

9A. PAPHIOPEDILUM BARBIGERUM VAR. LOCKIANUM Aver.

Paphiopedilum barbigerum var. **lockianum** *Aver.* in Komarovia 2: 11 (2002). **Type:** Northern Vietnam, Prov. Son La, Distr. Moc Chau, Mun. Dong Sang, Ba Phach village, 3 Nov. 2001, *Phan Ke Loc et al.* P–10546 (holotype HN, isotype LE).

Paphiopedilum coccineum H.Perner & R.Herrmann in Die Orchidee 51, 5: 623 (2000). Type: Vietnam, "Cao Bang Prov.", *G.C. Thieu* s.n. (holotype ?).

DESCRIPTION. This variety differs from the typical one in its large flower which has a carmine or carmine-brown centre to the dorsal sepal. *Peduncle* 14–18 cm long, suberect, brown-purple, shortly light

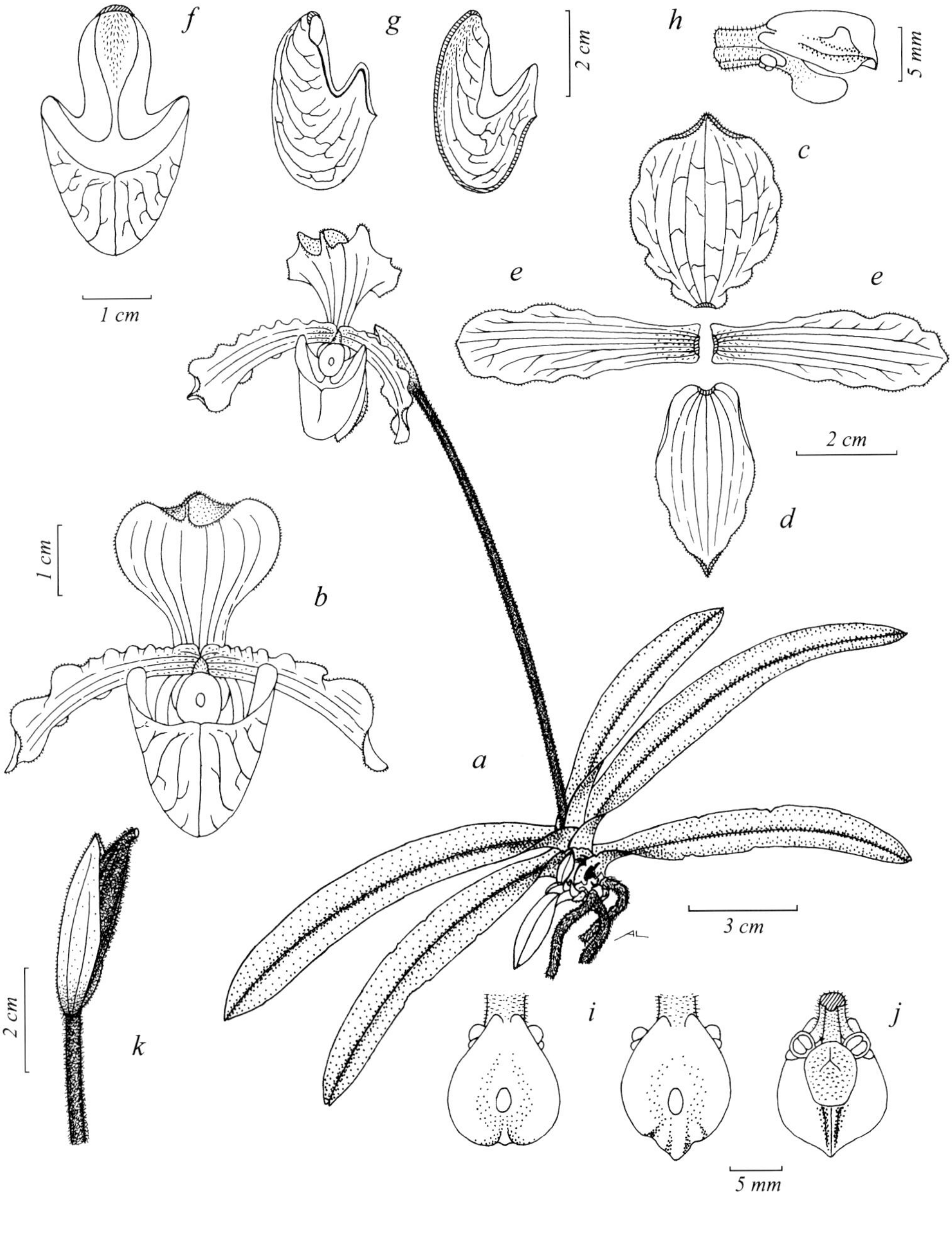

Fig. 127. *Paphiopedilum barbigerum* var. *lockianum*. (Drawing: L. Averyanov).

a — flowering plant; **b** — flower; **c** — median sepal; **d** — synsepal; **e** — petals; **f, g** — lip, frontal, side and section views; **h–j** — column, side, dorsal (normal and flattened) and ventral views; **k** — ovary and flower bract. (Drawn from *P.K.Loc* s.n., 29 September 2000).

brown pubescent; bract elliptic, obtuse, (2.5)3.5–4 cm long, shortly pubescent near the base. *Flower* about (5.7)9–10 cm across; dorsal sepal white with carmine or carmine-brown center; petals yellowish-brown to carmine-brown with dull yellow or greenish margin; lip yellowish-brown, yellow-pinkish or carmine-brown; staminode light yellow with orange umbo. *Pedicel and ovary* 3.1–4 cm long, shortly light brown pubescent. *Dorsal sepal* broadly ovate, obtuse, (3)4–5 cm long, (2.6)3.5–4.2 cm wide. *Synsepal* elliptic, obtuse, (2.4)4–5 cm long, (1.3)1.8–2.2 cm wide. *Petals* spathulate, obtuse, (3.6)4.5–5 cm long, (1)1.8–2.2 cm wide, sparsely hairy at the base, margin undulate and sparsely ciliate. *Lip* (3)3.5–4.5 cm long, 3–3.5 cm wide. *Staminode* transversely obovate to obovate-obcordate, 9–11 mm long, 8–10 mm wide. Figs. 127, 130.

DISTRIBUTION. Northern Vietnam (Cao Bang? and Lao Cai Province). Figs. 37, 128.

ECOLOGY. Evergreen, broad-leaved forests on highly eroded, crystalline, limestone mountains at elevations 600–1000 m. Fig. 129.

FLOWERING IN THE WILD. September–October.

IUCN STATUS. DD.

AFFINITIES. *Paphiopedilum barbigerum* is closely related to *P. insigne, P. helenae, P. henryanum* and *P. gratrixianum*. Among species of this group, *P. barbigerum* is distinctly reminiscent of a small *P. insigne*. It can be readily distinguished, however, by its small size, narrow leaves, short inflorescence, and flower with an unspotted dorsal sepal. The species represent a south-eastern vicariant of the *P. insigne* group with a small area of distribution in southern China and northern Vietnam.

HISTORY. *Paphiopedilum barbigerum* was described by the Chinese botanists T.Tang and F.T.Wang in 1940 based on a collection from Kweichow (now Guizhou) Province in south-west China made in 1930 by the French collectors Julian Cavalerie and Jean Joseph Fortunat. The species was largely forgotten until some living plants of *P. barbigerum* were introduced into cultivation from China a few years ago. They were first noticed around 1985 in a collection of *P. hirsutissimum* var. *esquirolei* coming from the border area of Guizhou and Guangxi Provinces (Mark, 1987; Fowlie & C.Z.Tang, 1987). Wild populations of this rare species have subsequently be found in Guangxi Province (Hu Song Hua, 1994; Cribb, 1998) and in Guizhou (Cribb, Luo & Siu, 2001). In China it is now known from Guangdong, Guangxi, Guizhou, and Yunnan Provinces (Hu Song Hua, 1999). For a long time this species was regarded as a Chinese endemic, confined to the subtropical south of the country north of the Vietnamese border.

In Vietnam the first flowering wild-collected samples of *Paphiopedilum barbigerum* var. *lockianum* was observed by Professor Phan Ke Loc of Hanoi National University in private nurseries of Sa Pa town in Lao Cai Province in September 2000. He carefully preserved a number of flowers in alcohol, made numerous photographs and mentioned that the flowers of Vietnamese plants differed from the type in having a carmine-purple centre to the central sepal and larger flowers. These plants obviously represent a distinct south-western geographical race of *P. barbigerum*, which

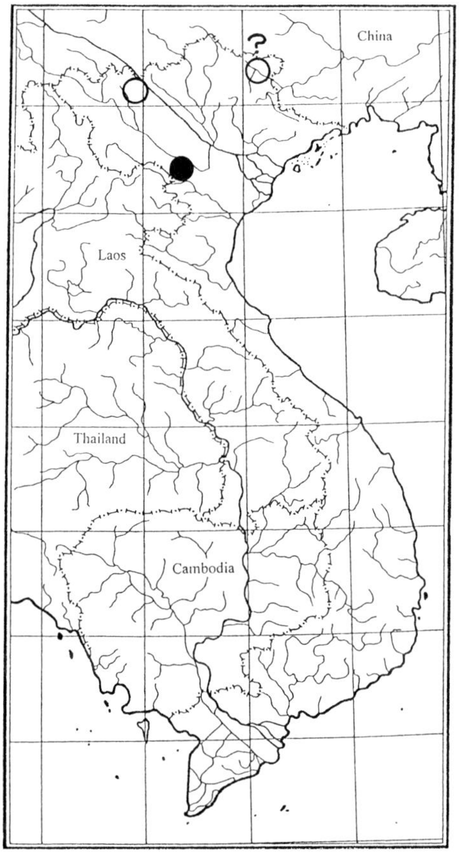

Fig. 128. Distribution of *Paphiopedilum barbigerum* var. *lockianum*. (Drawing: L. Averyanov).

Studied population is marked with a dot, open circles designate supposed localities according to local dealers.

Fig. 129. Type locality of *Paphiopedilum barbigerum* var. *lockianum*. (Photo.: P.K. Loc).

approaches *P. insigne* in its floral dimensions. The following year wild populations of *P. barbigerum* were discovered by Prof. Loc and his colleagues in limestone areas of Son La Province. Numerous plants in full flower were observed at the beginning of November.

At the same time photographs of a similar plant was taken by Rolf Herrmann, when he visited private nurseries in Hanoi. It was subsequently described on the basis of these photographs as a new species, *P. coccineum* (Perner & Herrmann, 2000). They reported that the photographed plants had originated from Cao Bang Province. The Vietnamese nurseryman who owned the plants did not, however, confirm this information (pers.comm.). The authors' citation of the new species states that the type is housed "in the Institute of Ecology and Biological Resources, Hanoi". No authentic samples of the species have been received at the Herbarium of the Institute of Ecology and Biological Resources of the National Center for Natural Sciences and Technology of Vietnam (HN). However, dried fragments of a slipper orchid flower which resembles the plant described by Perner and Hermann arrived in the winter of 2000 from Germany to the old office address of the Vietnamese botanist Dr. Nguyen Tien Hiep. The parcel lacked any sender address and contained no notes inside. Though the formal description of *P. coccineum* is quite valid, we prefer not to use this name for this variety.

HABITAT. In China *Paphiopedilum barbigerum* was observed growing on north-facing cliffs at 300–1200 m elevation on karst limestone mountains in heavy shade, e.g., of *Cinnamomum* trees and fan-palm *Guihaia*. Often the plants grow as a terrestrial and lithophyte in fissures of limestone, deriving nutrition from the soil and a thick layer of moss. The local environments may be humid from evaporation of water from shady gorges, caves and karst caverns. Orchids such as the terrestrial *Amitostigma guizhouensis, Habenaria* sp. and lithophytic *Bulbophyllum* spp., *Coelogyne* sp., *Eria* spp., *Pholidota chinensis* and *Podochilus* sp. grow with it (Hu Song Hua, 1994, 1999; Cribb, 1998).

In Vietnam, colonies of var. *lockii* were observed growing in cracks or narrow shelves of lightly

Fig. 130. Specimen, flower and staminode of the type of *Paphiopedilum barbigerum* var. *lockianum*. (Photos: P.K. Loc).

shaded vertical cliffs near the tops of highly eroded ridges composed of crystalline limestone at elevations of 1000–1100 m, mainly on slopes of south-east exposition. Observed plants were scattered or clustered in small colonies in heavily logged, evergreen, tropical, seasonal, broad-leaved, submontane forest with dominants such as *Acer* (Aceraceae), *Lithocarpus*, *Quercus* (Fagaceae), *Polyalthia* (Annonaceae), *Pistacia weinmannifolia* (Anacardiaceae), *Schefflera pesavis* (Araliaceae), *Phyllanthus* (Euphorbiaceae) and Sapotaceae. Orchids were not as abundant as usual on limestone mountains in the area. The commonest ferns and herbs of the forest floor are *Cyrtomium hemionitis, C. pachyphyllum, Polystichum* sp., *Pteris deltodon, Paphiopedilum dianthum*, species of *Begonia, Carex phankei* and numerous representatives of Urticaceae (Phan Ke Loc, pers. comm.). The climate in this area is a tropical monsoon one with a dry and cold winter period of 4–5 months that is typical for most territories of the northern part of the country (Nguyen Khanh Van *et al*., 2000). Data from the Moc Chau weather station (20°51'N, 104°38'E, elev. 958 m), that is presented in Figures 61, 116 and 201, gives quite a reliable picture of climate conditions in the natural habitats of *P. barbigerum* var. *lockianum* in Vietnam.

It is likely that the distribution of *Paphiopedilum barbigerum* in Vietnam is very limited. Regrettably the only known population of this rare plant has recently been found by commercial collectors. Since then most of plants have been collected for export abroad mainly via China. Without urgent effective protection this species may become extinct in Vietnam in the very near future.

VARIATION. *Paphiopedilum barbigerum* is a highly variable species particularly in the size of the plant, leaves and flower. The flower of this species is typically about 6 cm across and the leaves usually not wider than 1.3 cm. However, in Vietnamese plants the flower often reaches 10 and even more cm across with leaves commonly 1.3–1.5 cm wide (Fig. 127).

Most plants from Chinese populations have flowers with a white dorsal sepal with a basal green mark and sometimes a central brown or brown-purple vein, ochre petals with greenish tips, a reddish brown lip and a yellow staminode. The increased level of anthocyanin pigmentation in the flowers in Vietnamese populations is the major distinguishing feature of *P. barbigerum* var. *lockianum* (Fig. 129, 130). It usually has a white dorsal sepal with a carmine or carmine-brown central area and yellowish-brown to carmine-brown petals with a dull yellow or greenish margin. Specimens with darkish dots on the dorsal sepal were also occasionally seen in the wild in Vietnam. Such colouration may possibly be explained as a result of introgressive hybridization of this species with closely related slipper orchid species of *P. insigne* group having spotted flowers.

The form of the staminode in *Paphiopedilum barbigerum* may be transversely elliptic, ovate, obovate, obcordate and elliptic. The apex of the staminode is often strongly decurved, appearing in front view as apically bi- or tri-dentate (Figs. 127 i, 130 c).

An albino form of *Paphiopedilum barbigerum* in which all purple and brown pigments are absent was described as *P. barbigerum* var. *aureum* Hua in Orchids. Mag. Amer. Orch. Soc. 68, 3: 242 (1999)*.

Paphiopedilum barbigerum is closely allied to *P. insigne, P.* × *herrmannii, P. henryanum, P. tranlienianum* and *P. helenae*. The relationships of these species needs further investigation. Natural hybrid of *P. barbigerum* var. *lockianum* and *P. henryanum* was recently discovered in Vietnam and described under the name *P.* × *aspersum* (Averyanov, 2002 b).

Without any doubt *Paphiopedilum barbigerum* var. *lockianum* is one of the rarest slipper orchids in Vietnam. Detailed study of its natural populations, distribution and natural habitat is very urgently needed for its protection in the country.

10. PAPHIOPEDILUM TRANLIENIANUM Gruss & Perner

Paphiopedilum tranlienianum *Gruss & Perner* in Caesiana 11: 66 (1998); Braem, C. Baker & M. Baker, Gen. Paphiopedilum 2: 222 (1999). **Type:** Vietnam, Bac Thai Province, October 1998, fl. in cult. November 1998, leg. O.Gruss (holotype RO).
Paphiopedilum caobangense Nguyen Thien Tich in Hoa Canh 1999, 1: 14, (1999), *nom illeg.*

DESCRIPTION. Small *lithophytic herb* with 3–6 distichous leaves, often forming clumps. *Leaves* linear-ligulate, obtuse unequally bilobulate or minutely tridenticulate at apex, up to 18 cm long, 1.7 cm wide, glossy green above with a distinct paler margin, paler green on underside, sharply keeled beneath on mid-vein. *Inflorescence* suberect to arcuate, 1-flowered (very rarely with 2 flowers), 10–18 cm tall; peduncle 7–15 cm long, bright green, with dense, short yellow-brown, brown or purple-brown hairs; bract conduplicate, ovate, acute or obtuse, 1.7–2.5(3) cm long, up to 1.6 cm wide, green, spotted with maroon at base, with short hairs on the mid-vein and apical margin. *Flower* 5.5–6 cm across, 5–5.8 cm tall, shortly pubescent on outer surface of sepals; dorsal sepal white with a green base and purple veins; synsepal pale green; petals green flushed and veined glossy reddish brown; lip pale green heavily flushed with dark reddish brown; staminode yellow, with a olive-green tip of the umbo; pedicel and ovary 2.9–4(5) cm long, green, densely covered with short light brown, brown or purple-brown hairs. *Dorsal sepal* suborbiculate, shortly acute or obtuse, (2.5)3–3.2 cm long, (2.8)3–3.4 cm wide, somewhat hooded above but with reflexed basal margins. *Synsepal* more or less adpressed to ovary, narrowly ovate or ovate, subacute, 2.4–2.6 cm long, 1.1–1.8 cm wide, both margins strongly incurved. *Petals* spreading-slightly deflexed, oblong, obtuse at apex, 3–3.4 cm long, 0.7–0.9 cm wide, glossy, white-ciliate, slightly brown-purple hairy at base. *Lip* urceolate, obtuse, with a prominent claw-like base, 3.7–3.9 cm long, up to 1.6 cm wide; side lobes semi-elliptic, not confluent, glossy. *Staminode* obcordate to obovate, recurved down toward the apex, with short obtuse tip, 8–10 mm long and 7–9 mm wide, with a central glossy umbo, sides somewhat incurved and verrucose. Fig. 132.

DISTRIBUTION. Northern Vietnam (Bac Can?, Cao Bang?, Thai Nguyen and Tuyen Quang Provinces). Fig. 133.

ECOLOGY. Primary, broad-leaved, evergreen forests on highly eroded crystalline and stratified limestone mountains at elevations of 400–750 m. Fig. 131.

* **Type:** China, Yunnan, Malipo, Oct. 1997, *J.C.Lo* 1256 (holotype SCBI). This plant was discovered near Malipo town in south-eastern Yunnan near the border with Vietnam. It has light pale yellowish-green flowers with wide white margin of the dorsal sepal. Such plants have not yet been reported from Vietnam.

Fig. 131. Shady limestone cliffs in primary broad-leaved evergreen forests are a typical habitat of *Paphiopedilum tranlienianum*.

N. Vietnam, Thai Nguyen Prov. (Photo.: L. Averyanov).

FLOWERING IN THE WILD. September–November.

IUCN STATUS. EN.

AFFINITIES. *Paphiopedilum tranlienianum* is an unusual miniature slipper orchid of the *P. insigne* group most closely related to *P. barbigerum* and *P. × herrmannii.* The interrelations of these species need further investigation. This species has an elegant flower with a white dorsal sepal striped with purple-brown, strongly undulate purple-brown petals, a brown-purple lip and a bright yellow staminode. It is a local endemic in Vietnam distributed in the limestone lowlands of central provinces of the northern part of the country (Fig. 133).

HISTORY. The discovery of numerous new slipper orchids in the limestone regions of northern Vietnam around 1996 (Averyanov, 1996a, 1997b, 1998a; Averyanov *et al.*, 1996a, 1997a,b) and the political opening of the country has stimulated the illegal exploitation of slipper orchids. Very low prices for wild-collected plants, the absence of any control of the private exploration of natural plant resources and easy export has attracted international orchid dealers to the country. In recent years, a slipper orchid which resembles a large form of *Paphiopedilum helenae* when not in flower, was collected for sale in large quantities from the limestone regions of the northern Vietnam. It entered trade as "big helenae" or a new variety of *P. herrmannii.* Under the later name, plants soon reached Japan and Germany (Gruss and Perner, 1998).

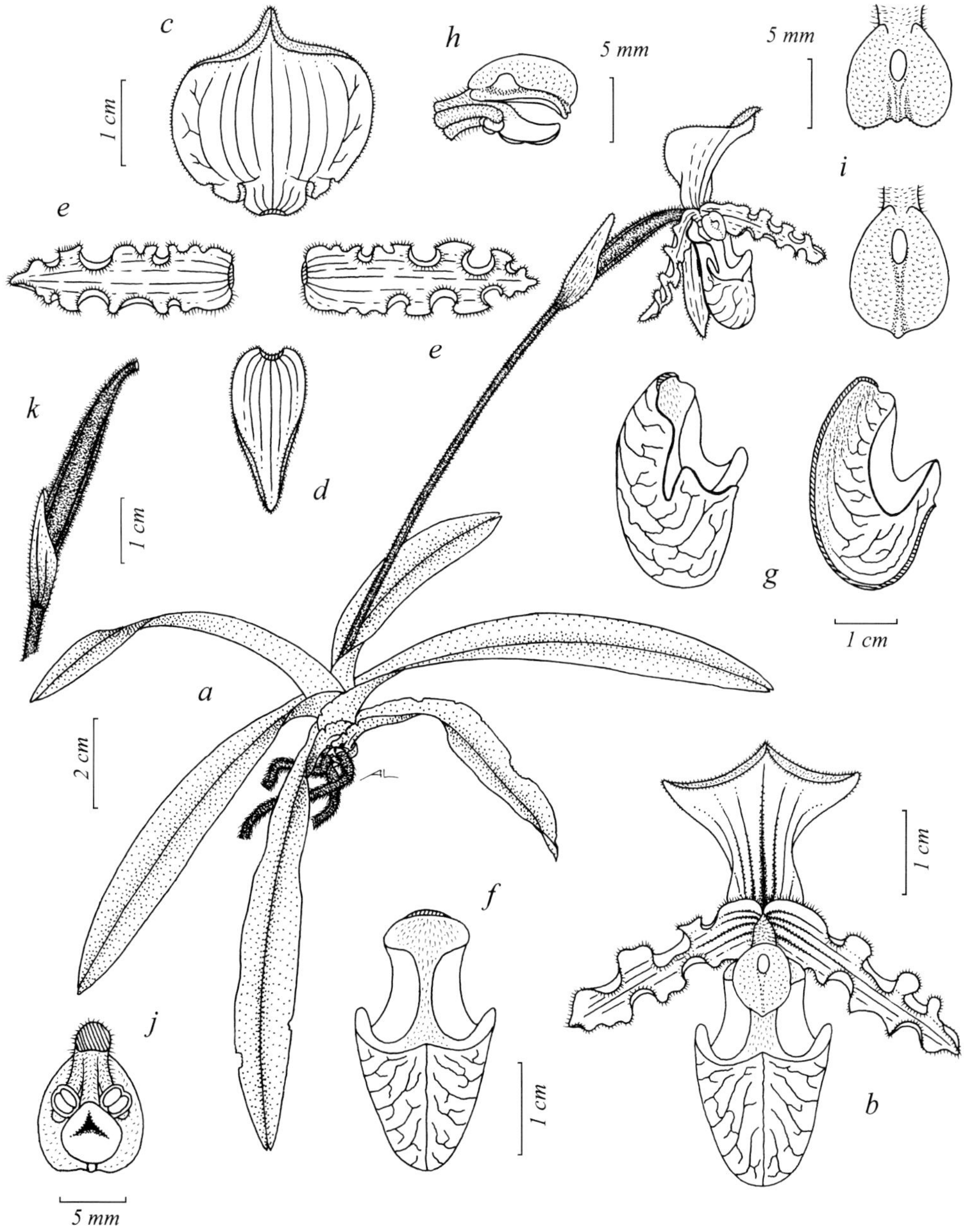

Fig. 132. *Paphiopedilum tranlienianum.* (Drawing: L. Averyanov).

a — flowering plant; **b** — flower; **c** — median sepal; **d** — synsepal; **e** — petals; **f, g** — lip, frontal, side and section views; **h–j** — column, side, dorsal (normal and flattened) and ventral views; **k** — ovary and flower bract. (Drawn from HAL 30).

Paphiopedilum tranlienianum was described by Gruss and Perner (1998) based upon a plant freshly imported from Vietnam, which flowered in Germany within a month of arrival. The species was named after the Vietnamese exporter, Mrs. Tran Ngo Lien. It was reported that the original collections, which certainly changed hands many times, were made in Bac Can or Thai Nguyen Provinces (formerly named Bac Thai Province) at elevations of about 500 m.

A little later *Paphiopedilum tranlienianum* was described once again, under the name *P. caobangense* (but without a Latin diagnosis), by the Vietnamese orchid grower Nguyen Thien Tich (1999) in Hoa Canh, the popular magazine of the Vietnamese orchid society. He reported that it originated from Cao Bang Province, but again gave no information about its precise provenance and ecology.

Wild populations of *Paphiopedilum tranlienianum* were first found and studied in spring 2001 by Averyanov and Phan Ke Loc in Thai Nguyen and Tuyen Quang Provinces of northern Vietnam during field work funded by the U.S. National Geographic Society, U.S. National Science Foundation and American Orchid Society. It was concluded that this Vietnamese endemic was very local and restricted to an area of low rocky limestone in the centre of northern Vietnam to the north of Hanoi (Fig. 133).

HABITAT. *Paphiopedilum tranlienianum* inhabits an area with a monsoon tropical climate having a cool dry winter and a wet warm summer. Temperature regimes in the area have a strong seasonality. The more or less dry winter lasts here from November to February or March. Common mean temperatures at this time lie usually between 12–16°C (Fig. 134). An absolute minimum temperature reported from the area is -2°C. Heavy cold fogs and persistent drizzling rains borne on cold NE winds are very typical in this area from late February until the end of March. Summer in the area is wet, with the peak rainfall in June, July and August (Fig. 134). Mean summer temperatures are about 25–28°C with observed maxima about 35–40°C. The mean annual rainfall commonly varies between 1500–1700 mm. However, on more humid south- and south-east-facing slopes, annual precipitation may reach 2200 mm and more.

Paphiopedilum tranlienianum inhabits primary, broad-leaved, evergreen and mixed forests that grow on very steep rocky slopes of highly eroded limestone mountains at elevations between 400–750 m.

The largest trees in these woods often reach 25–30 m tall and include such species as *Burretiodendron hsienmu* (Tiliaceae), *Choerospondias axillaris, Allospondias lakonensis* and *Dracontomelon duperreanum* (Anacardiaceae), *Chukrasia tabularis* (Meliaceae), *Cinnamomum* sp., *Persea* sp. (Lauraceae), *Diospyros* sp. (Ebenaceae), *Fernandoa* sp. (Bignoniaceae), *Ficus* spp. (Moraceae), *Lithocarpus* sp., *Quercus* sp. (Fagaceae) and *Pterospermum truncatolobatum* (Sterculiaceae). Sometimes conifers such as *Dacrycarpus elatus, Dacrydium elatum* and *Nageia* sp. (Podocarpaceae), are found in the forest especially along tops of ridges and near mountain peaks.

The undergrowth commonly includes a great number of medium-sized tropical trees from numerous families, particularly Araliaceae (*Schefflera*, *Brassaiopsis glomerulata*), Clusiaceae (*Calophyllum*), Elaeocarpaceae (*Elaeocarpus*), Euphorbiaceae (*Cleidion*, *Sauropus*), Lauraceae

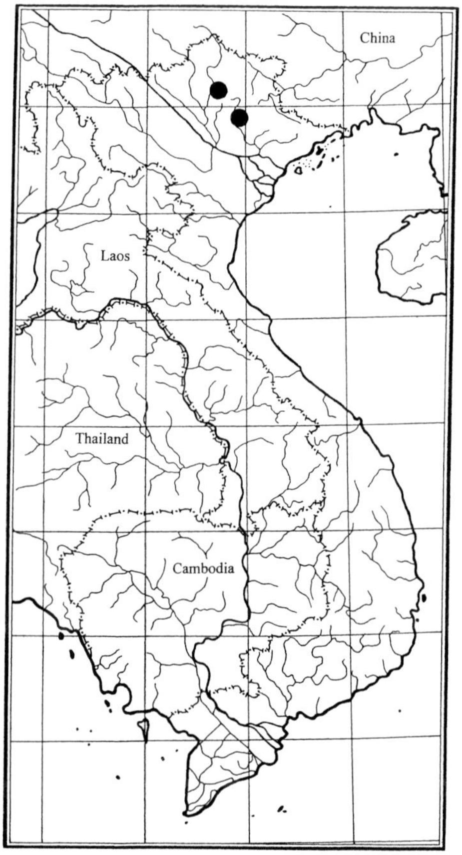

Fig. 133. Distribution of *Paphiopedilum tranlienianum*. (Drawing: L. Averyanov).

(*Cryptocarya, Litsea*), Moraceae (*Ficus, Streblus tonkinensis, S. ilicifolius*), Myrtaceae (*Syzygium*), Rubiaceae, Rutaceae (*Micromelum, Murraya*), Sapindaceae (*Dimocarpus longan*), Staphyleaceae (*Turpinia*), Sterculiaceae (*Sterculia*) families. Large palms include *Arenga pinnata, Livistona chinensis,* and a species of *Licuala*. Woody plants, such as *Dracaena cambodiana* and *Pandanus*, are commonly seen in these forests.

Shrubs that are commonly seen include *Ardisia* (Myrsinaceae), *Ilex* (Aquifoliaceae), *Leea indica* (Leeaceae) and *Lysimachia insignis* (Primulaceae). Numerous species of terrestrial and lithophytic herbs are an integral component of these habitats, especially representatives of families such as Acanthaceae, Araceae (*Aglaonema, Alocasia, Amorphophallus paeonifolius* and other species, *Arisaema, Colocasia*), Aristolochiaceae (*Asarum*), Begoniaceae (*Begonia*), Costaceae (*Costus speciosus*), Convallariaceae (species of *Aspidistra, Colonia, Ophiopogon, Polygonatum*), Gesneriaceae (species of *Chirita, Paraboea*), Hypoxydaceae (*Curculigo*), Maranthaceae (*Phrynium*), Rubiaceae (*Ophiorrhiza*), Urticaceae (species of *Elatostema, Pellionia, Procris*) and Zingiberaceae (species of *Alpinia, Amomum*). However, sedges such as *Carex indica* and *Scleria* (Cyperaceae) appear regularly as dominants of herb communities on drier slopes and cliffs. Many lithophytic ferns also occur here, among them *Adiantum caudatum, A. gravesii, Antrophyum, Asplenium belangeri, A. griffithianum, A. tenuifolium, A. unilaterale, Colysis, Cyclosorus, Nephrolepis cordifolia, Polypodium, Pyrrosia, Tectaria decurrens, T. deflexa* and *subpedata.* A few species of *Selaginella* (Selaginellaceae) and some short tall rather xerophytic mosses are also common in habitats of *P. tranlienianum.*

On rocky steep slopes, large examples of the nest-like ferns *Asplenium anthrophioides* and *A. nidus* and numerous epiphytic and lithophytic vines, such as *Acacia, Bauhinia, Entada phaseoloides, Pterolobium microphyllum* (Fabaceae), *Calamus* (Arecaceae), *Dischidia, Hoya* (Asclepiadaceae), *Embelia* (Myrsinaceae), *Jasminum* (Oleaceae), *Beaumontia grandiflora, Melodinus* (Apocynaceae), *Combretum* (Combretaceae), *Pothos repens, Pothos* sp., *Raphidophora* (Araceae) and *Tetrastigma* (Vitaceae) may commonly be seen.

The common epiphytes in these forests are orchids, particularly *Callostylis rigida, Cleisostoma fuersteinbergianum, C. rostratum, C. striatum, Coelogyne fimbriata, Dendrobium acinaciforme, D. lindleyi, Eria pubescens, E. sutepensis, Luisia* sp., *Oberonia* sp., *Phalaenopsis mannii* and *Thelasis khasiana.* However lithophytic orchids are more abundant and diverse, among them *Acampe rigida, Bulbophyllum purpureifolium, Calanthe alismaefolia, C. triplicata, Cheirostylis yunnanensis, Cleisostoma rostratum, Coelogyne fimbriata, Corymborkis veratrifolia, Cymbidium aloifolium, C.*

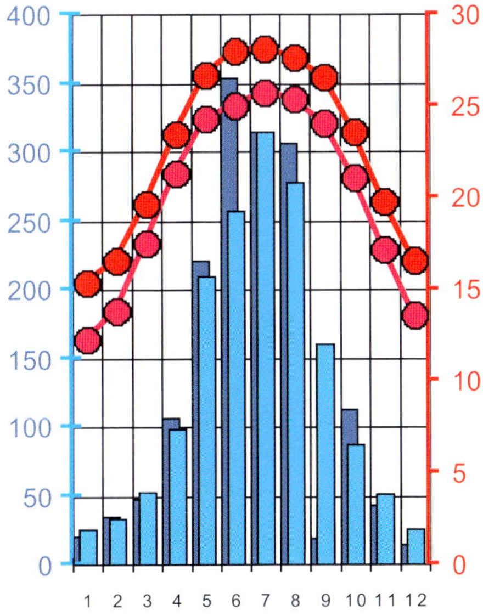

Fig. 134. Climatological data typical for the area of *Paphiopedilum tranlienianum.*

Black figures at the chart bottom are months. Blue axis represents atmospheric precipitation in mm (columns designate mean monthly rainfalls); red axis represents mean monthly temperature in °C. The deep blue and red are designated data from the Dinh Hoa weather station, 21°54'N 105°38'E, elev. 220 m; light blue and purple are data from the Ngan Son weather station, 22°26'N 105°59'E, 566 m, all according to Nguyen Khanh Van *et al.*, 2000.

Fig. 135. *Paphiopedilum tranlienianum* in its natural habitat.
N. Vietnam, Thai Nguyen Prov. (Photo.: L. Averyanov).

lancifolium, Dendrobium chrysanthum, Flickingeria sp., Hemipilia calophylla, Liparis mannii, L. nervosa, L. paradoxa, Liparis sp., *Neuwiedia* sp., *Oberonia ensiformis, Pholidota rubra, Podochilus microphyllus, Tropidia angulosa, T. curculigoides* and *Vandopsis gigantea.* In some places, *P. tranlienianum* is found growing in association with other *Paphiopedilum* species, especially *P. concolor, P. hangianum, P. henryanum, P. hirsutissimum* and *P. vietnamense*.

In habitats of *Paphiopedilum tranlienianum* that have been studied it typically grows as a lithophyte usually on high vertical shady cliffs hanging under deep karst cavities and gorges with its roots clinging in narrow crevices and small holes (Fig. 135). In places with intact primary vegetation, it occurs under forest shade on slopes of any exposition between 400–750 m. It usually occurs sporadically and does not form large colonies. The plants commonly inhabit the middle part of the slopes and rocky cliffs near the summit of highly eroded ridges and mountains composed of hard crystalline or stratified more or less soft limestone. It flowers in the nature in September–November.

Fig. 136. *Paphiopedilum tranlienianum*. In private nursery, Hanoi. (Photo.: L. Averyanov).

Paphiopedilum tranlienianum is highly sought after on both the local and international orchid markets and is presently collected in large quantities for sale. This rare slipper orchid is a local endemic with a very limited distribution in a densely populated part of central northern Vietnam. According to available data, it is now restricted to a very few small populations which may easily become extinct in the near future due to commercial collecting.

VARIATION. Plants from populations of *Paphiopedilum tranlienianum* that have been studied have rather uniform flowers with only slight variation in flower dimensions. Colouration of the flowers is also more or less standard and does not differ essentially from the type (Fig. 136). However, wider studies may reveal a greater variation in the flowers of *P. tranlienianum*, especially in areas where it grows with other closely related *Paphiopedilum* species. In culture, wild-collected samples which are somewhat intermediate between *P. tranlienianum*, *P. barbigerum* and *P.* × *herrmannii* appear from time to time. However, their provenance, as usual, is uncertain.

11. PAPHIOPEDILUM HELENAE Aver.

Paphiopedilum helenae *Aver.* in Bot. Journ. (Leningrad) 81, 9: 108 (September 1996); Aver. et al. in Orchids 65, 10: 1064 (October 1996); Cribb, Gen. Paphiopedilum, ed. 2: 238 (1998); Braem, C.Baker & M.Baker, Gen. Paphiopedilum 2: 220 (1999). **Type:** N. Vietnam, Cao Bang Prov., Tra Linh Distr., Quoc Toan Subdistr., 850–900 m, near Thang Hen Lake, environs of Thang Hen and Lung Tao villages, 21 October 1995, *L.Averyanov, N.T.Hiep & D.D.Huyen* CB 012 (holotype LE!, isotype HN!).

Paphiopedilum helenae forma *aureum* O.Gruss & J.Röth in Die Orchidee 50, 1: 3 (1999).

Paphiopedilum delicatum Z.J.Liu & J.Y.Zhang in Acta Phytotax. Sin. 39, 1: 78 (2001), **syn. nov. Type:** Vietnam, without exact loc., *Z.J. Liu* 20920 (holotype, Herb. Shenzhen City Wutonshan Nurseries).

DESCRIPTION. Very small *lithophytic herb* with 3–4(5) distichous leaves and clustered growths. *Leaves* coriaceous-fleshy to rigid, rather succulent, oblanceolate to oblong-elliptic, tridenticulate at obtuse apex, (3)4–9(12) cm long, (0.6)0.8–1.6(2) cm wide, green with paler yellowish white margin, with a few purple-violet marks beneath near the base. *Inflorescence* arching to almost horizontal or slightly hanging, one-flowered; peduncle slender, 4–7 cm long, shortly and densely black-purple pubescent; bract conduplicate, ovate to elliptic, obtuse, 8–12 mm long, 3–4 mm wide. *Flower* large for plant size, (2.5)5–6 cm across, pubescent on outer surface of sepals; dorsal sepal bright golden yellow with a white margin to pale dull yellowish; synsepal white; petals orange or ochreous yellow to dull yellowish with darker brown-orange veins; lip bright brownish-orange to dull yellowish-pink or dull pale pink; staminode pale yellow with a bright yellow or green shiny central umbo; pedicel and ovary 2–3.5 cm long, densely black-purple pubescent with short hairs. *Dorsal sepal* obovate-subcircular to circular, obtuse or shortly emarginate at apex, 18–35 mm long, 15–30 mm wide, side margins reflexed, apical margin somewhat slightly incurved. *Synsepal* ovate-elliptic to broadly ovate, obtuse or shortly bidentate at the apex, 15–25 mm long, 8–15 mm wide. *Petals* straight to somewhat incurved, ligulate to subspathulate, obtuse or truncate at apex, ciliate, slightly pubescent at base on inner surface, (15)25–35 mm long, (3)4–8 mm wide. *Lip* deeply saccate, not tapering much in apical part, 20–30 mm long, 15–20 mm wide, glabrous on outer surface; side lobes inflexed, not warty. *Column* 5–6 mm long; staminode broadly obovate, circular or indistinctly rhombic to transversely broadly elliptic, 7–8 mm long and wide, papillose with a central raised hemispherical umbo. Fig. 139.

DISTRIBUTION. Northern Vietnam (Bac Can and Cao Bang Provinces). Figs. 37, 140.

ECOLOGY. Primary, mixed and coniferous, open, mossy, dry forests (with *Pseudotsuga brevifolia* and *Pinus fenzeliana*) on tops of highly eroded limestone mountains at elev. (500)600–900(1000) m. Figs. 137, 138, 141.

FLOWERING IN THE WILD. September–November.

IUCN STATUS. EN.

Fig. 137. Fragments of primary coniferous forests with *Pseudotsuga brevifolia* and *Pinus fenzeliana* on the tops of remnant limestone mesas are the home of *Paphiopedilum helenae*.
N. Vietnam, Cao BangProv. (Photo.: L. Averyanov).

AFFINITIES. *Paphiopedilum helenae* is one of the prettiest of all recent slipper orchid discoveries. This surprisingly tiny species is certainly the smallest slipper orchid in the genus. Flowering samples in the wild often have leaves only 4–5 cm long with a flower 4–6 cm across (Fig. 142). The combination of a dwarf plant with large flowers makes this species particularly desirable for ornamental cultivation and the production of miniature slipper-orchid hybrids.

Paphiopedilum helenae is closely allied to *P. barbigerum* from which differs in having rather broader, short, subsucculent leaves, predominantly yellow flowers with or without a narrower white margin to the dorsal sepal, petals with straight rather than undulate margins, a somewhat more inflated

Fig. 138. Primary dry open coniferous forests with *Pseudotsuga brevifolia* supported large population of *Paphiopedilum helenae* in its locus classicus before its commercial exploration.

N. Vietnam, Cao Bang Prov. (Photo.: L. Averyanov).

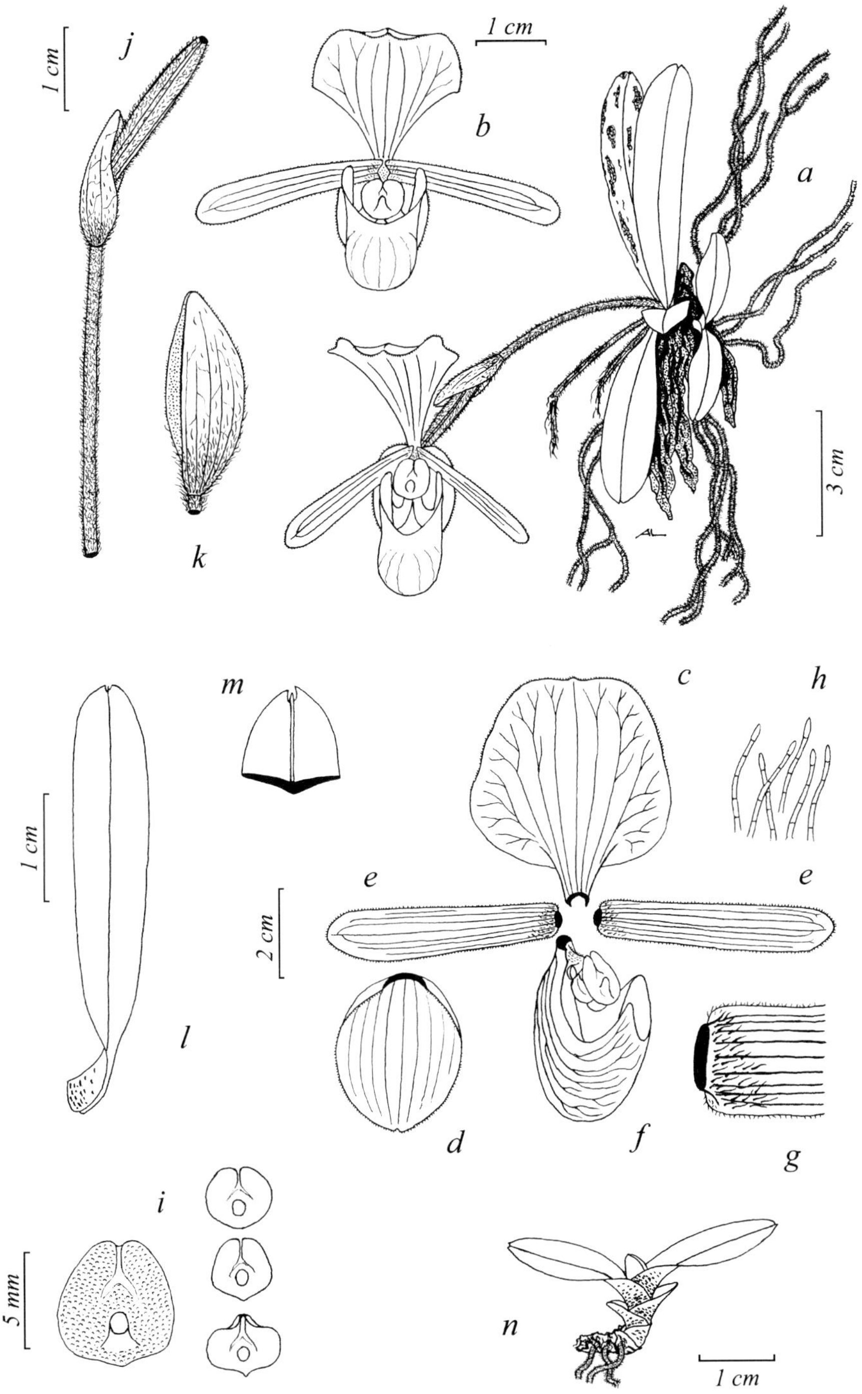

Fig. 139. *Paphiopedilum helenae.* (Drawing: L. Averyanov).

a — flowering plant; **b** — flower; **c** — median sepal; **d** — synsepal; **e** — petals; **f** — side view; **g** — base of lateral petal; **h** — hairs of flower hairiness; **i** — most common and other variants of form of staminode; **j** — inflorescence stalk and ovary; **k** — flower bract; **l** — leaf; **m** — leaf apex; **n** — seedling. (Drawn from the type).

lip, and a suborbicular staminode. These features seem sufficient to maintain *P. helenae* as distinct at specific rank, although investigation of further populations may link it more closely with *P. barbigerum*. The species is a local endemic with a very restricted distribution in northeastern limestone regions of Vietnam (Fig. 140).

HISTORY. *Paphiopedilum helenae* was discovered in October 1995 by Leonid Averyanov, N.T.Hiep and D.D.Huyen during field-work supported by the American Orchid Society. The plants were found in a remote area in Cao Bang Province of Vietnam just south of the Chinese border. It is named in honor of Averyanov's wife.

The discovery and publication of *Paphiopedilum helenae* attracted immense interest in orchid circles and on the black market. Although information on its geographical distribution was purposefully reduced on publication (Averyanov, 1996a; Averyanov *et al.*, 1996a), orchid dealers quickly traced the area of its discovery. According to information from local people, numerous orchid dealers from Hanoi and Ho Chi Minh City (Saigon) appeared in the region within two months of the publication of *P. helenae* in the journal of the American Orchid Society (Averyanov *et al.*, 1996a). Using copies of the photographs in that publication, they engaged local villagers to collect as many plants as possible. Over the following few weeks populations of *P. helenae* in the area were totally exterminated. Plants were sold for about 2–3 USD per kilo by local collectors. According to our estimate, in a very short time, 15000–20000 growths of *P. helenae* were collected. Most of these plants were immediately exported from the country and appeared on the international market shortly afterwards. It is likely that this illegal importation into Germany provided the basis for the description of *P. helenae* forma *aureum* O.Gruss & J.Röth in *Die Orchidee* 50, 1: 3 (1999)*. This plant differs slightly from the type in having a dorsal sepal that is entirely golden in colour and lacks a white margin. According to our observations such colour forms were not rare in populations of the species in the area of the type locality before commercial exploitation. Shortly afterwards, many hundreds of *P. helenae* plants were illegally collected by staff of a commercial trading company during its extensive exploitation of natural plant resources of the country (Tanaka, 1998–1999).

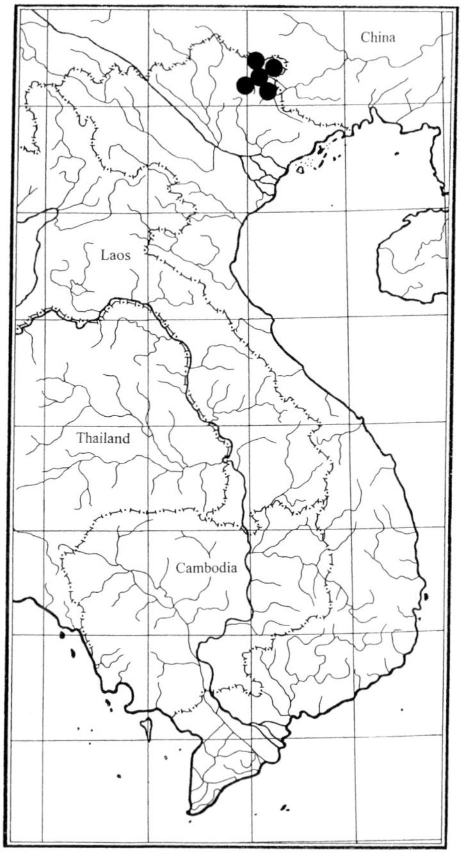

Fig. 140. Distribution of *Paphiopedilum helenae*. (Drawing: L. Averyanov).

HABITAT. *Paphiopedilum helenae* has a very limited distribution in the Cao Bang Province and adjacent regions of Bac Can Province of northern Vietnam (Fig. 140). It is a vicariant of the *P. insigne* group, closely related to *P. barbigerum*, *P. henryanum* and *P. tranlienianum* but having very specific ecological requirements.

The very picturesque rocky landscape of the region where *Paphiopedilum helenae* is found is composed of crowded, highly eroded, limestone ridges and mesas between 600–1000 m in elevation and emerging from relatively flat alluvial valleys at about 300–600 m (Fig. 137). Local depressions in the

* **Type:** North Vietnam, Kao Bang (Cao Bang), near to the Guangxi border, c. 850 m 12.10.1997, *Ngo Diem* s.n., cult. J.Röth & O.Gruss (holotype HAL 078518).

valley, formed by numerous karst craters, usually fill with water during the rainy season. At this time, a number of small or medium-sized lakes, which develop between the bluffs of the limestone ridges, make it difficult to travel in this country and provide additional humidity during the rainy season. Toward the end of the dry season, these lakes, however, usually disappear. Numerous aquatic organisms survive this period in underground caves. The limestone mountains of this region are covered by relatively dry, primary, evergreen forests that are replaced toward the more or less isolated peaks by rather open, mixed and coniferous forests rich in local endemics that are dominated by *Pseudotsuga brevifolia* and *Pinus fenzeliana* (Fig. 137, 138, 141).

Ecologically, *Paphiopedilum helenae* is closely associated with coniferous and mixed forests with *Pseudotsuga brevifolia* and *Pinus fenzeliana*, which cover the tops of the limestone table mountains that usually have vertical slopes about 100–150 m high. Old (about 200 to 250 years) gnarled trees 8–12 m high of *Pseudotsuga brevifolia* form the canopy of a unique, rather open forest. The subcanopy in these forests is formed of numerous kinds of broad-leaved, evergreen and semi-deciduous trees around 4–8 m high, including *Acer* sp. (Aceraceae), *Burretiodendron hsienmu* (Tiliaceae), *Illicium* sp. (Illiciaceae), *Myrsine kwangsiense* (Myrsinaceae), *Pistacia weinmannifolia* (Anacardiaceae), *Platycarya strobilacea* (Juglandaceae), *Quercus* sp. (Fagaceae), *Nageia fleuryi* (Podocarpaceae), *Schefflera pesavis* (Araliaceae), *Tirpitzia sinensis* (Linaceae), *Ulmus* sp. (Ulmaceae), *Pinus fenzeliana* (Pinaceae) and *Podocarpus neriifolius* (Podocarpaceae). Beneath this are a number of small xeromorphic trees and shrubs such as *Campylotropis bonii, Sophora glauca* (Fabaceae), and species of *Euonymus* (Celastraceae), *Mahonia* (Berberidaceae), *Pittosporum* (Pittosporaceae), *Rhapis* (Arecaceae) and *Wikstroemia* (Thymellaeaceae). Some vines, from genera such as *Bauhinia* sp. (Fabaceae), *Jasminum* sp. (Oleaceae), *Dioscorea* spp. (Dioscoreaceae) and *Tetrastigma* sp. (Vitaceae), are usually common in these woods.

Numerous lithophytic ferns and orchids inhabit the forests of *Pseudotsuga brevifolia* and *Pinus fenzeliana*. The commonest epiphytic orchids are *Aerides odorata, A. rosea, Bulbophyllum ambrosia, B. macraei, B. purpureifolium, Ceratostylis himalaica, Cleisostoma rostratum, Coelogyne fimbriata, Dendrobium acinaciforme, D. lindleyi Epigeneium amplum, E. chapaense, Eria globulifera, E. rhombodalis, E. siamensis, E. sutepensis, Hygrochilus parishii, Liparis*

Fig. 141. Open dry primary coniferous forest on narrow shelves above high mountain cliffs is the typical habitat of *Paphiopedilum helenae*.
N. Vietnam, Cao Bang Prov. (Photo.: L. Averyanov).

Fig. 142. The type specimen of *Paphiopedilum helenae* in natural habitat.
N. Vietnam, Prov. Cao Bang Prov. (Photo: L. Averyanov).

balansae, L. distans, L. mannii, Luisia morsei, Thrixspermum centipeda, T. calceolus and *Trichotosia dasyphylla*. Epiphytes are particularly abundant on small but old trees of tropical oaks (*Quercus* sp.).

Mossy rocks in the forest are covered in lithophytic orchids, including *Acampe rigida, Bulbophyllum purpureifolium, Cymbidium aloifolium, Coelogyne fimbriata, Coelogyne* spp., *Cleisostoma paniculatum, Dendrobium chrysanthum, D. fimbriatum, D. loddigesii, Epigeneium amplum, Eria coronaria, E. rhombodalis, Liparis mannii, L. viridiflora, Monomeria barbata, Oberonia cavaleriei, O. ensiformis, Pelatantheria insectifera, Phalaenopsis mannii, Pholidota leveilleana, P. roseans, P. rubra, P. yunnanensis, Podochilus microphyllus* and *Vanda concolor.* The rich orchid flora of the area is highly specific and a number of local endemics are closely connected with these unique forests, amongst them *Bulbophyllum arcuatilabium, B. purpureifolium, Cheirostylis marmorifolia, Eria crassifolia, Gastrochilus minutiflorus, Liparis averyanoviana, L. conopea, Luisia appressifolia, Phaius longicruris,* and *Renanthera citrina*. In some habitats *P. helenae* was observed growing in close association with other slipper orchids, such as *P. hirsutissimum* var. *esquirolei, P.* × *herrmannii* and *P. micranthum.*

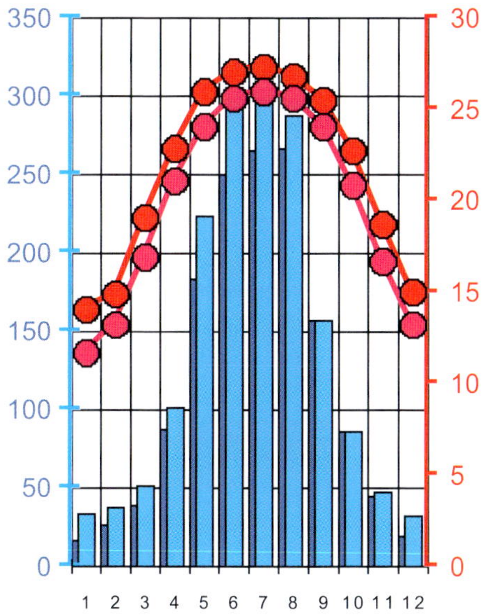

Fig. 143. Climatological data typical for the area of *Paphiopedilum helenae.*

Black figures at the chart bottom are months. Blue axis represents atmosphere precipitation in mm (columns designate mean monthly rainfalls); red axis represents mean monthly temperature in °C. The deep blue and red are designated data from the Cao Bang weather station, 22°39'N 106°14'E, elev. 258 m; light blue and purple are data from the Trung Khanh weather station, 22°50'N 106°31'E, 520 m, all according to Nguyen Khanh Van *et al.*, 2000.

Fig. 144. *Paphiopedilum helenae* in typical habitats. (Photos: L. Averyanov).

a, b — N. Vietnam, Cao Bang Prov.

Local conditions on the tops of karst limestone ridges and mesas are severe because of extreme drought. As a result many plants here, including epiphytic and lithophytic orchids, have xeromorphic adaptations. The winter in this area is moderate and dry, but the summer is hot and wet. Usual winter temperatures lie around 10–16°C with a recorded absolute minima about -3.4°C. Common temperatures during the summer are 20–28°C, however they may sometimes reach 35–40°C. The peak rainfall is between May-August (Fig. 143). Mean annual precipitation is relatively low and varies from 1400 to 1650 mm.

Populations of *Paphiopedilum helenae* are commonly found on slightly shaded or open vertical ledges with a northern exposure towards the top of vertical cliffs. They are found on karst limestone mountains and mesas mainly between 600–900 m. However, plants can occasionally be seen lower down at about 500 m, their upper altitudinal limit lies at about 950–1000 m. The rocks are usually

Fig. 145. Most common variants of *Paphiopedilum helenae* in natural populations. (Photos: L. Averyanov).
a — N. Vietnam, Cao Bang Prov., type;
b, c — N. Vietnam, Cao Bang Prov.

highly eroded, ancient (supposedly late Paleozoic), crystalline limestone. The most typical habitats are narrow shelves below overhanging bluffs and rock crevices often at the base of small trees or shrubs. It is not unusual, however, to see some specimens on vertical limestone walls without cracks where the plants are attached closely to the more or less smooth rock surface (Fig. 144). Convective air flow along vertical cliffs warmed during the day is probably very favourable for plants that settle down in these places. The flowering period in nature is from late September to early November. Very poor fruit production is usually observed in natural populations of the species.

Being associated with the highly restricted coniferous forests dominated by *Pseudotsuga brevifolia, Paphiopedilum helenae* is one of the rarest Vietnamese slipper orchids. The limestone mountain *Pseudotsuga* forests found in northern Vietnam represent a geographically very restricted and endangered primary plant community that includes numerous unique plant species in addition to orchids (Nguyen Tien Hiep, Averyanov & Phan Ke Loc, 1998; Phan Ke Loc, Nguyen Tien Hiep & Averyanov 1999a,b,c; Nguyen Tien Hiep, Phan Ke Loc & Averyanov, 2000). These forests are under great pressure nowadays from logging and burning. As a rule, populations of *P. helenae* are situated near the peaks on almost inaccessible vertical limestone cliffs that are difficult to reach. However, local collectors engaged by orchid dealers collect plants for sale in large quantities. Frequent forest fire, provoked by widespread forest cutting, is another factor that often completely destroys the habitats of *P. helenae*, even on the inaccessible tops of limestone ridges and mesas. We have never observed any regeneration of *P. helenae* populations after the primary vegetation has been burned. There is little doubt that this species once had a wider distribution in Cao Bang and Bac Can Provinces. Unfortunately, the present-day almost total deforestation in the rocky limestone landscapes of northern Vietnam makes the survival of this species in nature doubtful.

VARIATION. *Paphiopedilum helenae* is very variable species in both plant dimensions and flower colouration. Plants in natural habitats rarely have leaves longer than 6–8 cm. However in shady, occasionally more humid conditions and in cultivation plants may become larger and develop leaves up to 10–12 cm long and up to 1.5–2 cm wide. Leaves are usually deep green with a narrow yellowish line along the margin.

Flowers in natural populations vary from very small, not more than 2.5 cm wide, to relatively large up to 6–7 cm across. The dorsal sepal is commonly bright yellow or pale yellow with or without white margin. The lateral petals vary from bright orange or ochreous yellow to light dull yellowish with brownish or orange veins. The lip is usually bright orange or brownish-orange, but in plants from some populations the lip may be light yellowish-brown or dull pale pink. The staminode is light yellow, with a bright yellow, yellowish-green or green shiny umbo. The form of the staminode varies from broadly obovate or circular, to indistinctly rhombic or transversely broadly elliptic (Fig. 145).

Natural hybridization of *P. helenae* with sympatric slipper orchid species needs further field investigation. Not uncommonly, for example, where populations of *P. helenae* and *P. hirsutissimum* grow together plants can be found that may represent their natural hybrid. The plant described as *P. herrmannii* may be one of these. Some unusual wild-collected plants with transitional morphological features between *P. helenae* and *P. × herrmannii*, that appear time to time on sale, may have arisen from the hybridization of these taxa.

12. PAPHIOPEDILUM HENRYANUM Braem

Paphiopedilum henryanum *Braem* in Schlechteriana 1, 1: 4 (3 Sept. 1987); Seidenfaden, Orch. Indochina: 19 (1992); Aver., Identif. Guide. Vietnam. Orch.: 20 (1994); Cribb, Gen. Paphiopedilum, ed. 2: 246 (1998); Braem, C. Baker & M. Baker, Gen. Paphiopedilum 2: 204 (1999); Pham Hoang Ho, Ill. Fl. Viet. 3: 764 (2000). **Type:** China, border region to Vietnam, 1987, *Azadehdel* s.n. (holotype SCHLE).

Paphiopedilum dollii Lückel in Die Orchidee 38, 5: 266 (18 Sept. 1987). **Type:** ex China, cult. H. Doll (holotype K!).

Paphiopedilum chaoi Hua in Die Orchidee 50, 5: 495 (1999), **syn. nov. Type:** China, Guangxi Province, Napo County 10. May 1998, *M.H.Chao* 5322 (holotype SCBI).

Fig. 146. Mixed semideciduous and coniferous primary forests on karst limestone in Ha Giang Province is typical habitat of *Paphiopedilum henryanum* in northern Vietnam.

N. Vietnam, Ha Giang Prov. (Photo.: L. Averyanov).

DESCRIPTION. Small *lithophytic herb* with 3–6 distichous leaves, and clustered growths. *Leaves* lanceolate or linear-ligulate, obtuse, unequally bilobulate or minutely tridenticulate at apex, 10–17 cm long, 1.2–1.6 cm wide, glossy dark green, with a distinct yellowish line along margin above, paler green on underside, sharply keeled beneath on mid-vein. *Inflorescence* suberect, arcuate or arcuate-pendent, 1-flowered; peduncle 12–15 cm long, green to purple-brown, purple-brown pubescent; bract 2–3.6 cm long, 0.6–2 cm wide, dark green with a purplish base. *Flower* 4–5 cm high, 4–6 cm across, densely pubescent with short hairs on outside of sepals; dorsal sepal light yellow to dull yellowish-green with deep purple-brown maroon spots; synsepal pale yellow or light yellowish-green with maroon spotting; petals dull yellowish to pink-violet, spotted with dark maroon in basal half; lip rose-purple to light pink-violet with a yellow rim; staminode yellow with a purple or purple-violet tint, which changes color to orange as the flower ages. *Pedicel and ovary* 3.6–4 cm long, light green with deep purple-brown hairs. *Dorsal sepal* elliptic-subcircular to broadly-ovate, obtuse or shortly acuminate, 3.3–3.6 cm long, 2.9–3.5 cm wide, undulate on margins, often with recurved lower margins. *Synsepal* concave, elliptic-ovate, acute, 2.5–2.8 cm long, 1.6 cm wide. *Petals* spreading, sub-horizontal, oblong-spathulate to narrowly-obovate, obscurely tridentate or obtuse at apex, 3.4–3.6 cm long, 1.4–1.6 cm wide, with undulate margins, shortly ciliate. *Lip* deeply saccate, with a broad mouth, 3.7–4.3 cm long, 1.9–2.1 cm wide. *Column* short, about 9 mm long; staminode broadly obovate, obscurely tridentate or shortly acuminate, 7–10 mm long, 8–9 mm wide, papillose, with small low central umbo. Fig. 147.

DISTRIBUTION. Southern China (southeastern Yunnan, southwestern Guangxi), northern Vietnam (Ha Giang Province). Figs. 37, 148.

ECOLOGY. Primary, broad-leaved, mixed and coniferous, mossy, open, evergreen or semi-deciduous forests (commonly dominated by *Pseudotsuga brevifolia* and *Tsuga chinensis*) on highly eroded crystalline limestone mountains at (700)900–1400 m. Fig. 146.

FLOWERING IN THE WILD. September–November.

IUCN STATUS. VU.

AFFINITIES. *Paphiopedilum henryanum* is a small pretty slipper orchid with a distinctively coloured, boldly marked and relatively thick-textured flower. In many ways it resembles a miniature complex hybrid. This attractive orchid is closely related to *P. insigne, P. villosum* and their allies in sect. *Paphiopedilum,* which are distinguished by their green leaves, one-flowered inflorescences, usually spathulate incurving petals, large dorsal sepal and usually obovate staminode with a rough surface and central glossy umbo. It differs from other allied species in the region in its distinctive flower colour and size, particularly the boldly spotted dorsal sepal, the spotting at the base of the petals, and the pink lip. The only other Vietnamese species with a similarly spotted dorsal sepal are *P. gratrixianum* and *P. villosum* var. *boxalli*, but both have larger flowers, more spathulate petals, and neither is found growing

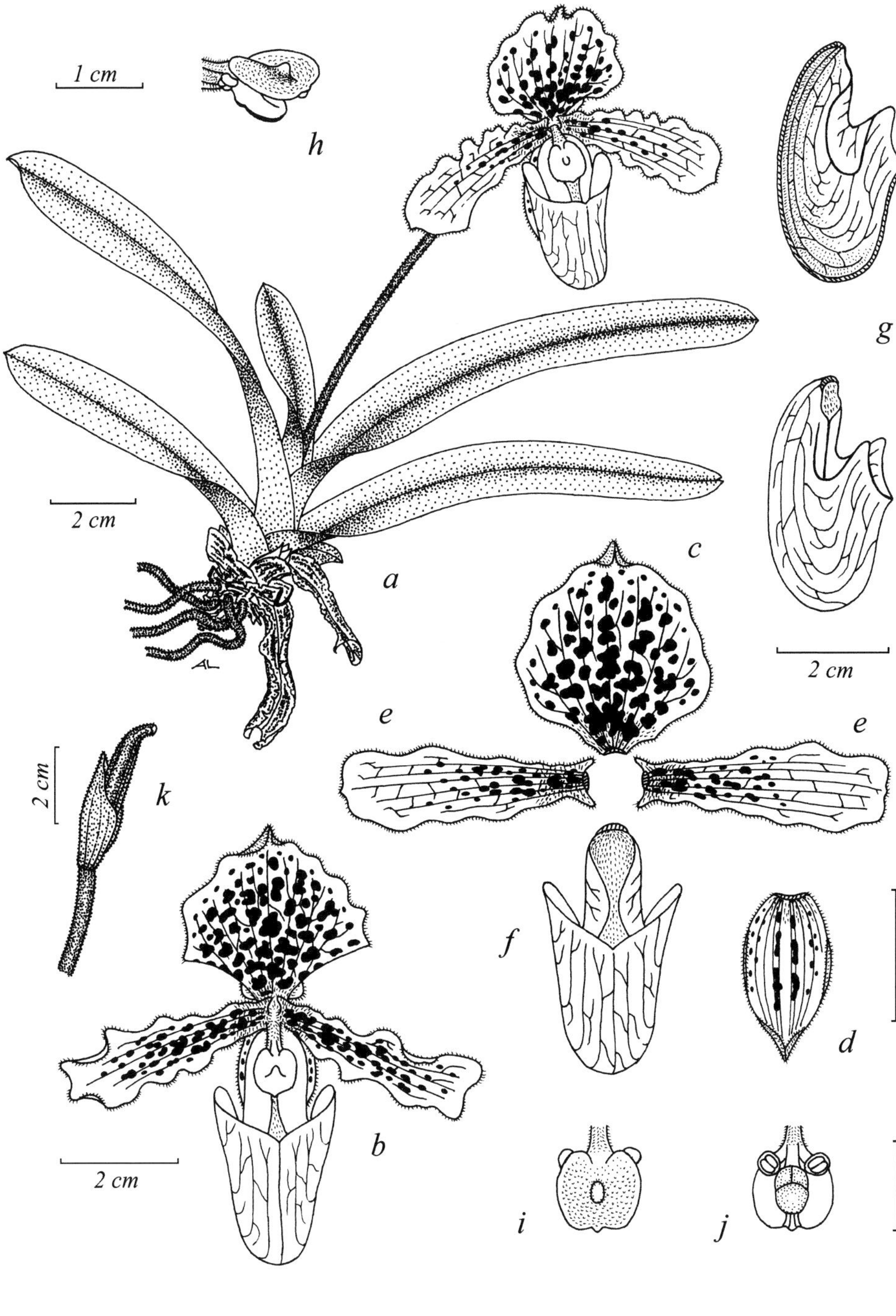

Fig. 147. *Paphiopedilum henryanum.* (Drawing: L. Averyanov).

a — flowering plant; **b** — flower; **c** — median sepal; **d** — synsepal; **e** — petals; **f, g** — lip, frontal, side and section views; **h–j** — column, side, dorsal and ventral views; **k** — ovary and flower bract. (Drawn from NTH 3590).

on limestone. *Paphiopedilum henryanum* has a very limited distribution in a restricted limestone area which lies along the border of southeastern Yunnan, southwestern Guangxi and Ha Giang Province of northern Vietnam (Fig. 37, 148). It is a vicariant of the *P. insigne* complex closely related to its calcicolous allies, such as *P. barbigerum, P. helenae, P. tranlienianum* and *P. × herrmannii.*

HISTORY. *Paphiopedilum henryanum* was described by Braem (1987) based on a wild-collected specimen imported from China that flowered in Germany. The name *P. henryanum* has formal priority, just pipping *P. dollii*, a name published by Lückel two weeks later, also based on a recently imported plant.

The extensive commercial collecting of slipper orchids in Vietnam in recent years produced the first evidence of *Paphiopedilum henryanum* in the country towards the end of 1998. Quantities of *Paphiopedilum henryanum* were received from the wild during 1998 by a commercial trading company (Tanaka, 1998).

Wild populations of *Paphiopedilum henryanum* were discovered in Vietnam by L. Averyanov and P.K. Loc in the spring of 1999 during field work supported by the National Geographic Society (Averyanov & Averyanova, 2000). In China, the species was found in the southwestern extremity of Guangxi (Hu Song Hua, 1999) and in southeastern Yunnan, but field work in Yunnan suggests that it may already be extinct there in all its known localities (Tsi *et al.*, 1999). Obviously, *P. henryanum* is one of the rarest species in the genus with a very restricted natural distribution.

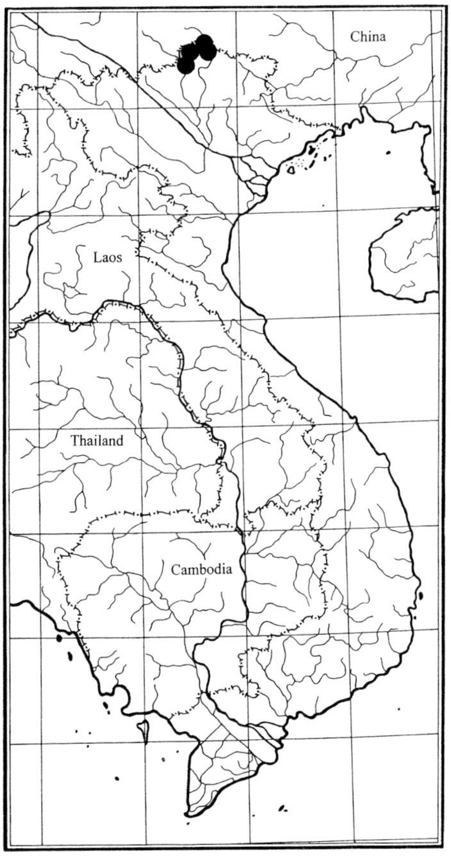

Fig. 148. Distribution of *Paphiopedilum henryanum* in Vietnam. (Drawing: L. Averyanov).

HABITAT. *Paphiopedilum henryanum* is restricted to a very small limestone area in the border triangle of Yunnan, Guangxi and Ha Giang Province of northern Vietnam (Fig. 37, 148). It typically grows as a lithophyte in small colonies on steep slopes and cliffs on karst, highly eroded limestone mainly between 900 to 1400 m (Fig. 149, 150). Very rarely, it may be found as low as 700 m.

The cold relatively dry northern winds in the area where it grows are replaced by the wet warm southwestern monsoon at the end of April. The flowering of the species coincides with end of the summer monsoon in late September and early October. At this time of the year the weather is cooler and drier as autumn begins. The moisture in autumn, winter and spring is the result of rare showers but more often mists and fogs. In the winter, mean air temperatures hover around 12–14°C but can drop to near freezing, although in the forest it always remains a few degrees higher. Summer temperatures seldom rise above 30°C and usually range betweend 22–28°C. Maximum rainfall is observed during May-August (Fig. 151). In most regions mean annual precipitation in the area varies between 1800 and 2500 mm, however locally on southeast facing slopes annual rainfall may reach 4800 mm, the highest rainfall in all of northern Indochina (Nguyen Khanh Van *at al.*, 2000).

Fig. 149 (opposite). *Paphiopedilum henryanum* may be found in south-facing slopes in intact primary forest where they flower better.

N. Vietnam, Ha Giang Prov. (Photo.: P. Cribb).

Fig. 150. Specimens of *Paphiopedilum henryanum* survive only on cliffs of northern exposition especially in deep and narrow gorges after felling of the primary forest.
N. Vietnam, Prov. Ha Giang Prov. (Photo.: P.J. Cribb).

In Vietnam, *Paphiopedilum henryanum* commonly inhabits primary, broad-leaved, mixed and coniferous, wet mossy, more or less open, evergreen or semi-deciduous forests on highly eroded narrow ridges and mesas, between (700) 900 and 1400 m, composed of crystalline limestone (Fig. 146). It is usually found in mixed and coniferous forests dominated by *Pseudotsuga brevifolia* and *Tsuga chinensis* (Cribb, Averyanov & Nguyen Tien Hiep, 2000). The centre of this slipper orchid distribution coincides with range of the unique and very rare Vietnamese endemic conifer, *Xanthocyparis vietnamensis* (Cupressaceae).

Mixed and coniferous forest is the climax primary vegetation of the peaks and ridge-tops of this karst limestone area in northern Vietnam. It is unlike the gloomy coniferous forests of northern temperate regions but is more like a narrow eyebrow of woodland along the ridge tops with the silhouettes of the individual trees clearly defined and the floor of the forest a tangle of rocks, crevices, small cliffs and plants.

The summit and ridge-top forest trees include *Pseudotsuga brevifolia*, *Tsuga chinensis*, *Podocarpus neriifolius, Nageia wallichiana,* and occasionally *Taxus chinensis* and *Xanthocyparis vietnamensis*. Broad-leaved species in the canopy here are represented mainly by deciduous or semi-deciduous trees in genera such as *Acer* (Aceraceae), *Carpinus* (Betulaceae), *Cinnamomum* (Lauraceae), *Lithocarpus, Quercus* (Fagaceae) and *Ulmus* (Ulmaceae), *Pistacia weinmannifolia* (Anacardiaceae) and *Platycarya strobilacea* (Juglandaceae). The tallest trees in this tier may reach 15–20 m in high (Fig. 146). Beneath the canopy are found numerous species of small trees, such as *Amentotaxus argotaenia*

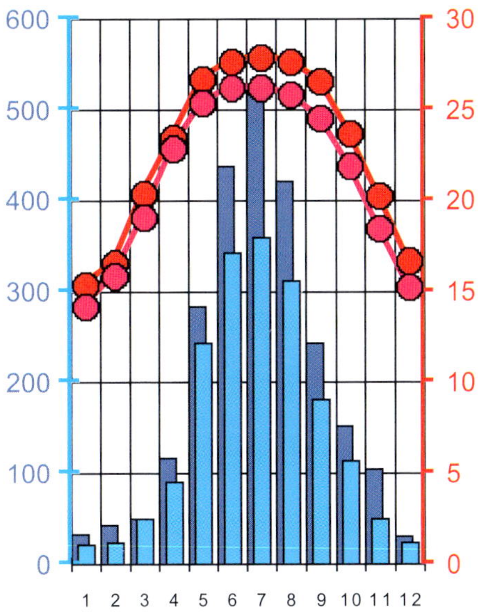

Fig. 151. Climatological data typical for the area of *Paphiopedilum henryanum* in Vietnam.

Black figures at the chart bottom are months. Blue axis represents atmosphere precipitation in mm (columns designate mean monthly rainfalls); red axis represents mean monthly temperature in °C. The deep blue and red are data from the Ha Giang weather station, 22°49'N 104°59'E, elev. 118 m; light blue and purple are data from the Hoang Su Phi weather station, 22°45'N 104°40'E, 553 m, all according to Nguyen Khanh Van *et al.*, 2000.

(Cephalotaxaceae), species of *Elaeocarpus* (Elaeocarpaceae), *Eriobotrya*, *Sorbus* (Rosaceae), *Myrsine kwangsiense* (Myrsinaceae), *Schefflera pesavis, Schefflera* sp. (Araliaceae) and *Tirpitzia sinensis* (Linaceae). Numerous shrubby species are also found here, the commonest being lithophytic species of *Ardisia* (Myrsinaceae), *Pittosporum* (Pittosporaceae), *Polygala* (Polygalaceae), *Rhododendron, Vaccinium* (Ericaceae) and *Spiraea* (Rosaceae). Some dwarf palms, such as *Guihaia* sp., are also very typical of these open forests.

Primary forests in the area support a very rich flora of lithophytic and epiphytic bryophytes, ferns and herbs. The herbs include Acanthaceae, Begoniaceae (*Begonia* spp.), Convallariaceae (species of *Aspidistra*, *Ophiopogon*, *Peliosanthes*, *Tupistra*), Gesneriaceae, Rubiaceae (species of *Ophiorrhiza*, *Hedyotis*) and Urticaceae (species of *Elatostema*, *Pellea*, *Pellionia*), and numerous sedges (species of *Carex*, *Scleria*) and selaginellas (*Selaginella* spp.). The commonest ferns in these forests are species of *Asplenium, Cyrtomium, Polypodium, Polystichum* and *Pyrrosia.*

Orchids are outstandingly diverse and abundant in these habitats and numerous lithophytic species are found there, including *Bulbophyllum ambrosia, B. macraei, B. maculosum, B. purpureifolium, Coelogyne fimbriata, C. lockii, C. malipoense, C. schultesii, Cleisostoma paniculatum, Dendrobium chrysanthum, D. nobile, D. thyrsiflorum, Eria coronaria, E. globulifera, E. sutepensis, E. siamensis, Epigeneium amplum, Flickingeria* sp., *Liparis averyanoviana, L. distans, L. mannii, L. stricklandiana, Monomeria barbata, Pholidota gracilis, P. roseans, P. yunnanensis,* and *Podochilus microphyllus*. In shady places and in deep karst depressions numerous terrestrial orchids may be found, such as *Anoectochilus calcareus, Calanthe alismaefolia, C. argenteostriata, Cheirostylis bipunctata, Goodyera hispida, Nephelaphyllum*

Fig. 152. *Paphiopedilum henryanum* is a typical lithophyte growing usually on vertical limestone walls.

N. Vietnam, Ha Giang Prov. (Photo.: P. Cribb).

tenuiflorum, Phaius flavus, Ph. longicruris and *Rhomboda petelotii*. Epiphytic orchid species also very common here, including species such as *Agrostophyllum callosum, Bulbophyllum hirtum, B. reptans, Ceratostylis himalaica, Cleisostoma striatum, Coelogyne schultesii, Epigeneium amplum, E. chapaense, Eria pannea, Luisia appressifolia, Ornithochilus difformis, Pholidota missionariorum, P. yunnanensis, Sunipia scariosa, Thelasis khasiana, Thrixspermum calceolus* and *Renanthera coccinea*. It is remarkable that this assemblage of many very rare local endemic orchids, *Anoectochilus calcareus, Bulbophyllum purpureifolium, Cheirostylis bipunctata, Coelogyne lockii, C. malipoense, Dendrobium chittimae, Holcoglossum wangii, Liparis averyanoviana, Luisia appressifolia, Pholidota gracilis, P. missionariorum, P. yunnanensis, Phaius longicruris* and *Rhomboda petelotii*, occurs in close association with *P. henryanum*. In such conditions, it sometimes grows with other slipper orchids such as *Paphiopedilum dianthum, P. hirsutissimum* var. *chiwuanum, P. malipoense* and *P. micranthum*.

Paphiopedilum henryanum often grows on the sunnier, but still well shaded, south-facing steep rocks that fall away in a series of low cliffs just below the summit. The plants usually occur on sheer cliffs with their roots in the cracks (Fig. 152). Like *P. hirsutissimum, P. helenae* and *P. dianthum*, it

Fig. 153. Most common variants of the flowers in Vietnamese populations of *Paphiopedilum henryanum*. **a** — N. Vietnam, Prov. Ha Giang (Photo.: L. Averyanov); **b–d** — N. Vietnam, Ha Giang Prov. (Photos: P. Cribb).

grows with its roots in intimate contact with the rock. Sometimes the roots form small platforms onto which leaf-litter falls and decomposes to provide nutrients for the plant. The soil and leaf-litter gathered from around the roots is commonly friable and dark brown. The pH of the soil on the roots is about 6.4, although the base rock was hard limestone (Cribb, Averyanov & Nguyen Tien Hiep, 2000).

The fact that only few localities have been found for *Paphiopedilum henryanum* in northern Vietnam, despite extensive searches over many years, suggests that it is a highly endangered species. The rate of logging in the region and the continuing collection of slipper orchids for sale make its future very precarious indeed. Most populations have been seriously damaged by forest cutting and commercial collecting. After destruction of the primary vegetation, single plants may survive for some time on shady mossy cliffs of north exposition. They usually eventually disappear due to progressive degradation and the rise of aridity in the habitats.

VARIATION. *Paphiopedilum henryanum* is polymorphous species with flowers that are very variable in size and in spotting, especially on the dorsal sepal. In Vietnamese populations plants occur with very different flowers (Fig. 153). In populations we have examined the median sepal varies in shape from narrowly elliptic and ovate to broadly ovate and even transversely elliptic. It may have dense heavy spotting or very few small marks only in the basal part. Also the lateral petals may be densely spotted or bear only a few small marks.

A remarkable variant where the heavy spotting on the dorsal sepal and petals is replaced by fewer faint dark streaks and dots was recognized in a bulk importation to Germany. The flowers of this plant had a light yellowish-green dorsal sepal and petals that contrasted boldly with the pink lip. It was described by Braem (in Schlechteriana 2, 4: 160, 1991) as var. *christae* *. The type of this semi-albino form was introduced, according to Braem, in the same consignment as the type of *P. henryanum*, but flowered some time afterwards in the collection of Helmut and Christa Sang of Essen, Germany. Presently, it is best considered as a form — *P. henryanum* f. *christae* (Braem) Gruss & Röth (Gruss & Röth, 1999).

The recently described *P. chaoi* (Hua in Die Orchidee 50, 5: 495, 1999) is illustrated by two photographs. These show a flower which is typical of *P. henryanum* except that it lacks the heavily spotted dorsal sepal while the spotting on the petals is rather dense, pale and widespread. The type was collected near Napo in Guangxi Province of southern China, an area very close to both the known Chinese and Vietnamese localities for *P. henryanum*. On the existing evidence, we do not believe that this plant merits recognition at specific rank. The small differences in floral marking suggests that, at best, it may warrant varietal or formal rank within *P. henryanum*.

13. PAPHIOPEDILUM GRATRIXIANUM (Masters) Rolfe

Paphiopedilum gratrixianum *Masters ex Rolfe* in Orchid Rev. 13: 63 (1905); Guillaumin in Bull. Soc. Bot. France 24: 556 (1924); id., Fl. Gen. Indo-Chine 6: 640 (1934); Karasawa, Saito in Bull. Hiroshima Bot. Gard. 5: 59 (1982); Cribb, Gen. Paphiopedilum: 146 (1987); id., Gen. Paphiopedilum, ed. 2: 249 (1998); Seidenfaden, Orch. Indochina: 17 (1992); Aver., Identif. Guide. Vietnam. Orch.: 20 (1994); Braem, C.Baker & M.Baker, Gen. Paphiopedilum 2: 236 (1999); Pham Hoang Ho, Ill. Fl. Viet. 3: 764 (2000). **Type:** Annam, *Micholitz* s.n., cult. Sander (holotype K).

Cypripedium gratrixianum Masters in Gard. Chron. ser. 3, 37: 76, fig. 35 (1905), *nom. inval. non* Williams (1897).

Cordula gratrixiana (Masters ex Rolfe) Rolfe in Orchid Rev. 20, 1: 2 (1912).

Paphiopedilum villosum (Lindl.) Stein var. *gratrixianum* (Masters ex Rolfe) Braem, Paphiopedilum: 119 (1988).

* **Type**: Border region China/Vietnam, 1987, *Azadehdel* s.n. (holotype SCHLE).

DESCRIPTION. *Lithophytic* or *terrestrial herb* often forming clumps. *Leaves* distichous, suberect, 4–7, linear to lanceolate, broadly lanceolate and oblong, acute at minutely tridentate apex, up to 30 cm long, 2–2.3(4) cm wide, green, usually spotted with dirty purple-violet on underside near the base. *Inflorescence* 1-flowered; peduncle erect, up to 25 cm long, green, covered with brownish-purple hairs; flower bract narrowly oblong-lanceolate to obovate, acute to obtuse, 3–4.5 cm long, 1.2–1.2 cm wide, green, sometimes sparsely spotted with purple-violet, densely covered with short hairs. *Flower* 7–8 cm across; dorsal sepal white, pale brownish-green at the base, more or less heavily spotted with brown-purple in basal two-thirds; synsepal pale green; petals glossy, yellow-brown, flushed and veined purple-brown; lip glossy, yellow flushed brown; staminode yellow; pedicel and ovary triangular in cross-section, 4–5.5 cm long, pale green, covered with short brown-purple hairs. *Dorsal sepal* ovate, broadly ovate to subcircular, obtuse, 4.8–5.2 cm long, 4.4–4.6 cm wide, shortly white-pubescent on outside. *Synsepal* ovate-elliptic, acute, 3.5–5 cm long, 2.5 cm wide, shortly white-pubescent on outside. *Petals* spathulate,

Fig. 154. Primary wet broad-leaved evergreen forests on very steep slopes of Dam Dao ridge at 950–1000 m support populations of *Paphiopedilum gratrixianum* in the northern Vietnam. (Photo.: L. Averyanov).

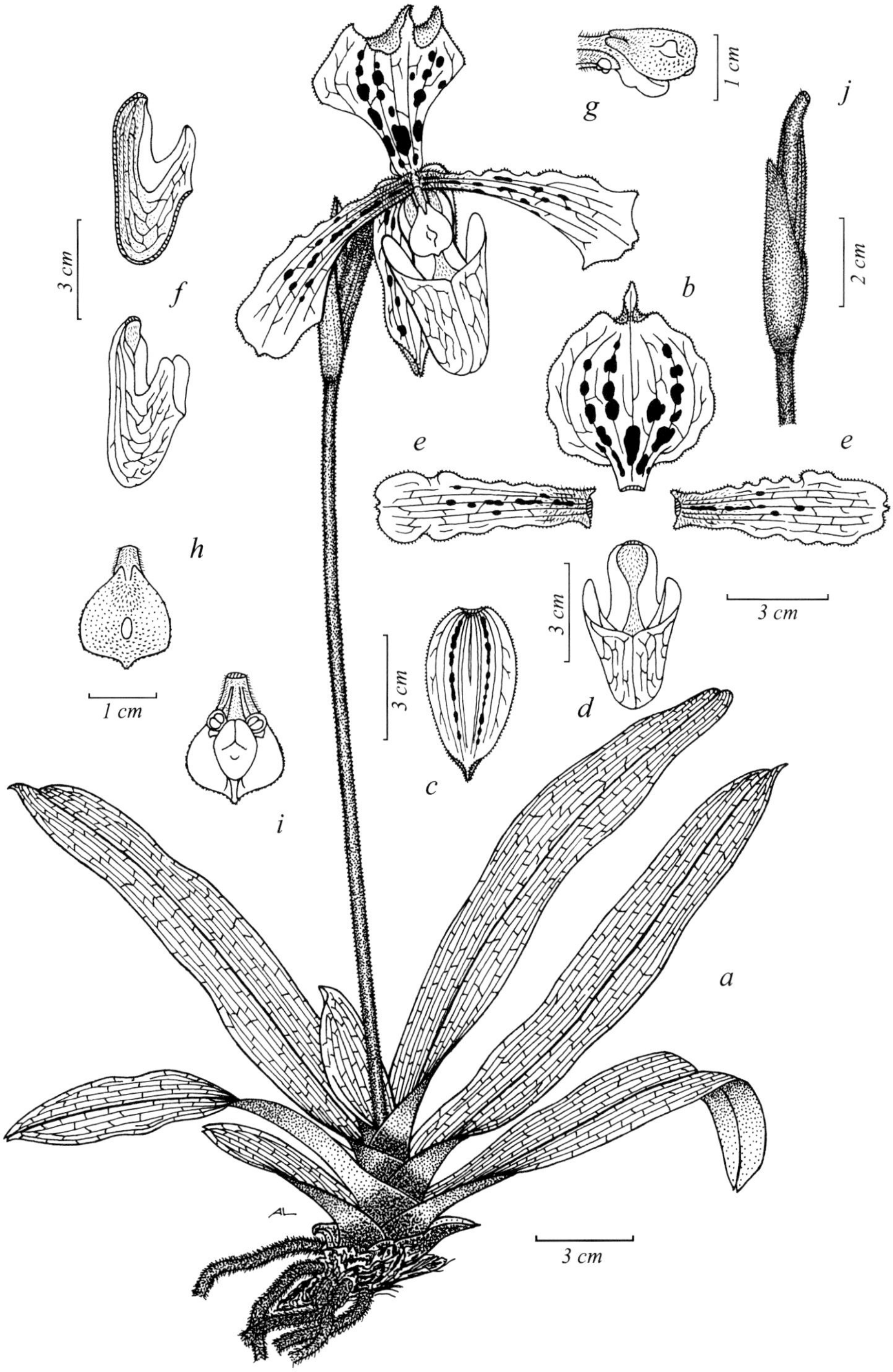

Fig. 155. *Paphiopedilum gratrixianum.* (Drawing: L. Averyanov).

a — flowering plant; **b** — median sepal; **c** — synsepal; **d** — lip; **e** — petals; **f** — lip, side and section views; **g–i** — column, side, dorsal and ventral views; **j** — ovary and flower bract. (Drawn from *L. Averyanov and N.T.Hiep* s.n., 10 Nov. 1998).

obtuse, 4.5–5.2 cm long, 1.8–2.5 cm wide, minutely ciliate, side margins undulate or reflexed. *Lip* tapering to apex, 4–4.2 cm long, 2.4–2.8 cm wide. *Staminode* obcordate, 10–12 mm long and wide, shortly acute at the apex, papillose, purple-violet hairy in basal half, with a central raised knob. 2n = 26. Fig. 32, 155.

DISTRIBUTION. Southeastern Laos (Attopeu Province) and northern Vietnam (Lao Cai?, Lai Chau?, Vinh Phuc and Thai Nguyen Provinces). Figs. 37, 156.

ECOLOGY. Primary, wet, broad-leaved, evergreen forest on steep mountain slopes composed of granite, gneiss and rhyolite at 900–1100(2000?) m. Fig. 154.

FLOWERING IN THE WILD. October–December.

IUCN STATUS. EN.

AFFINITIES. *Paphiopedilum gratrixianum* is a showy species with a large, elegant, long-lasting flower. It is most closely related to northeast Indian *P. insigne* and to *P. exul* from peninsular Thailand. This plant represents an eastern vicariant in the *P. insigne* complex growing on acidic calcium-free substrates. Its known distribution comprises two relatively small disjunct areas in the south-eastern extremity of Laos and in the central part of northern Vietnam.

HISTORY. Maxwell Masters described *Cypripedium gratrixianum* from plants exhibited by Messrs. Sander & Sons of St Albans who had imported them from Annam (supposedly from the region situated in southeastern extremity of Laos in Attopeu Province) where they had been collected by Wilhelm Micholitz. Specimens were cultivated in numerous private collections in Vietnam for many years and this suggested that *P. gratrixianum* occurred in the country. However, this rare species was only discovered in the wild in Vietnam in 1985, during field work by a Russo-Vietnamese botanical expedition exploring the central part of the northern Vietnam (Averyanov, 1988d). *Paphiopedilum gratrixianum* was found on the Dam Dao mountain massif that spreads along the border of Vinh Phuc and Thai Nguyen Provinces about 50 km to the north-west of Hanoi. This was a surprising discovery more than 1400 km from the original collection of the species.

HABITAT. In Vietnam, *Paphiopedilum gratrixianum* is only known from a single locality. A few populations were found on very steep slopes and cliffs on the Tam Dao ridge at elevation from 950 to 1100 m (Averyanov, 1988d, Averyanov, 2000b). The base rocks here are composed of rhyolite of late Mesozoic age. Many samples of *P. gratrixianum* have been cultivated for a many years in numerous private nurseries in Sa Pa town of Lao Cai Province near the northwestern border of Vietnam. Most of these were certainly collected from the wild near the city. According to local collectors, *P. gratrixianum* was collected on very steep granite slopes and cliffs of the Hoang Lien Son range in the peak area of Fan Si Pan at elevations of 1400–2000 m. However, this information has not been verified. Tanaka (1998) reported *P.*

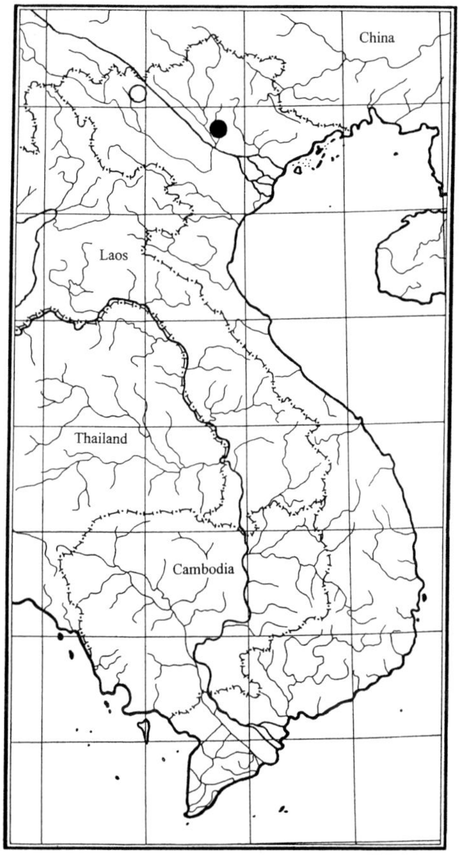

Fig. 156. Distribution of *Paphiopedilum gratrixianum* in Vietnam. (Drawing: L. Averyanov).

Studied population marked with a dot, open circle designates a locality according to information from local collectors.

gratrixianum to be a common species in the north of Vietnam. This is probably based on the observation that numerous specimens were delivered to a commercial trading company by local dealers. Indirect data accompanying the plants coming into local trade gave the impression that *P. gratrixianum* was extinct in southern Laos or had become very rare there. All plants on local sale probably originate now from northwestern Vietnam and northern Laos. We have little doubt that *P. gratrixianum* exists in several localities in northern Vietnam, but no exact data is yet available to confirm this.

In northern Vietnam, *Paphiopedilum gratrixianum* occurs on very steep wet mountain slopes on outcrops composed of silicate rocks such as granite, gneiss and especially rhyolite. It grows at elevations of 900–1100 m but may probably be found as high as 1500–2000 m. It occurs in very humid conditions in mountain ranges that receive at least 2500–3000 mm mean annual precipitation. A monsoon tropical climate is associated with these mountains. They lack a dry period due to the constant formation of heavy clouds and mists during winter and early spring that is usually the dry season in the northern Vietnam. Atmosphere humidity at this time never falls below 100%. Regular winter temperatures in the area vary from 8–12°C (Fig. 157). The coldest months are December and January when night temperatures occasionally fall to below zero. Wet night snow is not uncommonly observed above 1570 m. The absolute recorded temperature minima in the area of *P. gratrixianum* lie between zero and -3°C. The summer here is very wet with maximum rainfall from June to September. Temperatures during this period are rather moderate, commonly ranging between 18 and 25°C (Fig. 157).

Paphiopedilum gratrixianum grows in areas of very humid, broad-leaved, closed forests that develop because of the very high level of precipitation and high moisture content of silicate soils. Due to moderate summer and cool winter temperatures these forests lack tropical trees typical of the Vietnamese lowlands. The commonest forest dominants here are in the Aceraceae (*Acer*), Clusiaceae (*Garcinia*), Elaeocarpaceae (*Elaeocarpus*), Fagaceae (*Castanopsis, Lithocarpus, Quercus*), Hamamelidaceae (*Altingia, Rhodoleia championii, Symingtonia populnea, S. tonkinensis*), Juglandaceae (*Engelhardtia*), Lauraceae (*Actinodaphne, Cinnamomum, Cryptocarya, Lindera, Litsea, Machilus, Neolitsea, Notaphoebe, Phoebe*), Magnoliaceae (*Magnolia, Michelia, Talauma*), Myrtaceae (*Syzygium*), Proteaceae (*Helicia*), Rosaceae (*Pyrus, Sorbus, Pygeum*), Sterculiaceae (*Reevesia*), Theaceae (*Eurya, Pyrenaria, Schima, Gordonia*), and Ulmaceae (*Gironniera*). Woody and semi-woody species of Euphorbiaceae, Melastomataceae, Rubiaceae, and Sterculiaceae dominate the undergrowth . Large woody ferns, such as *Cyathea* and *Cibotium* which may reach 4–6 m in high in humid rivers canyons, are very common here.

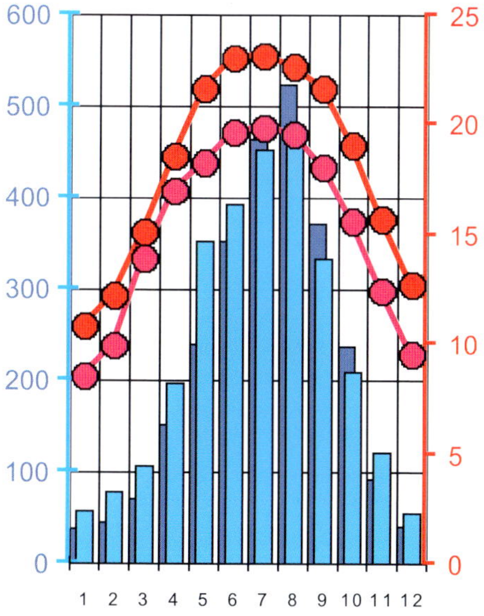

Fig. 157. Climatological data typical for the area of *Paphiopedilum gratrixianum* in Vietnam.

Black figures at the chart bottom are months. Blue axis represents atmosphere precipitation in mm (columns designate mean monthly rainfalls); red axis represents mean monthly temperature in °C. The deep blue and red are data from the Tam Dao weather station, 21°27'N 105°38'E, elev. 897 m; light blue and purple are data from the Hoang Sa Pa weather station, 22°20'N 103°50'E, 1570 m, all according to Nguyen Khanh Van *et al.*, 2000.

On the wet rocky slopes and cliffs where *P. gratrixianum* grows, communities dominated by small trees and shrubs, such species of *Ardisia* (Myrsinaceae), *Blastus* (Melastomataceae), *Desmos* (Annonaceae), *Rhododendron*, *Vaccinium* (Ericaceae), *Schefflera*, *Trevesia palmata* (Araliaceae), *Cinnamomum*, *Eugenia* (Lauraceae), *Elaeocarpus* (Elaeocarpaceae), *Ficus* (Moraceae), *Ixora*, *Pavetta*, *Lasianthus* (Rubiaceae), *Euonymus* (Celastraceae) and *Pandanus* (Pandanaceae), may commonly be seen.

Sometimes colonies of *P. gratrixianum* may be found on wet, open, rocky, south-facing cliffs in association with other lithophytic herbs, such as species of *Begonia* (Begoniaceae), *Curculigo* (Hypoxidaceae), *Impatiens* (Balsaminaceae), *Ophiorrhiza* (Rubiaceae), *Amorphophallus*, *Arisaema* (Araceae), *Alpinia*, *Amomum* (Zingiberaceae), *Sonerila*, *Phyllagathis* (Melastomataceae), *Aspidistra*, *Ophiopogon* (Convallariaceae), *Viola* (Violaceae), and species of Acanthaceae, Gesneriaceae, moisture-loving sedges and grasses. In shady places numerous lithophytic and epiphytic ferns, selaginellas and mosses are very common. The shady wet vertical rocks, kept constantly wet by cloudy mists and fogs, are often covered with numerous species of thin-leaved, moss-like, creeping ferns, especially *Hymenophyllum* and *Trichomanes* (Hymenophyllaceae).

Many species of orchid occur in close association with *Paphiopedilum gratrixianum*, among them terrestrial and lithophytic species such as *Acanthephippium simplex, A. striatum, Anoectochilus elwesii, A. daoensis, A. lanceolatus, A. roxburghii, A. sikkimensis, A. tridentatus, Apostasia wallichii, Calanthe herbacea, Cephalantheropsis obcordata, Collabium assamicum, C. chinense, Dendrobium henryi, Eria gagnepainii, E. pusilla, Habenaria rhodocheila, Mischobulbon macranthum, Nephelaphyllum tenuiflorum, Pholidota rubra, Platanthera angustata,* and *Podochilus microphyllus*. However, epiphytic orchids are the largest and most diverse group found here, including *Bulbophyllum ambrosia, Callostylis*

Fig. 158. Wet closed evergreen cloud forest on vertical north-facing rhyolite cliffs at 950–1000 m supports *Paphiopedilum gratrixianum* populations in northern Vietnam, Thai Nguyan Prov. (Photo: L. Averyanov).

Fig. 159. *Paphiopedilum gratrixianum* on inaccessible shelves of high vertical mountain cliffs. N. Vietnam, Thai Nguyen Prov. (Photo.: L. Averyanov).

rigida, Ceratostylis himalaica, C. radiata, Cleisostoma birmanicum, C. striatum, Coelogyne assamica, C. fimbriata, Dendrobium aduncum, D. faulhaberianum, D. hercoglossum, D. linguella, D. longicornu, D. nobile, Epigeneium amplum, E. chapaense, Eria foetida, E. pannea, E. thao, Flickingeria sp., *Liparis bootanensis, L. chapaensis, L. latilabris, L. stricklandiana, Oberonia quadridentata, Pholidota chinensis, Podochilus microphyllus* and *Schoenorchis gemmata*.

Colonies of *P. gratrixianum* commonly grow in rather wet conditions on shelves of rhyolite cliffs in a thin humic soil (Figs. 158–160). Elsewhere, plants grow in small crevices on vertical rocky walls of high mountain chasms (Fig. 161). Water seepages from overhanging ledges are common here during rainy season. The area has abundant small streams and waterfalls. Soils, leaf litter and rocks on the forest floor maintain a constant humidity.

Paphiopedilum gratrixianum is one of the most endangered of the Vietnamese slipper orchids. All known populations of this handsome species are situated on the Tam Dao ridge near the Tam Dao tourist resort which has expanded during recent years into a major tourist centre visited by a great number of people throughout the year. In recent years all colonies of *P. gratrixianum* have been seriously depleted. It is quite probable that *P. gratrixianum* may become extinct in this region in the very near future due to over-collecting and uncontrolled depredation by tourists.

VARIATION. In his description, Maxwell Masters quite reasonably compared *Paphiopedilum gratrixianum* with *P. insigne* and *P. exul* stating that the "flowers have so much of the appearance of those of some forms of *P. insigne* or of *P. exul,* if that be considered as distinct, that its position as a separate species is a matter of opinion ...".

The clones of *Paphiopedilum gratrixianum* in cultivation generally have smaller flowers than typical *P. insigne* with shorter petals and an elliptic dorsal sepal with fewer, but bolder dark spots. The

Fig. 160. Colonies of *Paphiopedilum gratrixianum* on wet open south-facing rocky outcrops at 950–1000 m
N. Vietnam, Vinh Phuc Prov. (Photo.: L. Averyanov).

leaves of the plants are suberect and narrow with purple spots at the base. However, Masters described the type collection as having 4 cm wide leaves. Broad leaves, 2–3 cm wide, are not uncommon in Vietnamese plants. The dorsal sepal in Vietnamese plants is usually white with deep purple-brown spots (Fig. 162), more or less flushed with purple, greenish or purple-greenish at the base or along the mid-vein, rarely pure white.

The morphological differences between *Paphiopedilum gratrixianum* and *P. insigne* are small but consistent. From *P. exul,* it differs rather more and can be readily distinguished by its habit, foliage, the dorsal sepal which tapers to the base and the lip which is larger and with longer side-lobes.

Paphiopedilum gratrixianum is also closely related to *P. villosum* and is sympatric with that species in Laos and Vietnam. However, it can be distinguished by its smaller flower, which is boldly marked with dark maroon raised spots, its narrower glossy ochreous petals and smaller lip. The leaves and long erect peduncle are not

Fig. 161. *Paphiopedilum gratrixianum* often grows as a lithophyte on high shady vertical cliffs.
N. Vietnam, Vinh Phuc Prov. (Photo.: L. Averyanov).

unlike those of *P. insigne*, while the flower has the shortly pubescent ovary of *P. insigne* but an overall similarity of its floral segments to *P. villosum*.

The relationships and geographic connections of all these species needs further study which will become increasingly difficult as populations of all of them are rapidly disappearing. According to available information, we suspect that this group is still actively evolving and may not have stabilized. Seidenfaden (1992) suggested that natural hybridisation may confuse the picture. *Paphiopedilum affine* De Wild., which regarded in the past as a possible synonym of *P. gratrixianum* (Cribb, 1987), may actually represent a natural hybrid of *P. villosum* and *P. appletonianum* (Cribb, 1998).

Fig. 162. Common variants of *Paphiopedilum gratrixianum* in wild Vietnamese populations.

a, b — N. Vietnam, Thai Nguyen Prov.; **c, d** — N. Vietnam, Vinh Phuc Prov. (Photos: L. Averyanov).

14. PAPHIOPEDILUM VILLOSUM (Lindl.) Stein

Paphiopedilum villosum *(Lindl.) Stein*, Orchideenbuch: 490 (1892); Pfitzer in Bot. Jahrb. Syst. 19: 41 (1894); id. in Engler, Pflanzenr. Orch. Pleon.: 72 (1903); Guillamin in Bull. Soc. Bot. France 71: 557 (1924); id. Fl. Gen. Indo-Chine 6: 637 (1934); Karasawa & Saito in Bull. Hiroshima Bot. Gard. 5: 41 (1982); Cribb, Gen. Paphiopedilum: 150 (1987); id., Gen. Paphiopedilum, ed. 2: 254 (1998); Braem, Paphiopedilum: 115 (1988); Seidenfaden, Orch. Indochina: 19 (1992); Aver., Identif. Guide. Vietnam. Orch.: 19 (1994); Braem, C. Baker & M. Baker, Gen. Paphiopedilum 1–2: 242 (1998–1999); Pham Hoang Ho, Ill. Fl. Vietnam 3: 765 (2000). **Type:** Burma, *Lobb* s.n. (holotype K-LINDL!).
Cypripedium villosum Lindl. in Gard. Chron.1854: 135 (1854).
Cordula villosa (Lindl.) Rolfe in Orchid Rev. 20: 2 (1912).

DESCRIPTION. *Epiphytic* or rarely *lithophytic herb*, usually forming more or less large clumps. *Leaves* 4–5, linear-ligulate, acuminate to acute at unequally bilobulate apex, 14–42 cm long, 2.5–4 cm wide, light green on beneath, purple-spotted at the base, basal margins ciliate. *Inflorescence* suberect to arcuate or pendent, 1-flowered; peduncle 7–24 cm long, green, commonly spotted with purple, with long villose whitish to deep purple hairs; bract elliptic, obtuse, 3.7–6.5 cm long, 3–3.8 cm wide, green, spotted with maroon, glabrous. *Flower* 7.5–13.5 cm across; dorsal sepal white or greenish toward the center, with a glossy deep maroon central area; synsepal pale green; petals glossy reddish-brown, often with a central maroon stripe, commonly yellow-brown along lower portion; lip ochre, flushed with pink or reddish, with maroon veins; staminode yellow, with a yellow or green central umbo; pedicel and ovary triangular in cross-section, 3–6 cm long, light-green coloured, densely covered with whitish to purple villose hairs. *Dorsal sepal* obovate, obtuse, 4.5–7 cm long, 3–4.6 cm wide, with reflexed basal margins. *Synsepal* adpressed to ovary, narrowly ovate, subacute, 3.8–7.6 cm long, 1.8–2.6 cm wide. *Petals* incurved, obovate-spathulate, retuse-emarginate, or broadly tridentate, rounded at apex, 4.7–8.6 cm long, 2.5–4.6 cm wide, glossy, ciliate, purple-villose at base. *Lip* tapering to apex, 4–6.8 cm long, 3–3.8 cm wide. *Staminode* obcordate-obovate, truncate, about 16 mm long, 14 mm wide, verrucose, hirsute, with a central glossy umbo. $2n = 26$. Fig. 164.

DISTRIBUTION. Northeast India, Myanmar, southern China, Laos, northern and southern Vietnam (Lai Chau, Lao Cai; Lam Dong, Khanh Hoa, Kontum Provinces), Thailand. (Fig. 165 a, b).

ECOLOGY. Primary, evergreen, wet, closed, mixed and broad-leaved, montane, cloud forests at elev. 1300–2000 m. Fig. 163.

FLOWERING IN THE WILD. April–May.

IUCN STATUS. VU.

AFFINITIES. *Paphiopedilum villosum* is a widespread and variable species with large attractive flowers. Its closest allies are *P. gratrixianum* and *P. insigne*. From both it can be distinguished by usually epiphytic habit, its regularly shorter scape, larger bract and larger flower which is distinctively coloured and has more spathulate broader petals, an obovate dorsal sepal, a villose ovary, triangular in cross-section and longer lip.

Three varieties of *Paphiopedilum villosum* have been recognized. The typical variety ranges from northeastern India, northern Burma and southwestern Yunnan to northern Thailand. *Paphiopedilum villosum* var. *boxallii* is distributed in Burma, northern Laos, southwestern Yunnan and northwestern Vietnam. The third variety, var. *annamense*, is found in Indochina and southern China.
Two varieties of *Paphiopedilum villosum*, var. *annamense* and var. *boxallii,* occur in Vietnam.

HISTORY. Thomas Lobb discovered *Paphiopedilum villosum* in 1853 growing at 1300–1600 m in the mountains near Moulmein in southeast Burma and it was introduced into cultivation in the same year by Messrs. Veitch & Sons. This well known species is nowadays widely distributed in cultivation. *Paphiopedilum villosum* was one of the first slipper orchids to be discovered in Vietnam by French botanists at the beginning of the last century. For many years, wild-collected samples of this species were commonly on sale in Dalat town in the southern Vietnam. In the past, this slipper orchid was very common in the Dalat area, especially on Lang Bian Mountain.

Fig. 163. Primary evergreen broad-leaved cloud forests at 1800 m a.s.l. support populations of *Paphiopedilum villosum* var. *annamense* in southern Vietnam.

S. Vietnam, Lam Dong Prov. (Photo.: L. Averyanov).

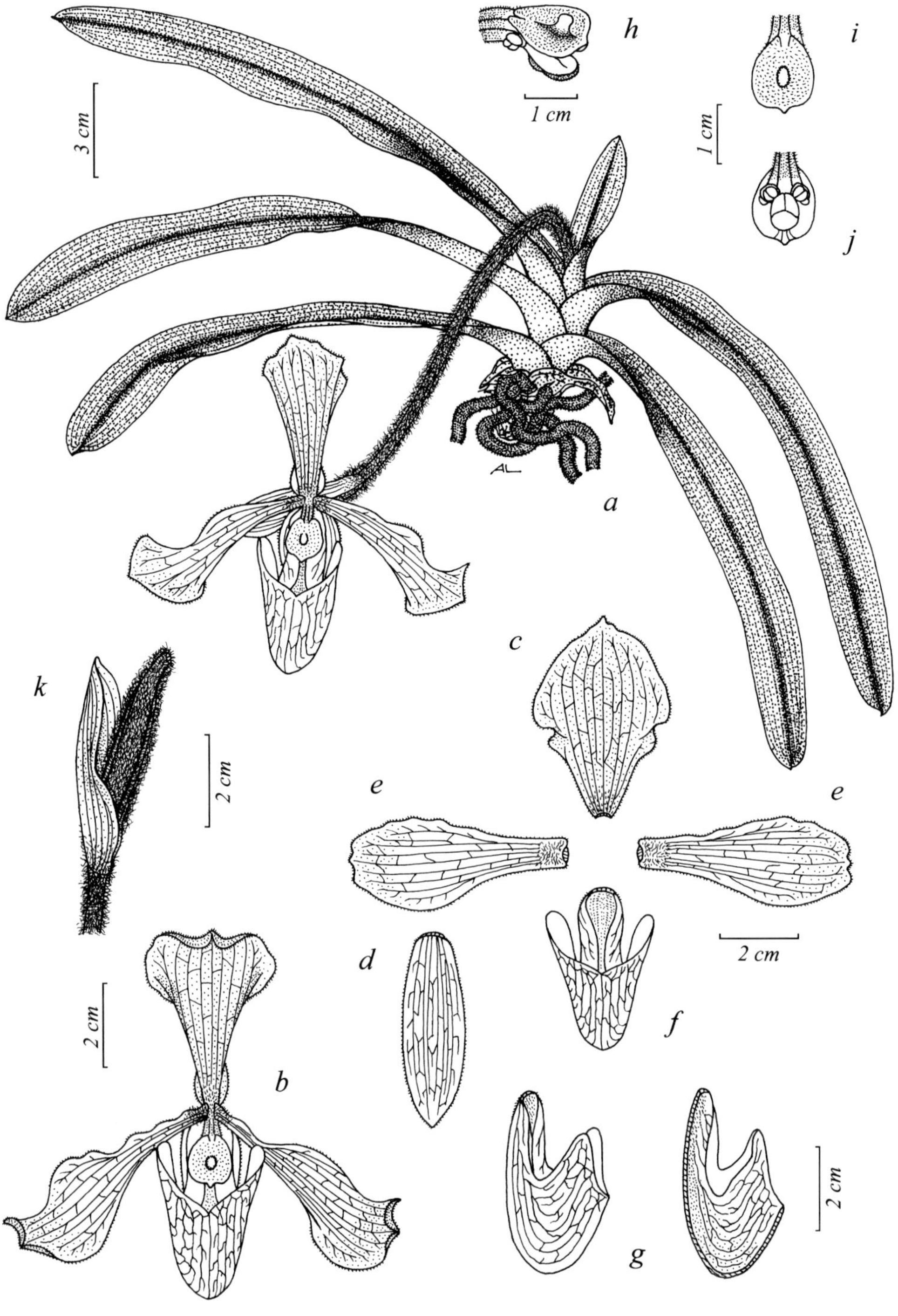

Fig. 164. *Paphiopedilum villosum* var. *annamense.* (Drawing: L. Averyanov).

a — flowering plant; **b** — flower; **c** — dorsal sepal; **d** — synsepal; **e** — petals; **f, g** — lip, frontal, side and section views; **h–j** — column, side, dorsal and ventral views; **k** — ovary and flower bract. (Drawn from VH 3333).

HABITAT. The widespread distribution of *Paphiopedilum villosum* in mainland southeastern Asia may be linked to its usually epiphytic habit. However, this species may also occasionally be found as a lithophyte. The preferred host trees in northern Thailand are *Eugenia angkae, Podocarpus neriifolius, Nyssa javanica* and *Quercus rex.* Some large trees of *Eugenia angkae* were observed in Thailand with up to 30 clumps of this slipper orchid (Bänziger, 1996).

In Vietnam, *Paphiopedilum villosum* inhabits wet, primary, broad-leaved and mixed, montane forests commonly at elevations of 1400–2000 m (Fig. 163). It is usually found as a humus epiphyte on high trees along mountain ridges in mossy cloud forests (Fig. 166). In areas inaccessible to commercial collectors, this species may also occasionally be found as a lithophyte growing on granite, dacite, gneiss or other acidic siliceous rocks (Fig. 167).

The climate where *Paphiopedilum villosum* is found is a monsoon tropical one associated with mountains. Such weather conditions are typical of mountain areas of Vietnam which lie above 1400–1500 m. This climate is associated with highland areas of large mountain ranges and high mountains. Mean annual precipitation in these regions ranges between 1800–3900 mm and certainly reaches higher levels in cloudy highland forests on mountain peaks and summits. The rains usually last from May to September but, in the south of the country, they often shift to August-October. Frequent heavy foggy mists are very typical for these areas which lack a dry season.

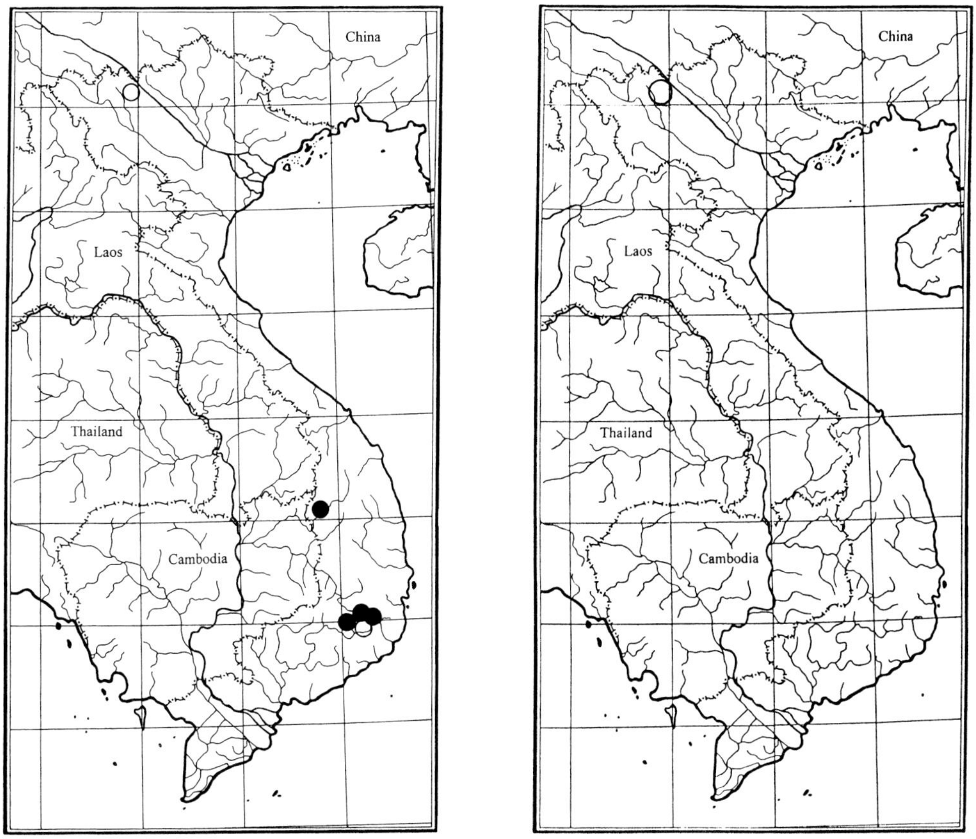

Fig. 165. Distribution of *Paphiopedilum villosum* var. *annamense* (above left) and var. *boxallii* (above right) in Vietnam. (Drawing: L. Averyanov).

Studied populations are marked with dots, open circles designate localities from local collectors.

Fig. 166. *Paphiopedilum villosum* var. *annamense* growing as a humus epiphyte in a typical habitat. S. Vietnam, Lam Dong Prov. (Photo.: L. Averyanov).

Mean summer temperatures in this mountain zone vary around 18–20°C both in the south and the north, but winter temperatures in the mountains of northern Vietnam are much lower than in the south (Fig. 168). Minimum winter temperatures may fall to 0° in the south and -3°C in the north of Vietnam. Local people living in these mountain regions are familiar with frost and ice. The highest temperature reported for the area is 32°C.

In Vietnam, *Paphiopedilum villosum* is found in humid, evergreen, closed, primary, montane forests developed on acidic silicate rocks at 1300–2000 m (Fig. 163). The canopy of these forests is dominated by cool-loving, evergreen, broad-leaved and coniferous trees. Broad-leaved woods commonly cover the slopes of mountain ridges and include numerous subdominant tree species, including Aceraceae (*Acer*), Betulaceae (*Alnus, Betula, Carpinus*), Elaeocarpaceae (*Elaeocarpus*), Ericaceae (*Rhododendron*), Fagaceae (*Castanopsis, Lithocarpus, Quercus*), Hamamelidaceae (*Rhodoleia, Symingtonia*), Illiciaceae (*Illicium*), Juglandaceae (*Engelhardtia*), Lauraceae (*Actinodaphne, Cinnamomum, Cryptocarya, Lindera, Litsea, Machilus, Neolitsea, Notaphoebe, Phoebe*), Magnoliaceae (*Magnolia, Michelia, Talauma*), Moraceae (*Ficus*), Myrtaceae (*Syzygium*), Rosaceae (*Eriobotrya, Pyrus, Sorbus, Photinia, Pygeum*), Sabiaceae (*Meliosma*), Theaceae (*Eurya, Pyrenaria, Schima, Gordonia*) and Ulmaceae (*Gironniera*).

Some gymnosperms are regularly found in these forests along ridges and mountain peaks, including *Amentotaxus* sp., *Cephalotaxus mannii, Dacrycarpus imbricatus, Dacrydium elatum, Fokienia hodginsii, Keteleeria evelyniana, Nageia fleuryi, N. wallichiana, Pinus kesiya, P. krempfii, P. dalatensis, P. wangii, Podocarpus neriifolius,* and *Taxus wallichiana*. In Vietnam, *P. villosum* exhibit a preference for habitats with coniferous and mixed forests of this type, but conifers are never a host tree for this orchid.

The subcanopy layer of montane forests commonly includes shrubby species of Araliaceae (*Brassaiopsis, Macropanax, Trevesia, Tupidanthus, Schefflera*), Chloranthaceae (*Chloranthus*), Ericaceae (*Agapethes, Gaultheria, Craibiodendron, Lyonia, Rhododendron, Vaccinium*),

Grossulariaceae (*Polyosma*), Melastomataceae (*Blastus, Poilannammia, Pseudodissochaeta*), Myrsinaceae (*Ardisia*), Oleaceae (*Fraxinus*), Rubiaceae (*Ixora, Pavetta, Lasianthus*), Saxifragaceae (*Dichroa*), Symplocaceae (*Symplocos*), and Theaceae (*Adinandra, Anneslea, Eurya, Ternstroemia*).

Common forest floor herbs are species of *Argostemma, Ophiorrhiza* (Rubiaceae), *Begonia* (Begoniaceae), *Curculigo* (Hypoxidaceae), *Dicliptera, Justicia, Nelsonia, Peristrophe, Phlogacanthus, Staurogyne, Strobilanthes* (Acanthaceae), *Disporum, Ophiopogon, Peliosanthes* (Convallariaceae), *Impatiens* (Balsaminaceae), *Pellionia, Pellea* (Urticaceae), *Pentaphragma* (Pentaphragmaceae), *Phyllagathis* and *Sonerila* (Melastomataceae). A very rich, terrestrial, lithophytic and epiphytic fern flora is typical with a great abundance of moisture-loving species such as *Angiopteris, Asplenium, Belvisia, Coniogramme, Crypsinus, Diplasium, Histiopteris, Humata, Leptochilus, Lindsaea, Marattia, Microlepia, Microsorum, Paragramma, Polystichum, Prosaptia, Stegnogramma, Tectaria, Thelypteris, Vittaria* and *Woodwardia.* Mosses and moss-like, creeping, thin-leaved fern species of *Hymenophyllum, Macroglena, Mecodium, Trichomanes* and *Vandenboschia* (Hymenophyllaceae) cover stems and branches of trees in these highland montane forests with a thick, lush, continuous cover. Woody ferns such as *Cyathea, Ctenitopsis,* and *Cibotium* reach 10 and more metres tall here.

Fig. 167. *Paphiopedilum villosum* var. *annamense* growing as a lithophyte on granite/gneiss bluffs by a waterfall in a deep narrow canyon.

S. Vietnam, Khanh Hoa Prov. (Photo.: L. Averyanov).

Paphiopedilum villosum is commonly found as an epiphyte about 10–15 m above the ground on the tallest trees (Fig. 166). Host trees belong to many species, particularly Magnoliaceae and Fagaceae. In intact primary forest *P. villosum* forms large clumps with 10–20 or even more growths. Here, *it* commonly settles in the forks of large old tree branches and grows as a humus epiphyte (Fig. 169) in close association with numerous epiphytic mosses, lichens, ferns and other epiphytes such as species of *Peperomia, Piper* (Piperaceae), *Aeschynanthus* (Gesneriaceae) and *Vaccinium* (Ericaceae). The large woody vine, *Schisandra coccinea* (Schisandraceae), is very common in these forests

Epiphytic orchids that are commonly associated with *Paphiopedilum villosum* in Vietnam are *Agrostophyllum callosum, Appendicula gracilis, Bulbophyllum crassiusculifolium, B. hymenanthum, B. longiflorum, B. luanii, B. ngoclinhensis, B. odoratissimum, B. reptans, B. retusiusculum, Callostylis rigida, Ceratostylis himalaica, C. radiata, C. subulata, Cleisostoma birmanicum, C. duplicilobum, C. striatum, C. williamsonii, Coelogyne calcicola, C. cycnoches, C. eberhardtii, C. fimbriata, C. mooreana, C. sanderae, Cymbidium eburneum, C. schroederi, Dendrobium bellatulum, D. cariniferum, D. crystallinum, D. gratiosissimum, D. khanhoaense, D. langbianense, D. nobile, D. virgineum, D. wattii, Epigeneium amplum, E. cacuminis, E. chapaense, Eria apertiflora, E. bidupensis, E. globifera, E. lactiflora, E. paniculata, E. pannea, E. siamensis, E. thao, Flickingeria angustifolia, F. ritaeana, Gastrochilus calceolaris, G. pseudodistichus, G. simplicilabius, Holcoglossum subulifolium, Hygrochilus parishii, Liparis cespitosa, L. compressa, L. elliptica, L. luteola, Monomeria dichroma, Oberonia acaulis, O. anthropophora, O. caulescens, O. dalatensis, O. dolichocaulis, O. langbianensis, Otochilus albus, O. fuscus, Panisea albiflora, Pholidota*

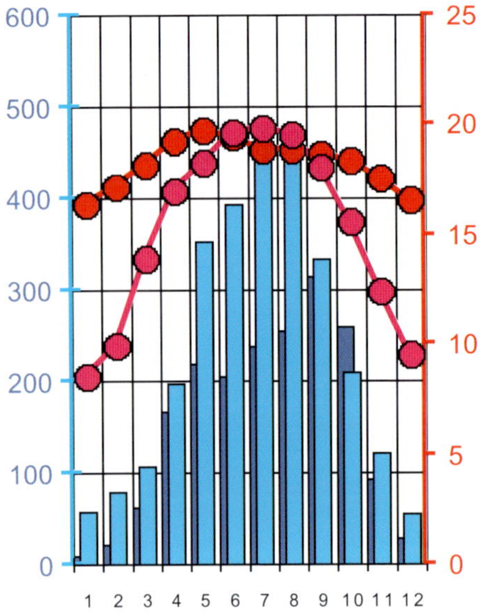

Fig. 168. Climatological data for the area of *Paphiopedilum villosum* in southern and northern Vietnam on the basis of data from Da Lat (11°57'N 108°26'E, elev. 1513 m) and Sa Pa weather stations (22°20'N 103°50'E, elev. 5 m).

Black figures at the chart bottom are months. Blue axis represents atmosphere precipitation in mm (columns designate mean monthly rainfall); red axis represents mean monthly temperature in °C. The deep blue and red are data from the Da Lat weather station; light blue and purple are data from the Sa Pa weather station, all according to Nguyen Khanh Van *et al.*, 2000.

articulata, P. chinensis, P. convallariae, P. leveilleana, P. recurva, Polystachya concreta, Schoenorchis eberhardtii, S. gemmata, Sunipia annamensis, S. pallida, S. scariosa, Trichotosia dasyphylla, T. microphylla, T. pulvinata, and *T. velutina. Paphiopedilum villosum* may be occasionally found in Vietnam as a lithophyte growing on vertical bluffs of mountain rivers canyons or on waterfall flanks (Fig. 167). Rocks and soils here are kept constantly wet by mists along fast-flowing streams. In some areas in the south, *P. villosum* grows in close association with *P. appletonianum, P. callosum* and *P. delenatii.* All these occur regularly on granite, gneiss, dacite, rhyolite and other related siliceous rocks. Numerous lithophytic mosses, ferns and herbs are associated with *P. villosum* in its lithophytic habitats. Common orchids here include terrestrial or lithophytic species such as *Anoectochilus acalcaratus, A. lylei, A. roxburghii, Apostasia wallichii, Appendicula cornuta, A. hexandra, A. reflexa, Calanthe clavata, C. densiflora, Cymbidium ensifolium, C. lancifolium, Eria gagnepainii, Goodyera foliosa, G. fumata, G. schlechtendaliana, Pristiglottis umbrosa, Tainia hookeriana, Bulbophyllum tixieri, Dendrobium alleizettii, D. fimbriatum, D. sociale, Flickingeria fimbriata, Liparis luteola, Monomeria barbata, Pholidota levelleana, P. rubra* and *Podochilus microphyllus.*

VARIATION. *Paphiopedilum villosum* is a widespread and variable species. Its morphological variation certainly needs further investigation based on studies of wild populations. It is presently considered that the species comprises three varieties, two of which — var. *annamense* and var. *boxallii* occur in Vietnam (Cribb, 1998). The typical variety has not been found in Vietnam.

KEY TO THE VARIETIES

1. Dorsal sepal broadly obovate to obovate, white with a wide maroon-brown central area; lateral petals narrowly obovate to broadly spatulate with broad basal part, brown-orange with a median purple-brown narrow stripe . . . . **14a. *P. villosum* var. *annamense***

+. Dorsal sepal narrowly obovate, light green, boldly spotted with large maroon marks and with a narrow white margin; lateral petals narrowly spatulate with a narrow stalk-like basal part, heavily striped with purple-brown . . . . **14b. *P. villosum* var. *boxallii***

14B. PAPHIOPEDILUM VILLOSUM VAR. ANNAMENSE Rolfe

Paphiopedilum villosum var. **annamense** *Rolfe* in Curtis's Bot. Mag. 133: t. 8126 (1907).
Lectotype: Annam, *Micholitz* s.n., cult. Sander (icon. in Curtis's Bot. Mag. 133: t. 8126).

DESCRIPTION. It differs from the typical variety in having narrower suberect *leaves*, 2.5 cm wide and in a smaller *flower* with a *dorsal sepal* which is white with a dark brown-purple central streak or blotch edged with pale green or yellow (Figs. 169, 170).

DISTRIBUTION. Southern China (Yunnan? and Guangxi Provinces?), Laos, Vietnam (Lai Chau, Lao Cai Provinces?, Lam Dong, Khanh Hoa and Kontum Provinces). Fig. 165 a.

ECOLOGY. Primary, evergreen, wet, closed, mixed and broad-leafed cloud, mountain forests at elev. of 1300–2000 m. (Fig. 163).

FLOWERING IN THE WILD. April–May.

IUCN STATUS. VU.

AFFINITIES, HISTORY AND VARIATION. *Paphiopedilum villosum* var. *annamense* is an eastern race of the species. It slightly differs from the typical variety in its small flowers and colouration of the dorsal sepal (Fig. 170). It was discovered by Wilhelm Micholitz, collecting for Messrs. Sander & Sons of St Albans, and was described by Robert Allen Rolfe in Curtis's Botanical Magazine in 1907. It should be mentioned that variation of the flower colouration of *P. villosum* var. *annamense* in Vietnamese populations is very wide and how they differ from the type is still unclear.

Numerous wild populations of var. *annamense* were recently found in the southern Vietnam (Averyanov, 1988d, 2000b). Before its widespread commercial collection, it was quite common in highland mountain areas in the south. It has yet to be found in northern Vietnam. It has, however, been

Fig. 169. *Paphiopedilum villosum* var. *annamense* growing as a humus epiphyte.
S. Vietnam, Lam Dong Prov. (Photo: L. Averyanov).

Fig. 170. Common variants of *Paphiopedilum villosum* var. *annamense*. (Photos: L. Averyanov).

a — S. Vietnam, Lam Dong Prov., Dalat, private collection;
b, c — S. Vietnam, Lam Dong Prov., wild plants.

cultivated in private collections in Sa Pa town for a many years and it seems likely that most of these plants were collected in the wild in the highlands of Hoang Lien Son range in Lai Chau and Lao Cai Provinces. Nevertheless, we have not yet any documentary evidence of its occurrence in the north.

An albino variant of the species was illustrated by Braem (1988, 1999) under the name *Paphiopedilum villosum* var. *aureum*. It was recognized among cultivated plants and is probably best treated as an albino form of the typical variety.

14B. PAPHIOPEDILUM VILLOSUM VAR. BOXALLII (Rchb.f.) Pfitzer

Paphiopedilum villosum var. **boxallii** *(Rchb.f.) Pfitzer* in Engler & Prantl, Nat. Pflanzenfam. 2, 6: 83 (1888); id., in Engler, Pflanzenr. Orch. Pleon.: 73 (1903). **Type:** Burma, Moulmein, *Boxall* s.n. (holotype W!).

Cypripedium boxallii Rchb.f. in Gard. Chron. n.s. 7: 367 (1877).

Cypripedium boxallii Rchb.f. var. *atratum* Masters in Gard. Chron. ser. 3, 1: 210 (1887).

Cypripedium dilectum Rchb.f. in Gard. Chron. ser. 3, 3: 330 (1888).

Cypripedium villosum Lindl. var. *boxallii* (Rchb.f.) Veitch, Man. 4: 54 (1889).

Cordula boxallii (Rchb.f.) Rolfe in Orchid Rev. 20: 2 (1912).

DESCRIPTION. Distinguished from the type in narrow, light green, boldly spotted *dorsal sepal* and narrow lateral *petals* heavily striped with purple-brown. Fig. 171.

DISTRIBUTION. Myanmar, northern Vietnam (Lai Chau and Lao Cai Provinces?). Fig. 165b.

ECOLOGY. Primary broad-leaved evergreen mountain forests?

FLOWERING IN THE WILD. ?

IUCN STATUS. DD.

AFFINITIES, HISTORY AND VARIATION. *Paphiopedilum villosum* var. *boxallii* is a remarkable local geographical variant readily distinguished by its brightly coloured showy flowers with a boldly spotted dorsal sepal. It was discovered in 1877 by William Boxall, collecting orchids for Messrs. Hugh Low & Co., probably from the Tongku District of Burma. Its distribution may extend to northern Vietnam. Numerous examples of this plant are cultivated in orchid nurseries in Sa Pa town. Some of them were said to have been collected on granite cliffs of the Hoang Lien Son range in Lai Chau and Lao Cai Provinces reportedly between 1400–2000 m. However, we have not yet been able to confirm this.

Paphiopedilum villosum var. *boxallii* differs from the typical variety in the colouring of its flowers, in particular it has a heavily spotted dorsal sepal. The synsepal is also more acuminate and the petals purple-marked. The floral segments are consistently narrower at the base (Fig. 171).

Fig. 171 (right). Flower of cultivated sample *Paphiopedilum villosum* var. *boxallii* of uncertain origin.

Plants with flowers of identical colouration may be collected on highlands of Hoang Lien Son range in northeastern Vietnam. Europe, private collection. (Photo.: P.J. Cribb).

Reichenbach, in describing this as a new species, suggested that further distinctions from *P. villosum* were to be found in its larger bract and purple-blotched peduncle covered with purple and colourless hairs. The variation found in typical *P. villosum* encompasses this. The type of *Cypripedium dilectum* is, in these respects, rather intermediate between the two varieties whilst retaining the spotting on the dorsal sepal of *C. boxallii*. *Cypripedium boxallii* var. *atratum,* on the other hand, is a more boldly marked form of var. *boxallii* and is here reduced to its synonymy.

Paphiopedilum sect. Barbata (Kraenzl.) V.A. Albert & B. Pettersen

Paphiopedilum sect. **Barbata** *(Kraenzl.) V.A. Albert & B. Pettersen* in Lindleyana 9, 2: 137 (1994). **Type:** *Paphiopedilum barbatum* (Lindl.) Pfitzer (syn. *Cypripedium barbatum* Lindl.).
Cypripedium sect. *Barbata* Kraenzl., Orch. Gen. Sp. 1: 53 (1897).
Paphiopedilum sect. *Spathopetalum* Pfitzer in Engler, Pflanzenr. Orch. Pleon.: 78 (1903).
Paphiopedilum sect. *Blepharopetalum* Pfitzer, l.c.: 82 (1903).
Paphiopedilum sect. *Phacopetalum* Pfitzer, l.c.: 87 (1903).
Paphiopedilum sect. *Sigmatopetalum* Hallier f. ex Karasawa & Saito in Bull. Hiroshima Bot. Gard. 5: 42 (1982).
Paphiopedilum sect. *Punctatum* Karasawa & Saito, l.c.: 49 (1982).
Paphiopedilum sect. *Planipetalum* Karasawa & Saito, l.c.: 54 (1982).

DESCRIPTION. *Leaves* deep green–light green tessellated, commonly uniformly light green, without purple-violet marks beneath. *Inflorescence* erect, single or rarely two-flowered. *Median sepal* and synsepal distinctly different from petals in color, size and shape. *Petals* oblong to narrowly obovate, much narrower than median sepal, spotted and warted along margin in most species. *Lip* with large, well developed incurved, usually warted side-lobes, its apical margin straight (not involute). *Staminode* usually lunate, with a three-toothed apex, or an obvious variation on that theme, rarely obcordate, with a more or less rugose surface, lacking a distinct central boss (umbo). 2n=28–44. 27 species distributed almost throughout the range of the genus (Fig. 38).

15. PAPHIOPEDILUM APPLETONIANUM (Gower) Rolfe

Paphiopedilum appletonianum *(Gower) Rolfe* in Orchid Rev. 4: 364 (1896); Pfitzer in Engler, Pflanzenr. Orch. Pleon.: 79 (1903); Guillaum. in Bull. Soc. Bot. France. 71: 550 (1924); id., Fl. Gen. Indo-Chine 6: 641 (1934); Karasawa & Saito in Bull. Hiroshima Bot. Gard. 5: 46 (1982); Cribb, Gen. Paphiopedilum: 161 (1987); id., Gen. Paphiopedilum, ed. 2: 280 (1998); Braem, Paphiopedilum: 149 (1988); Seidenfaden, Orch. Indochina: 19 (1992); Aver., Identif. Guide. Vietnam. Orch.: 21 (1994); Braem, C. Baker & M. Baker, Gen. Paphiopedilum 2: 292 (1999); Pham Hoang Ho, Ill. Fl. Viet. 3: 763 (2000). **Type:** hort. F. Appleton, April 1893 (holotype K!).
Cypripedium appletonianum Gower in The Garden 1893: 95 (1893).
Cypripedium bullenianum Rchb.f. var. *appletonianum* (Gower) Rolfe in Orchid Rev. 1: 135 (1893).
Cypripedium poyntzianum O'Brien in Gard. Chron. ser. 3, 15: 36 (1894); Rolfe in Orchid Rev. 3: 55 (1895); id., ibid. 4: 18 (1896).
Cypripedium wolterianum Kraenzl. in Gard. Chron. ser. 3, 17: 166 (1895) & in Xenia Orchid. 3: 142, t. 282 (1896).
Paphiopedilum appletonianum (Gower) Rolfe var. *poyntzianum* (O'Brien) Pfitzer in Engler, Pflanzenr. Orch. Pleon.: 79 (1903); Guillaum., Fl. Gen. Indo-Chine 6: 642 (1934).

Paphiopedilum wolterianum (Kraenzl.) Pfitzer in Engler, Pflanzenr. Orch. Pleon.: 79 (1903).
Cordula appletoniana (Gower) Rolfe in Orchid Rev. 20: 2 (1912).
Paphiopedilum hookerae (Rchb.f. ex Hook.f.) Stein subsp. *appletonianum* (Gower) M.W.Wood in Orchid Rev. 85, 1007: 11 (1977).
Paphiopedilum hainanense Fowlie in Orchid Dig. 51, 2: 69 (1987).
Paphiopedilum robinsonii forma *viride* Braem in Orchid. Cult. Protect. 36: 35–38 (1998).
Paphiopedilum cerveranum Braem in Orchid. Cult. Protect. 38: 28 (1999). **Type:** Indochina, Vietnam-Cambodian border, *De Raeve* s.n. (holotype SCHLE).
Paphiopedilum cerveranum forma *viride* Braem in Orchid. Cult. Protect. 38: 29 (1999).
Paphiopedilum appletonianum (Gower) Rolfe var. *hainanense* (Fowlie) Braem, C.O.Baker & M.L.Baker, Gen. Paphiopedilum 2: 295 (1999).
P. tridentatum S.C. Chen, Z.J. Liu & J.Y. Zhang in Acta Phytotax. Sinica 39 (5): 455 (2001), **syn. nov. Type:** Vietnam, without exact loc., cult. Shenzhen City Wutongshan Nurseries, 28 March 2001, *Liu* 20141 (holotype Herb. Shenzhen City Wutongshan Nurseries).
P. puberulum S.P. Lei & J.Y. Zhang in Acta Bot. Yunnanica 24 (3): 309, fig. (2002), **syn. nov. Type**: N. Vietnam, without exact loc., cult. 2 May 2002, Shenzhen City Wutongshan Nurseries (holotype Herb. Schenzhen City Wutongshan Nurseries).

DESCRIPTION. *Terrestrial, lithophytic* or occasionally humus *epiphytic herb* with 4–6(8) leaves. *Leaves* narrowly elliptic to oblong-elliptic, subacute to obtuse at tridenticulate apex, (7)15–25 cm long, 2–4 cm wide, sparsely ciliate at base, distinctly tessellated with pale and darker green, rarely purple-marked at base on lower side. *Inflorescence* erect, single- or rarely two-flowered; peduncle 20–50 cm long, purple, shortly pubescent; bract lanceolate, acute, 1.5–2.1 cm long, green, ciliate. *Flower* 6–10(12) cm across; sepals whitish green or pale green, longitudinally veined with green, dorsal sepal often olive-green or brown-purple striped toward the base; petals green below with darker green and olive-brown stripes and maroon-black spots in basal half, dull pink or purple above; lip olive-green, ochre, to pale purple with darker veins and a yellowish or greenish margin; pedicel and ovary dull purple-violet spotted, 3–6 cm long, pubescent. *Dorsal sepal* ovate, cordate at base, apiculate above, 2.7–4.5 cm long, 2–3.2 cm wide, with recurved basal margins. *Synsepal* elliptic-lanceolate, acuminate, 2–3 cm long, 1.1–1.5 cm wide. *Petals* spathulate, acute, 4.4–5.8(6) cm long, 1.2–1.8 cm wide, often half-twisted in apical half, with marginal maroon-black warts in basal part of upper margin. *Lip* 3.6–4.6 cm long; front margin flat, curved or more or less distinctly 2- or 6-dentate. *Staminode* dull yellowish, yellowish-green, emerald-green or olive-green with yellowish-white margin and central yellowish spot, from broadly obcordate to deeply transversely lunate, with 2-, 4- or 5-toothed apex, 7–9 mm long, 8–9 mm wide; lateral teeth often falcate, divergent, usually longer than middle tooth. 2n=38. Figs. 173, 174.

DISTRIBUTION. China (Hainan Island), Thailand, southern Vietnam (Dak Lac, Khanh Hoa, Kontum, Lam Dong, Ninh Thuan and Thua Thien Hue Provinces), Laos, and Cambodia. Figs. 38, 175.

ECOLOGY. Primary, wet, coniferous, mixed or broad-leaved, evergreen, mountain forest, usually on steep slopes composed of granite, gneiss and sandstone at elev. of (700)1000–1900(2000) m. Fig. 172.

FLOWERING IN THE WILD. March–May.

IUCN STATUS. VU.

AFFINITIES. In the past *Paphiopedilum appletonianum* was certainly one of the commonest slipper orchids, widely distributed in mountain regions of the southern part of Truong Son Range in southern Vietnam (Fig. 175). This remarkable plant, with boldly tessellated leaves, commonly grows in closed mountain forests as a terrestrial or lithophytic herb (Fig. 176–178) along damp river canyons or on wet shady flanks of waterfalls. It has a very long erect inflorescence stalk bearing one or rarely two flowers (Fig. 176, 181). Flowers are very variable in shape, size and colouration and in some populations they reach 10–12 cm across. *Paphiopedilum appletonianum* is obvious very closely related to *P. bullenianum,* which has a wide distribution in peninsular Malaysia and the western islands of the Malay archipelago (Fig. 38).

Fig. 172. Primary wet mixed and broad-leaved evergreen mountain forests on granite and gneiss along rivers support large populations of *Paphiopedilum appletonianum* in the southern part of the Truong Son range in the southern Vietnam.

S. Vietnam, Dak Lac Prov. (Photo.: L. Averyanov).

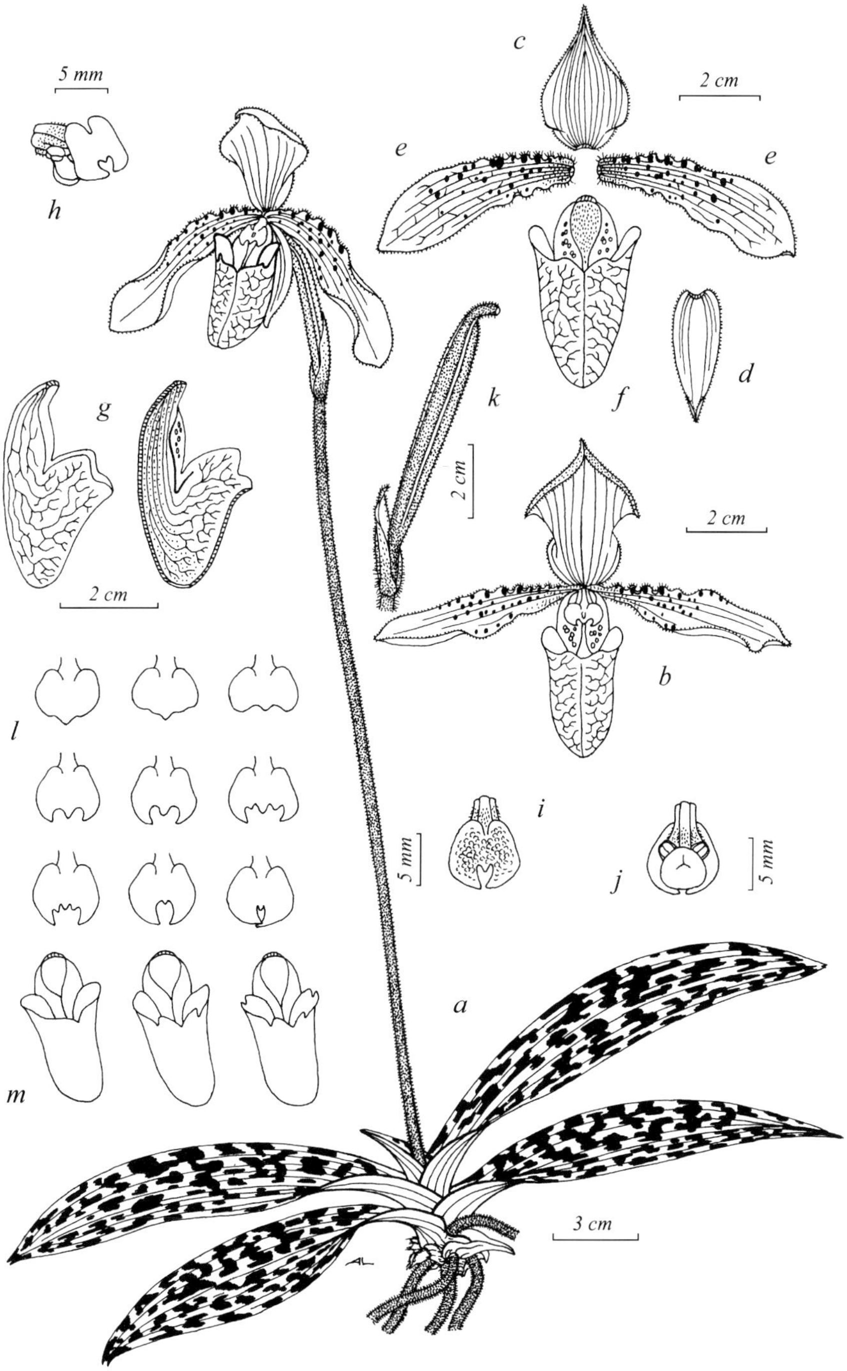

Fig. 173. *Paphiopedilum appletonianum.* (Drawing: L. Averyanov).

a — flowering plant; **b** — flower; **c** — dorsal sepal; **d** — synsepal; **e** — petals; **f, g** — lip, frontal, side and section views; **h–j** — column, side, dorsal and ventral views; **k** — ovary and flower bract; **l** — most common variants of the staminode shape observed in Vietnamese populations; **m** — most common variants of the frontal lip margin in Vietnamese populations of the species. (**a–k** — drawn from VH 5316).

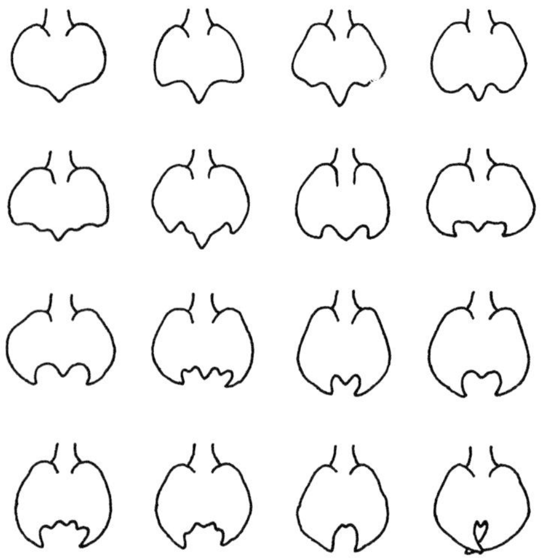

Fig. 174. Variations in staminode shape regularly observed in Vietnamese populations of *Paphiopedilum appletonianum.*

(Drawing: L. Averyanov).

HISTORY. Gower described this species, as *Cypripedium appletonianum,* in 1893 based on a plant of unknown provenance that was flowered by Mr. J. Appleton. Rolfe (1893) initially treated it as a variety of *C. bullenianum* but, later raised it to specific rank, distinguishing it from the latter by its obscurely tessellated leaves (Rolfe, 1896b). Pfitzer (1903) included it in his revision of the genus, distinguishing it on staminode shape. Leaf marking and staminode shape are generally useful in distinguishing *P. appletonianum* from *P. bullenianum*, but variation of both these features in populations of *P. appletonianum* in Vietnam are very wide and obscure the differences between these species. In *P. appletonianum*, the dorsal sepal is usually longer with a longer acuminate apex and reflexed basal margins giving it a distinctively obovate appearance. The petals are usually slightly longer than in *P. bullenianum* with rather acute tips, but spotting in the basal half of the petals is usually greater in *P. appletonianum.* The lip of *P. appletonianum* usually has raised oblique blunt teeth on its apical margin.

Paphiopedilum wolterianum, for long considered synonymous with *P. appletonianum,* has been resurrected by Fowlie (1966, 1969) and others based on Kraenzlin's rather poor illustration of it in Xenia Orchidacea (1896). Kraenzlin described this species in apparent ignorance of the existence of *P. appletonianum* and *P. bullenianum,* comparing it rather with the distinct *P. barbatum.* The differences in staminode shape which are said to exist between *P. appletonianum* and *P. wolterianum* by Fowlie (1969) are no more than those between

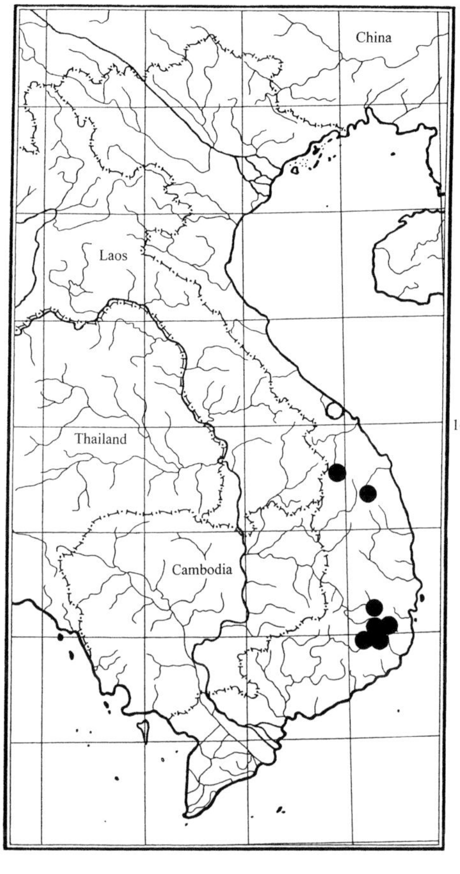

Fig. 175. Distribution of *Paphiopedilum appletonianum* in Vietnam. (Drawing: L. Averyanov).

Studied populations are marked with dots, a circle designates locality of the species from the literature.

Kraenzlin's two drawings of the staminode in the illustration in Xenia Orchidacea. Other differences of petal attitude and foliage variegation also fall within the range of variation seen in populations of *P. appletonianum* in Vietnam.

Cypripedium poyntzianum, described by O'Brien in 1894, is no more than a variant of *P. appletonianum* with more boldly marked leaves and large pale flowers, and we follow Rolfe (1895, 1896a) in reducing it to the synonymy of *P. appletonianum*.

Paphiopedilum hainanense was described by J.A.Fowlie in 1987 based upon a cultivated plant which originated in Hainan but was flowered in the Los Angeles Arboretum. It falls within the range of morphological variation of *P. appletonianum*, but the flower is more brightly colored and the leaves more boldly tessellated than specimens from mainland South-east Asia. Despite Fowlie's assertion that its staminode differs from that of *P. appletonianum* in ways "somewhat difficult to describe, but ascertainable with the kodachrome camera", the differences are too slight to warrant formal recognition of this entity.

Plants closely allied to *P. appletonianum* have been in cultivation for many years under the misapplied name *P. robinsonii*, but no one had any idea of their provenance. Cribb (1998) pointed out that the name *P. robinsonii*, of which the type came from peninsular Malaysia, could not be applied to these plants. The cultivated plants had a typical transversely reniform staminode rather than the broadly cordate or broadly lunate one of *P. appletonianum*.

A similar plant, collected in 1996 by De Raeve in southern Vietnam near the Cambodian border and flowered by Frank Cervera from Yonkers, New York, was described as *Paphiopedilum cerveranum*

Fig. 176. *Paphiopedilum appletonianum* commonly grows terrestrially on shady forest floors in deep leaf litter.
a — S. Vietnam, Ninh Thuan Prov.; **b** — S. Vietnam, Dak Lac Prov. (Photos: L. Averyanov).

(Braem, 1999). However, its morphological features fall within the natural variation range of *P. appletonianum*. *Paphiopedilum tridentatum* S.C. Chen, Z.J. Liu & J.Y. Zhang and *P. puberulum* S.P. Lei & J.Y. Zhang, both based on cultivated plants introduced into China from Vietnam, are certainly the same as *P. appletonianum*. Their description is probably the result of seeking commercial gain by trading them as "new species".

HABITAT. Fowlie (1987) described the habitat of *Paphiopedilum appletonianum* in Hainan as "sandstone mountains and flat expanses of original eroded plateau of southwest Hainan Island, 1500 and 2000 feet (600–800 m) elevation". In Vietnam *P. appletonianum* inhabits primary, humid, coniferous, mixed and broad-leaved, evergreen forests commonly on steep mountain slopes composed of granite, gneiss and sandstone between 700–1900 m (Fig. 172). It often grows along shady damp canyons of mountain streams as a lithophyte on wet mossy boulders or as a terrestrial in deep leaf litter (Figs. 176–178). It occasionally may also be found as a humus epiphyte on old mossy trees near the ground in very wet montane cloud forests (Fig. 179). In most cases *Paphiopedilum appletonianum* is found in similar situations to *P. villosum*. Both species may often be found growing together. The climate in these areas is a monsoon tropical one associated with mountains. Such weather conditions are also described for *P. villosum* (Fig. 168).

In the past *Paphiopedilum appletonianum* was one of the commonest slipper orchids in highland areas of the southern portion of Truong Son range in the southern Vietnam. Very large colonies could be seen in many inaccessible localities a few years ago before extensive commercial exploitation. In some places *P. appletonianum* was an integral component of the herbaceous flora of the forest floor along with other moisture-loving herbs (Fig. 179). Regrettably, nowadays this species has become much rarer due to commercial collecting.

VARIATION. *Paphiopedilum appletonianum* is one of the most variable slipper orchid species in Vietnam. The dorsal sepal and synsepal of the flower are usually pale or whitish-green, but may be bright emerald green, olive-green or greenish-brown with purple-brown longitudinal stripes in the basal portion. The petals are commonly green or olive-green at the base and pink, dull pink-purple or purple-violet toward the apex, often brown-purple striped with deep maroon spots and maroon margin warts.

Fig. 177. *Paphiopedilum appletonianum* may be often observed as a lithophyte on wet shady granite or gneiss bluffs along damp mountain canyons.

a — S. Vietnam, Khanh Hoa Prov.; **b** — S. Vietnam, Dak Lac Prov. (Photos: L. Averyanov).

Fig. 178. Large colonies of *Paphiopedilum appletonianum* can sometimes be found on the forest floor in virgin habitats inaccessible to commercial collectors .

S. Vietnam, Kontum Prov. (Photo. L. Averyanov).

The lip is olive-green, dull brown-pink or brown-purple with greenish venation and greenish margin. The front margin of the lip may be flat, curved, or with 2 to 6, more or less deep, distinct teeth. Form and colouration of the staminode is also extremely variable. The staminode of *P. appletonianum* in Vietnamese populations ranges from dull yellowish, yellowish-green, emerald-green to olive-green with a yellowish-white margin and a central yellowish spot. It shape varies from broadly obcordate to deeply transversely lunate, with a 2-, 4- or 5- toothed apex (Figs. 174, 180).

The inflorescence is usually one-flowered, but in some populations a few plants develop two-flowered inflorescences (Fig. 181). Flowers may reach 10–12 cm an even more across.

The shape, size and character of the leaf tessellation of *Paphiopedilum appletonianum* in Vietnamese populations are also very variable. Leaves varies from broadly elliptic to narrowly oblong or broadly lanceolate. Leaf tessellation may be bold with contrasting mottling of pale and deep green or, in contrast, dull and indistinctly mottled with a nearly uniformly green blade.

Albino variants of the species have recently been introduced into cultivation and been named as *Paphiopedilum appletonianum* var. *album* by Birk (1983), var. *immaculatum* (Braem, 1988) and *P. cerveranum* forma *viride* by Braem (1999). They are all probably best treated as forms. We have not yet seen such plants from Vietnam.

Fig. 179. *Paphiopedilum appletonianum* sometimes occurs as a humus epiphyte on mossy old trees near the ground in very wet conditions in mountain cloud forests.

S. Vietnam, Kontum Prov. (Photo.: L. Averyanov).

Fig. 180 (above and next page). Flower variation in *Paphiopedilum appletonianum* in Vietnam. (Photos: L. Averyanov).

a — S. Vietnam, Lam Dong Prov.; **b–d** — S. Vietnam, Kontum Prov.; **e** — S. Vietnam, Khanh Hoa Prov.; **f, g** — S. Vietnam, Ninh Thuan Prov.

Fig. 180 (contd.). Flower variation in *Paphiopedilum appletonianum* in Vietnam. (Photos: L. Averyanov).

h–m — S. Vietnam, Kontum Prov.; **n** — S. Vietnam, Dak Lac Prov.

Fig. 181. In some Vietnamese populations of *Paphiopedilum appletonianum* plants with two-flowered inflorescences may be seen.

S. Vietnam, Kontum Prov. (Photo.: L. Averyanov).

16. PAPHIOPEDILUM CALLOSUM (Rchb.f.) Stein

Paphiopedilum callosum *(Rchb.f.) Stein*, Orchideenbuch: 457 (1892); Pfitzer in Engler, Pflanzenr. Orch. Pleon.: 92 (1903); Guillaum. in Bull. Soc. Bot. France. 71: 551 (1924); id., Fl. Gen. Indo-Chine 6: 642 (1934); Karasawa & Saito in Bull. Hiroshima Bot. Gard. 5: 51 (1982); Cribb, Gen. Paphiopedilum: 184 (1987); id., Gen. Paphiopedilum, ed. 2: 331 (1998); Braem, Paphiopedilum: 197 (1988); Seidenfaden, Orch. Indochina: 20 (1992); Aver., Identif. Guide. Vietnam. Orch.: 21 (1994); Pham Hoang Ho, Ill. Fl. Viet. 3: 762 (2000). **Type:** Thailand, *Regnier* s.n. (holotype W!).

Cypripedium callosum Rchb.f. in Gard. Chron. ser. 2, 26: 326 (1886).

Cypripedium crossii E.Morren in Belg. Hort. 15: 226, t. 17 (1865), *nom. nud.*

Cypripedium schmidtianum Kraenzl. in Bot. Tidsskr. 24: 13 (1901); Fowlie in Orchid Digest 40: 164 (1976).

Paphiopedilum callosum var. *schmidtianum* (Kraenzl.) Pfitzer in Engler, Pflanzenr. Orch. Pleon.: 93 (1903).

Cordula callosa (Rchb.f.) Rolfe in Orchid Rev. 20: 2 (1912).

Paphiopedilum callosum var. *angustipetalum* Guillaum. in Bull. Soc. Bot. France ser. 4, 24: 551; id., 1934, Fl. Gen. Indo-Chine 6: 643 (1924).

Paphiopedilum crossii (Morren) Braem & Senghas in Sida 19, 2: 251 (2000), *comb. invalid.*

Paphiopedilum amabile sensu Guillaum. Bull. Mus. Paris 2 Ser. 9: 217 (1937); id. ibid., 2 Ser. 33, 3: 333 (1961); Gagnep. in Bull. Mus. Hist. Nat. Paris 2 ser. 22, 5: 628 (1950); Seidenf., Contrib. Revis. Orch. Fl. Cambod. Laos Vietnam: 88 (1975); Aver., Prelim. List Vietnam. Orch. 2: 31 (1988); Pham Hoang Ho, Ill. Fl. Viet. 3: 762 (2000), *non* Hallier.

DESCRIPTION. *Terrestrial* or occasionally *lithophytic herb* with 3–5(6) leaves. *Leaves* narrowly elliptic, oblong-elliptic or obovate, often tridenticulate at acute apex, 10–20 cm long, 3.2–4.8 cm wide, ciliate at base, tessellated pale and dark above, sometimes purple at the base on lower surface. *Inflorescence* 1- (or rarely 2-) flowered, 20–40 cm tall; peduncle purple, purple-pubescent, 12–25 cm long; bract ovate to elliptic, acute to subacute, 1.5–2.8 cm long, 1.5–2 cm wide, green, sometimes marked with purple, ciliate. *Flower* large, 8–11 cm across; sepals white flushed with few purple in lower half, veined with purple and green; dorsal sepal with small silver warts near the base; petals whitish, green or yellow-green with a purple apical third, spotted with maroon on upper margin and sometimes in basal half; lip dull brown-purple, light yellowish at the base, brown-purple warted on side lobes, often with olive-green venation; staminode white or light yellowish with dark green veins; pedicel and ovary 3–6.5 cm long, green, flushed with purple, purple-pubescent. *Dorsal sepal* broadly obovate to subcircular, apiculate, 4–5.5 cm long, 4.2–6 cm wide, with recurved ciliate margins. *Synsepal* concave, elliptic or broadly lanceolate, acute, 2.7–3.2 cm long, 1.6–2.5 cm wide. *Petals* sometimes reflexed, more or less sigmoid, ligulate, obtuse or rounded at the apex, 4.6–6.8 cm long, 1.2–1.8 cm wide, maroon-ciliated. *Lip* 2.5–4.4 cm long, 2–2.5 cm wide, warty on incurved side-lobes. *Staminode* lunate, 11 mm long, 7 mm wide, with narrow falcate lateral lobes. 2n=32. Fig. 183.

DISTRIBUTION. Thailand, Vietnam (Gia Lai, Lam Dong, Quang Nam, Quang Tri and Thua Thien Hue Provinces), Laos, Cambodia. Figs. 38, 184.

ECOLOGY. Primary, closed, evergreen, seasonal, broad-leaved, forests on granite and sandstone at elev. 300–1300 m. Fig. 182.

FLOWERING IN THE WILD. April–June.

IUCN STATUS. EN.

AFFINITIES. *Paphiopedilum callosum* is a native of Southeast Asia where it occurs in widely scattered localities in Thailand, Cambodia, Laos and Vietnam. It is closely allied to *P. barbatum* from Malaysia but differs in its larger dorsal sepal, usually more than 4 cm long and wide and flushed with purple across the center, and longer more or less sigmoid, almost pendent petals, more than 5 cm long and 1.2 cm broad.

Fig. 182. *Paphiopedilum callosum* commonly occurs in Vietnam in primary evergreen seasonal broad-leaved forests on granite along streams or small mountain rivers.

S. Vietnam, Lam Dong Prov. (Photo.: L. Averyanov).

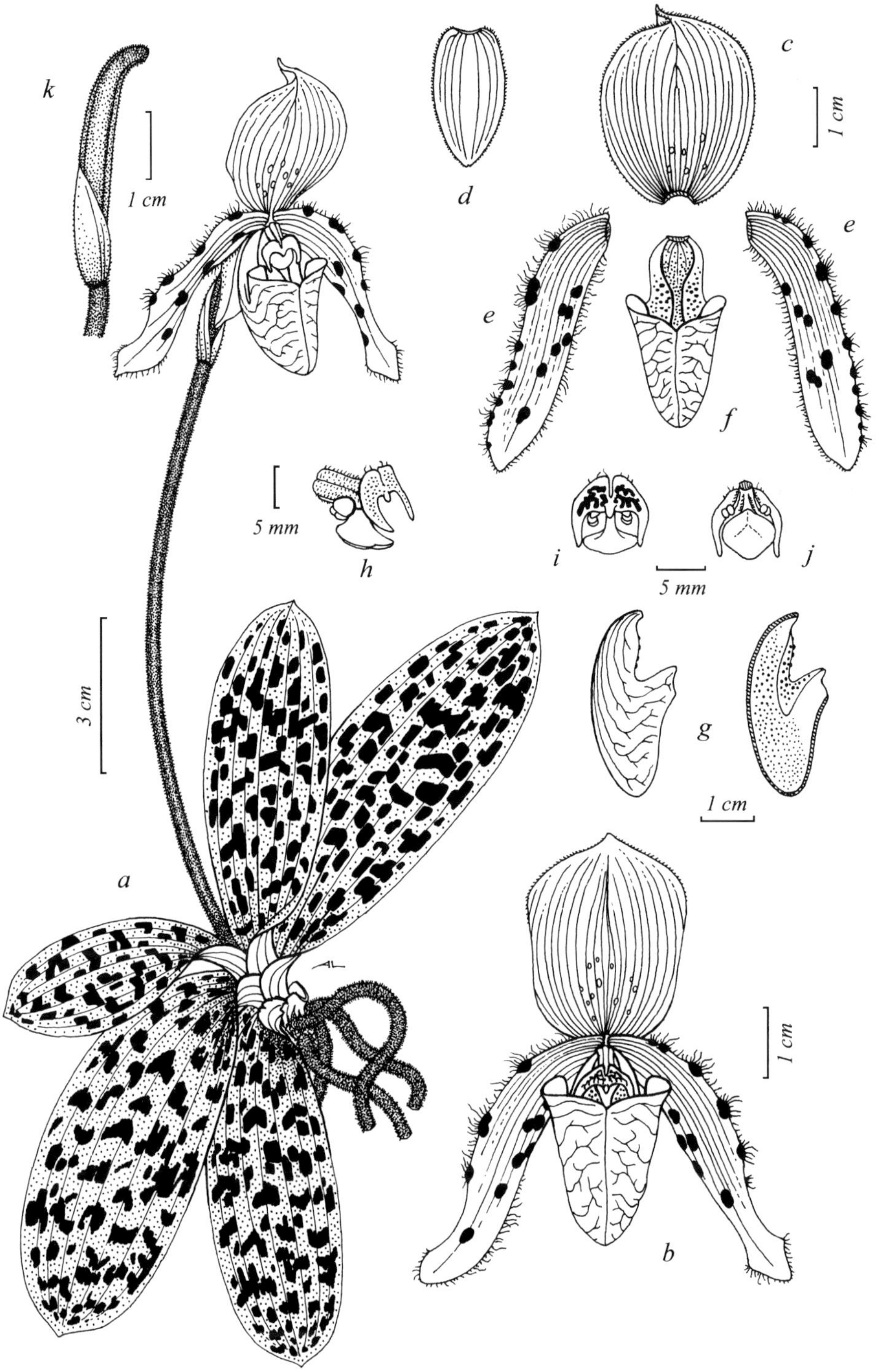

Fig. 183. *Paphiopedilum callosum.* (Drawing: L. Averyanov).

a — flowering plant; **b** — flower; **c** — dorsal sepal; **d** — synsepal; **e** — petals; **f, g** — lip, frontal, side and section views; **h–j** — column, side, dorsal and ventral views; **k** — ovary and flower bract. (Drawn from 2000, L. Averyanov, N.T.Hiep VH sine no).

HISTORY. H.G. Reichenbach described this species as *Cypripedium callosum* in 1886 based on a Thai plant sent to him by Alexandre Regnier. The illustration of it in his herbarium shows a plant with spotted petals. It seems likely that the plant illustrated as *C. crossii* by Morren in 1865 is the same species. However, Morren gave no description of the plant merely stating that Mr. Cross had discovered it in Peru, and as the plate contains no floral diagnosis, the name *C. crossii* must be treated as a *nomen nudum*.

The plant described and illustrated as *Cypripedium barbatum* var. *warnerianum* is likewise attributable to *Paphiopedilum callosum*, but the varietal epithet has never been used at specific level and can therefore be ignored for priority.

In 1901, Kraenzlin described *Cypripedium schmidtianum* based on a collection from Klong Son, Koh Chang (Elephant Island) off the south-east coast of Thailand. He compared it with *Paphiopedilum fairrieanum* and *P. dayanum* rather than with *P. callosum* and *P. barbatum* and stated that he described it as new "not without hesitation". Seidenfaden (1972) cited "*Schmidt* 789 (C)" as the holotype but Kraenzlin gave no collection number to the material he examined (three flowers in spirit) and, therefore, Seidenfaden's citation should be regarded as an effective lectotypification. This specimen and description agree closely with that of *P. callosum* that is found on the island (Fowlie, 1976). We have no hesitation, therefore, in including it in the synonymy of *P. callosum*.

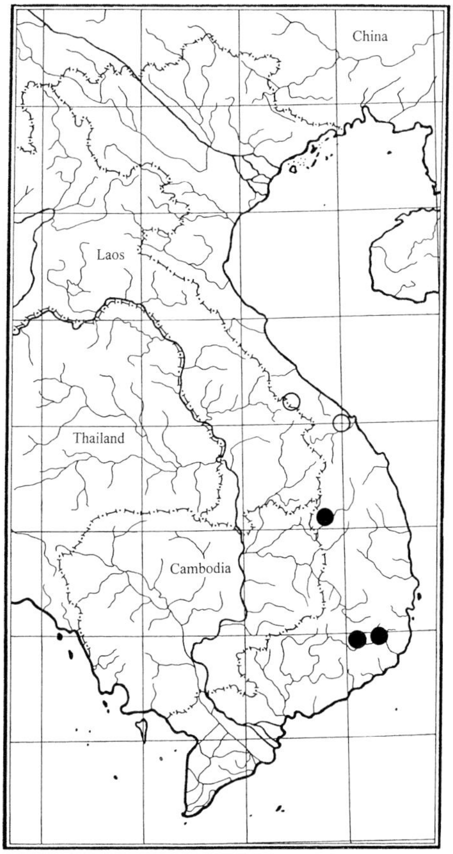

Fig. 184. Distribution of *Paphiopedilum callosum* in Vietnam. (Drawing: L. Averyanov).

Studied populations are marked with dots, circles designate localities from literature.

Paphiopedilum callosum is common in cultivation and the species has undoubtedly been severely depleted in the wild by commercial collection of thousands of plants at a time. This is made easy by the colonial habit of this species. It grows in nature in lower montane evergreen forest in leaf litter and on mossy rocks often along streams.

Many varieties of *P. callosum* have been described based on plants in cultivation. One of the finest of these is undoubtedly *Paphiopedilum callosum* var. *sanderae*, an albino form with a flower as large as the best typical *callosum*. Braem (1988) suggested that this should correctly be known as *P. callosum* var. *viridiflorum*, a name published a year earlier than var. *sanderae*. Others plants, such as varieties *grandiflorum*, *giganteum*, *rossianum* and *superbum* all refer to large-flowered plants which fall within the range of variation of the typical subspecies. All, including *sanderae* and *viridiflorum*, are best treated as cultivars of *P. callosum*.

Two distinct varieties of the species have been recognized from Thailand — var. *sublaeve* (Rchb.f.) Cribb and var. *potentianum* (Gruss & Röth) Cribb. The former is distinguished from the typical variety by its somewhat smaller flower with a smaller dorsal sepal, 3.4–4.2 cm long, 3–4.1 cm wide, and shorter relatively broader and less sigmoid petals, with warts only on the upper margin. The latter differs from the type in its petals which lack the marginal warts on the upper margin and in its narrower dorsal sepal, about 4 cm long and 2.6 cm wide. Both these plants have yet to be found in Vietnam.

HABITAT. Fowlie (1974) described the habitat of *Paphiopedilum callosum* on Gunong Jerai in northwest Malaya. It is found there growing in large scattered colonies between 700 and 950 m. The plants have their roots in leaf litter or fine sand in mossy *Podocarpus-Dacrydium* forest.

Fig. 185. *Paphiopedilum callosum* in Vietnam.
S. Vietnam, Lam Dong Prov. (Photo.: L. Averyanov).

In Vietnam, *Paphiopedilum callosum* inhabits primary more or less dry, tall, closed, evergreen, seasonal, broad-leaved, forests on granite and sandstone between 800–1100 m. (Fig. 182). The commonest dominants here are trees such as *Canarium* (Burseraceae), *Dipterocarpus*, *Shorea* (Dipterocarpaceae), *Gironniera subaequalis* (Ulmaceae), *Lithocarpus*, *Castanopsis* (Fagaceae), *Michelia* (Magnoliaceae), *Nephelium* (Sapindaceae), *Parkia sumatrana* (Fabaceae), *Rhodoleia championi* (Hamamelidaceae), *Schima wallichii* (Theaceae) and *Syzygium* (Myrtaceae). Usually, colonies of *P. callosum* can be found in rather open places along stream slopes where some species of bamboo play an important role in the forest undergrowth. *Paphiopedilum callosum* commonly grows as a terrestrial on the forest floor in deep leaf litter (Fig. 185), but sometimes it may be seen on mossy boulders. It grows in similar places to *P. appletonianum,* but it is usually found at lower elevations in drier conditions in areas with a distinctly seasonal climate.

A tropical monsoon climate with a long dry season and without any cold period is typical for the area where *Paphiopedilum callosum* is found in Vietnam. This is area of warm and relatively dry weather with insignificant winter/summer temperature differences. Median temperatures all the year round lie within the limits 19–25°C (Fig. 186). Maximum temperatures are observed in April-May and may occasionally reach 35–39°C. Minimum temperatures in the area very rarely approach 11–14°C during December-January. However, usual temperatures in winter time vary between 19–20°C. Annual mean rainfall does not exceed 2000 mm, but is usually lower between 1600 and 1900 mm. Heavy torrential rains fall from April to October. However, in winter-time, from November to the beginning of April, it is distinctly dry. Mean monthly precipitation during December-February here may be only 2–10 mm (Fig. 186).

Nowadays, *Paphiopedilum callosum* is very rare in Vietnam, populations having been seriously depleted by widespread commercial collecting. Growing in lower montane forests, it is an easy target for collectors and certainly needs urgent protection in all its known habitats. Based on our observations in Vietnam, most plants on sale in Vietnamese markets come from the lowlands of Laos and Cambodia where this species may presently be commoner.

VARIATION. Flowers of *Paphiopedilum callosum* in Vietnam are quite variable, especially in form of the dorsal sepal and petals. The former may be broadly obovate or circular and is usually white with a slight purple tint and green longitudinal stripes (Fig. 185b). The lateral petals are narrowly oblong, ligulate or broadly lanceolate. They may be reflexed, more or less sigmoid, or strongly decurved.

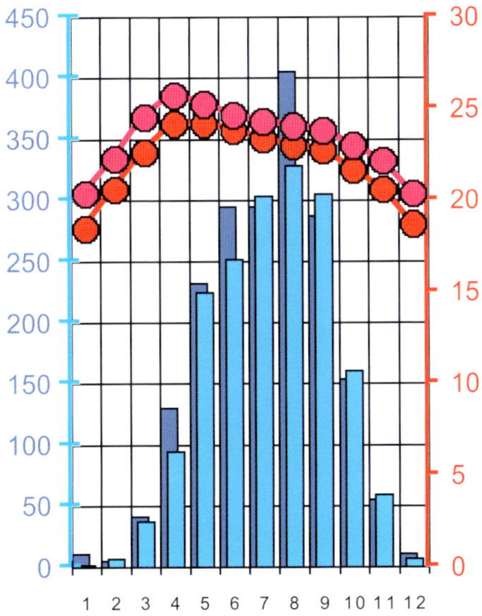

Fig. 186. Climatological data typical for the area of *Paphiopedilum callosum* in Vietnam.

Black figures at the chart bottom are months. Blue axis represents atmosphere precipitation in mm (columns designate mean monthly rainfalls); red axis represents mean monthly temperature in °C. The deep blue and red are data from the Dac To weather station, 14°42'N 107°49'E, elev. 620 m; light blue and purple are data from the Hoang Kon Tum weather station, 14°30'N 108°01'E, 536 m, all according to Nguyen Khanh Van *et al.*, 2000.

From time to time, albino forms, whose flowers lack anthocyanin pigmentation (Fig. 187), appear on the market in Vietnam. Similar forms are known in cultivation under the names var. *sanderae* (Hort.) Braem and var. *viridiflorum* (Hort.) Pfitzer.

According to Seidenfaden (1992) some plants, seen by him at the orchid market in Dalat city in southern Vietnam, may be natural hybrids between *P. callosum* and *P. appletonianum*. Natural hybridization of two species in Vietnam appear highly probable, but we have not yet seen such hybrids among Vietnamese plants.

Fig. 187. Albino form of *Paphiopedilum callosum* collected in southern Vietnam.

Dalat, private collection. (Photo.: Mr. Son Ha; digital correction: L. Averyanov).

17. PAPHIOPEDILUM PURPURATUM (Lindl.) Stein

Paphiopedilum purpuratum *(Lindl.) Stein*, Orchideenbuch: 481 (1892); Pfitzer in Bot. Jahrb. Syst. 19: 40 (1894); id., in Engler, Pflanzenr. Orch. Pleon.: 86 (1903); Karasawa & Saito in Bull. Hiroshima Bot. Gard. 5: 54 (1982); Cribb, Gen. Paphiopedilum: 205 (1987); id., Gen. Paphiopedilum, ed. 2: 380 (1998); Braem, Paphiopedilum: 179 (1988); Seidenfaden, Orch. Indochina: 20 (1992); Aver., Identif. Guide. Vietnam. Orch.: 21 (1994); Braem, C. Baker & M. Baker, Gen. Paphiopedilum 2: 336 (1999); Pham Hoang Ho, Ill. Fl. Viet. 3: 762 (2000). **Lectotype:** cult. Loddiges, Bot. Reg. 23: t. 1991 (1837), selected by Cribb (1998).

Cypripedium purpuratum Lindl. in Bot. Reg. 23: t. 1991 (1837); id., Gen. Spec. Orch. Pl.: 530 (1840); Hook. in Curtis's Bot. Mag. 82: t. 4901 (1856).

Cypripedium sinicum Hance ex Rchb.f., Walp. Ann. 3: 602 (1853). **Type:** Hong Kong, *Hance* s.n. (holotype W!).

Paphiopedilum sinicum (Hance) Stein, Orchideenbuch: 481 (1892).

Cordula purpurata (Lindl.) Rolfe in Orchid Rev. 20, 1: 2 (1912).

Paphiopedilum purpuratum (Lindl.) Stein var. *hainanense* F.Y.Liu & Perner in Orchidee 52, 1: 64 (2001) **syn. nov. Type:** China, Hainan, *M.E. Tang* 98003 (holotype KUN).

P. aestivum Z.J. Liu & J.Y. Zhang in Acta Phytotax. Sinica 39, 6: 568, fig. 1 (2001), **syn. nov. Type:** Yunnan, Wenshan, cult. Shenzhen City Wutongshan Nurseries, July 3rd 2000, *Z.J. Liu* 1525 (holotype, Herb. Shenzehen City Wutongshan Nurseries).

DESCRIPTION. *Terrestrial herb* with 4–6(8) leaves. *Leaves* elliptic to oblong-elliptic, acute or subacute at unequally trilobulate apex, 7–14(17) cm long, 2.3–4.2 cm wide, pale green below, tessellated light and darker green above, margins shortly ciliate towards base. *Inflorescence* erect, one-

or rarely two-flowered; peduncle (6)10–20 cm long, slender, purple-violet, deep purple-pubescent; bract narrowly ovate-elliptic, acuminate, 1.5–1.7 cm long, 0.7 cm wide, green, white-pubescent. *Flowers* 7–10 cm across; dorsal sepal white, greenish at the base, heavily veined with purple-maroon; synsepal green with darker veins, often with two brown-purple longitudinal stripes; petals glossy deep purple-maroon, light greenish near the base and yellowish-pink at the apex, heavily spotted with blackish maroon in lower two-thirds; lip purple-maroon; staminode pale ochre-yellowish, purple-tinted along margin, with green veins around whitish centre; pedicel and ovary 2.5–4 cm long, green, whitish pubescent; stigma yellowish-white; filaments green; pollen masses honey-yellow. *Dorsal sepal* broadly ovate-cordate, apiculate or acuminate, 2.5–3.5 cm long, 2.2–4.3 cm wide. *Synsepal* narrowly ovate, acuminate, 2–3.5 cm long, 1.2–1.6 cm wide. *Petals* more or less horizontal, narrowly elliptic to oblong, acute to obtuse, 3.5–4.6 cm long, 0.9–1.3 cm wide, long black ciliate. *Lip* 3.3–4.4 cm long, 2–2.7 cm wide, verrucose on incurved side-lobes. *Staminode* lunate, with narrow falcate acute side lobes, 8 mm long, 8–11 mm wide, pubescent, apical teeth acute. *Stamens* with elongate acute filament apices. 2n=40. Fig. 189.

DISTRIBUTION. China (Hong Kong, Guangdong, Hainan and possibly Guangxi), Vietnam (Cao Bang, Lao Cai?, Lam Dong?). Figs. 38, 190.

ECOLOGY. Primary, mixed and semi-deciduous, broad-leaved forests on highly eroded crystalline limestone at elev. 1200–1400 m. Fig. 188.

FLOWERING IN THE WILD. September–October.

IUCN STATUS. EN.

AFFINITIES. *Paphiopedilum purpuratum* is a distinctive species similar in its overall colouring and flower size to the Malaysian *P. barbatum*, but readily distinguished by its broader petals, which lack large warts on the upper margin, by its more lunate staminode, and relatively smaller ovate, acuminate dorsal sepal. The petals of *P. purpuratum* are also covered by small warts on the basal two-thirds of the inner side.

Fig. 188. *Paphiopedilum purpuratum* grows in primary semi-deciduous mixed and broad-leaved forests on limestone mountains in the northern Vietnam.

N. Vietnam, Cao Bang Prov. (Photo.: L. Averyanov).

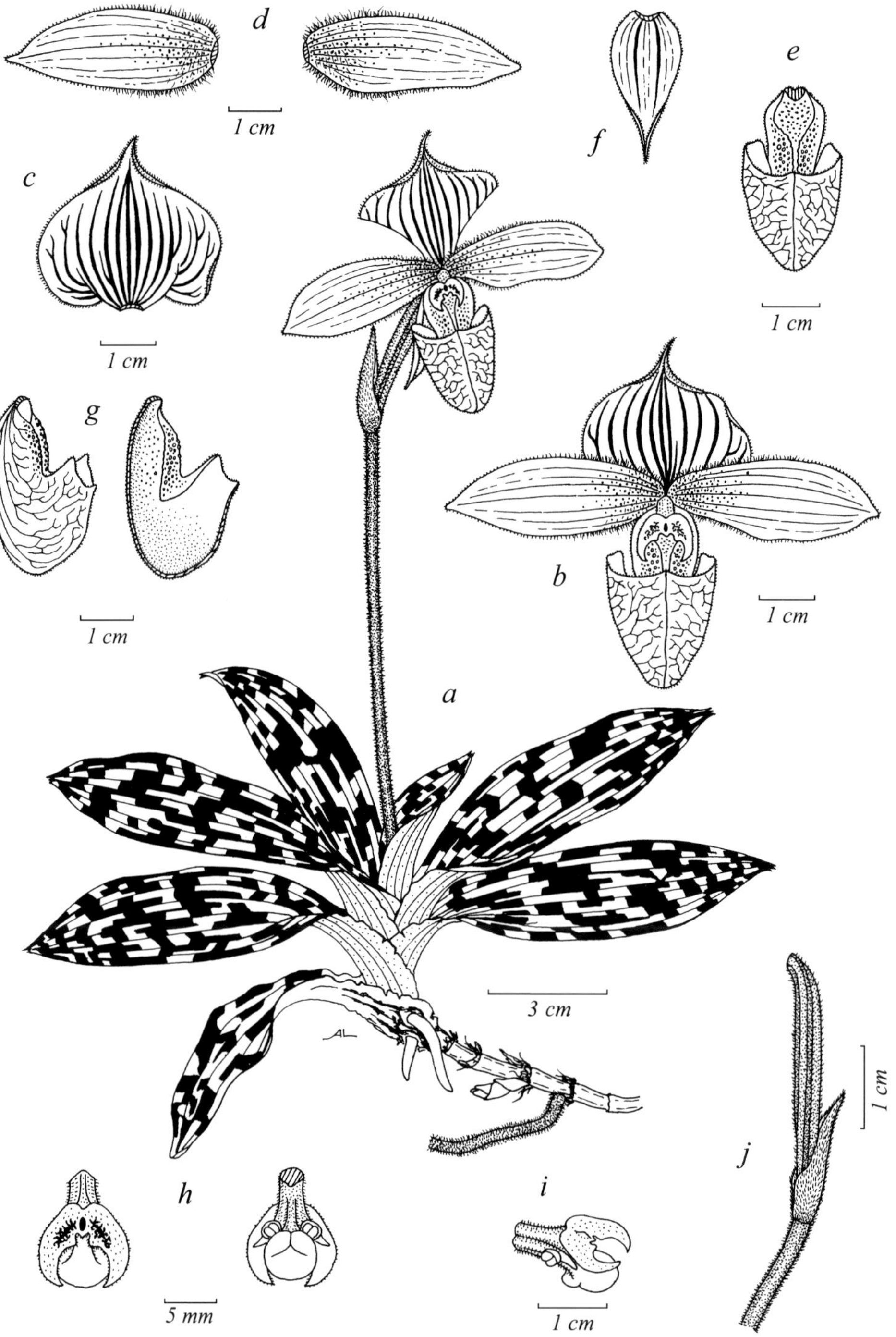

Fig. 189. *Paphiopedilum purpuratum.* (Drawing: L. Averyanov).

a — flowering plant; **b** — flower; **c** — dorsal sepal; **d** — petals; **e** — lip, frontal view; **f** — synsepal; **g** — lip, side and section views; **h, i** — column, dorsal, ventral and side views; **j** — ovary and flower bract. (Drawn from CBL 1458b).

HISTORY. *Paphiopedilum purpuratum* was originally described in 1837 by John Lindley who based his description on an illustration of a plant cultivated by Messrs. Loddiges of Hackney but imported by Messrs. Knight of King's Road Nursery, Chelsea. The provenance of this plant was reportedly the Malay Archipelago, an origin repeated by Hooker for his Curtis's Botanical Magazine account in 1856. However, most likely these plants came from Hong Kong (Seidenfaden, 1992). The plant illustrated as *Cypripedium purpuratum* by Wight (in his Icones Plantarum Indiae Occidentalis 5, t. 1760 (1851)) is almost certainly the Malaysian *P. barbatum*. This putative origin, no doubt, led to the description of *Cypripedium sinicum* from Hong Kong. However, Reichenbach obviously realized the conspecificity of this with *P. purpuratum* because he mounted the type of *C. sinicum* on the same sheet in his herbarium as a specimen clearly labeled "*Cypr. purpuratum* Lindl." in Lindley's handwriting.

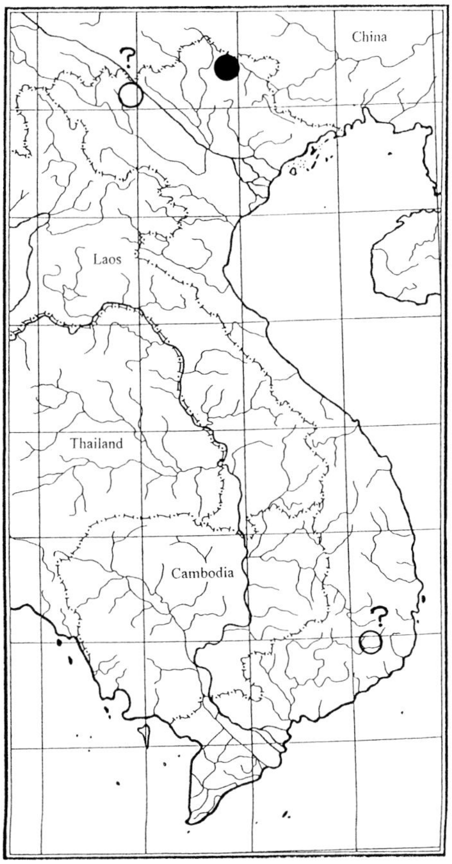

Fig. 190. Distribution of *Paphiopedilum purpuratum* in Vietnam. (Drawing: L. Averyanov).

Studied population is marked with a dot, circles designate localities according to local collectors and literature.

Paphiopedilum purpuratum was first recorded from Indochina by Wilhelm Micholitz in a letter to Frederick Sander. Cribb (1998) has studied the correspondence and has found that Micholitz in two letters from December 1903 reported collecting 1000 plants of *P. purpuratum* at Dankia, about 10–12 km NW of Dalat city in the southern Vietnam (Seidenfaden, 1992). This report still remains questionable. There are no other reliable reports of it from this well known locality. On the one hand, it is possible that Micholitz succeeded in exterminating the entire population of the species. On the other, *P. appletonianum* and *P. callosum* are common in Dalat area and Micholitz may have mistaken non-flowering specimens for *P. purpuratum*. Unfortunately, Micholitz's collection is not represented in any herbarium (Seidenfaden, 1992). The natural hybrid of *P. callosum* and *P. villosum*, described recently under the name *P.* × *dalatensis* (Averyanov, 2001), is superficially similar to *P. purpuratum*. This rare plant appears time to time, under the name *Paphiopedilum purpuratum*, in Dalat flower market. The first reliable report of *Paphiopedilum purpuratum* in Vietnam was provided by Averyanov *et al.* (2000a,b). They reported it from northern regions of the country near to the Chinese border. At the same time, specimens of *P. purpuratum* of Vietnamese origin have recently begun to appear in international trade.

The recently described *P. aestivum* Z.J. Liu & J.Y. Zhang is said to differ from *P. purpuratum* in flowering at a different time and in having rather clear veins on its leaves and a horizontal dorsal sepal. In the morphological characters it falls within the variation of *P. purpuratum* and we see no reason for recognising it as distinct at any rank. We also doubt the provenance of the type collection which is said to have come from Wenshan in SE Yunnan. Wenshan is a town near the border with Vietnam and Vietnamese orchids are traded on a large scale in its market. Many of the plants come from adjacent countries and from all over SW China.

The reported range of *P. purpuratum* outside Vietnam extends from Hong Kong into adjacent areas of Guangdong Province of south-east China, Hainan Island and Southeast Yunnan (Fig. 38).

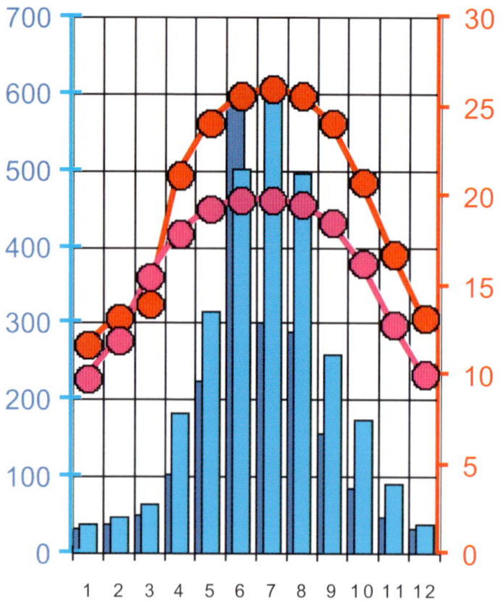

Fig. 191. Climatological data typical for the area of *Paphiopedilum purpuratum* in Vietnam.

Black figures at the chart bottom are months. Blue axis represents atmosphere precipitation in mm (columns designate mean monthly rainfall); red axis represents mean monthly temperature in °C. The deep blue and red are data from Trung Khanh weather station, 20°50'N 106°31'E, elev. 520 m; light blue and purple are data from the Sin Ho weather station, 22°21'N 103°15'E, 1529 m, all according to Nguyen Khanh Van, *et al.*, 2000.

HABITAT. In Hong Kong, *Paphiopedilum purpuratum* grows on north- and north-west-facing slopes exposed to the cold, dry north wind from China in the winter months when the temperature may drop to freezing (Lee, 1975). It is found there in small colonies on steep rocky slopes in light shade beneath a low cover of scrub at about 500 m. Graham (pers. comm.) reported it growing on the wooded slopes near streams flowing over large expanses of rock. The plants were growing in shade on moss-covered banks to within a 0.3 m of the stream where they were liable to occasional flooding. Elsewhere, they were growing in deep leaf litter amongst the roots of bamboo and between rocks, always with good shade and moisture. The plants consisted almost entirely of single growths. He also remarked on the wide range of variation to be found in the degree of leaf tessellation and the ground colouring which varied from pale green to bluish green in a single colony. Chau and Siu (pers. comm.) have found it in several places between 120 and 750 m elevation, mostly on north- and north-west-facing slopes, but also occasionally on south-west and south-east slopes along streams. Population size is low, ranging mostly between 20 to 30 plants.

Paphiopedilum purpuratum occurs in Vietnam in mountain regions at elevations of 1200–1500 m. The climate here is a monsoon

Fig. 192. *Paphiopedilum purpuratum* commonly occurs in Vietnam as a single scattered plants or forms small colonies.

N. Vietnam, Cao Bang Prov. (Photo.: L. Averyanov).

Fig. 193. Flowering sample of *Paphiopedilum purpuratum*.
N. Vietnam, Cao Bang Prov. (Photo.: L. Averyanov).

tropical one associated with mountains. An idea of the climate can be obtained from observations at Trung Khanh and Sin Ho weather stations (Fig. 191). The species grows in a distinctly seasonal climate with a dry cold winter and a wet summer when rather moderate temperatures are common. Median winter temperatures here are usually 9–13°C. The coldest months are December and January. Night temperatures during this time often fall to zero, with an absolute observed minimum of –4.5°C. The very dry season with almost no rain lasts from the end of September to the first days of April. On the contrary, air humidity regularly reaches 100% in February and March when cool humid north-east sea winds bring heavy cold fogs and mists to the area. The wet summer season starts in the middle of April and ends in September or October. Median summer temperatures range commonly around 20°C with maximum observed temperatures between 30–35°C. Mean annual rainfall in the area reaches 2500–2800 mm.

The largest populations of *Paphiopedilum purpuratum* seen in Vietnam were found on steep slopes of limestone mountains between 1200–1400 m (Fig. 192). It inhabits rather open, primary, evergreen and semi-deciduous, broad-leaved forests commonly near ridge tops (Averyanov *et al.*, 2000a). Commonly observed canopy trees in these woods include *Burretiodendron hsienmu* (Tiliaceae), *Gleditchia* sp. (Fabaceae), *Symingtonia populnea* (Hamamelidaceae), evergreen and deciduous species of tropical oaks — *Lithocarpus*, *Quercus* (Fagaceae), species of *Acer* (Aceraceae), *Alangium* (Alangiaceae), *Carpinus* (Betulaceae), *Ulmus* (Ulmaceae) and *Platycarya strobilacea* (Juglandaceae). Occasionally, conifers, such as *Pinus fenzeliana, Pseudotsuga brevifolia, Tsuga chinensis* (Pinaceae), *Amentotaxus argotaenia* (Cephalotaxaceae), *Fokienia hodginsii* (Cupressaceae), *Nageia fleuryi* and *Podocarpus neriifolius* (Podocarpaceae), may be seen here. Trees of the upper stratum in these mountain forests usually do not exceed 14–18 m tall.

Below these, small trees, such as *Garcinia* (Clusiaceae), *Illicium jiadifengpi* (Illiciaceae), *Morus* (Moraceae), *Pistacia weinmannifolia* (Anacardiaceae), *Schefflera pesavis* (Araliaceae), *Tirpitzia sinensis* (Linaceae), *Photinia, Sorbus* (Rosaceae) and *Madhuca* (Sapotaceae), are very common.

Fig. 194. *Paphiopedilum purpuratum* in Vietnam, typical flower.

N. Vietnam, Cao Bang Prov. (Photo.: P.J. Cribb).

The most typical shrubs in these forests are species of *Celastrus, Euonymus* (Celastraceae), *Lespedeza*, *Sophora glauca* (Fabaceae), *Pittosporum* (Pittosporaceae), *Wikstroemia* (Thymelaeaceae), and numerous species of *Rhododendron* and *Vaccinium* (Ericaceae).

The herbaceous flora commonly includes species of *Ainsleya* (Asteraceae), *Alpinia, Hedychium* (Zingiberaceae), *Arisaema* (Araceae), *Lysimachia* (Primulaceae), *Selaginella* (Selaginellaceae), *Begonia* (Begoniaceae), *Aspidistra, Disporum, Ophiopogon, Peliosanthes, Polygonatum* and *Tupistra* (Convallariaceae), *Elatostema, Pellea, Pellionia* (Urticaceae), *Carex* and *Scleria* (Cyperaceae). Many different kinds of terrestrial and lithophytic ferns and herbaceous species of Acanthaceae, Gesneriaceae and Melastomataceae also occur in these forests.

Many other orchids can be found here, including the terrestrial species *Anoectochilus calcareus, A. setaceus, Calanthe aleizettii, C. argenteostriata, Cymbidium ensifolium, C. lancifolium, Goodyera hispida, G. schlechtendaliana, G. viridiflora, Malaxis acuminata, Phaius flavus, P. longicruris* and *Rhomboda petelotii*. The rocky slopes and cliffs are rich in lithophytic orchids, including *Aerides rosea, Bulbophyllum ambrosia, B. macraei, B. purpureifolium, Cleisostoma paniculatum, C. rostratum, C. striatum, Coelogyne fimbriata, Cymbidium floribundum, C. suavissimum, Dendrobium chrysanthum, D. fimbriatum, D. loddigesii, Eria coronaria, E. pusilla, E. rhombodalis, Gastrochilus pseudodistichus, Hemipilia calophylla, Liparis balansae, L. bootanensis, L. distans, L. mannii, Monomeria barbata, Oberonia cavaleriei, Pelatantheria insectifera, Pholidota leveilleana, P. roseans, P. yunnanensis,* and *Podochilus microphyllus.* Many epiphytic orchids also occur in these forests, the most typical being *Aerides rosea, Callostylis rigida, Ceratostylis himalaica, Cleisostoma rostratum, C. striatum, Dendrobium acinaciforme, D. aduncum, D. lindleyi, D. longicornu, D. nobile, Epigeneium amplum, E. chapaense, Eria pannea, E. sutepensis, Holcoglossum wangii, Hygrochilus parishii, Pelatantheria insectifera, Pholidota articulata, P. missionariorum, P. recurva, Thelasis pygmaea, Thrixspermum calceolus, Vanda fuscoviridis, V. concolor,* and *Trichotosia velutina.*

Fig. 195. Occasionally specimens of *Paphiopedilum purpuratum* in Vietnam may develop two-flowered inflorescences.

N. Vietnam, Cao Bang Prov. (Photo.: L. Averyanov).

Paphiopedilum purpuratum sometimes occurs with some other slipper orchids, such as *P. dianthum, P. micranthum* and *P. hirsutissimum* var. *chiwuanum.*

Paphiopedilum purpuratum usually grows in Vietnam as a terrestrial on deep soils rich in humus and covered by thick leaf litter. The species commonly occurs as single scattered plants or forms small colonies on slopes of limestone hills or mountains of any exposition under a light canopy (Figs. 192, 193). In Vietnam, it is rare nowadays, having been extensively collected by commercial collectors. It needs urgent protection.

VARIATION. Considerable variation in leaf tessellation is typical for the plants from Vietnamese populations of *Paphiopedilum purpuratum.* As with Chinese plants, ground colouring varies from whitish or pale green to bluish green in a single colony. At the same time the colouration and form of the flower is relatively constant in all studied plants (Fig. 194). Commonly, plants develop a one-flowered inflorescence, but from time to time samples may be seen with two flowers on a stalk (Fig. 195).

Paphiopedilum sect. Pardalopetalum Hallier f. & Pfitzer

Paphiopedilum sect. **Pardalopetalum** *Hallier f. & Pfitzer* in Engler, Pflanzenr. Orch. Pleon.: 66 (1903). **Lectotype:** *Paphiopedilum lowii* (Lindl.) Stein.

Paphiopedilum sect. *Polyantha* Pfitzer in Bot. Jahrb. Syst. 19: 41 (1894), p.p.

Paphiopedilum sect. *Mystropetalum* Pfitzer in Engler, l.c.: 65 (1903); Karasawa & Saito, Bull. Hiroshima Bot. Gard. 5: 38 (1982).

DESCRIPTION. *Leaves* green, coriaceous, strap-shaped, without purple-violet marks beneath. *Inflorescence* pendent, arching, or suberect multi-flowered, with 2–10 flowers; flowers open simultaneously. *Median sepal* and synsepal very distinct from the petals in colour, size and shape. *Petals* narrowly lanceolate or linear, twisted, much narrower than median sepal, with clusters of black hairs along lower margin and on inflated apex. *Lip* relatively small with markedly incurved side-lobes. *Staminode* obcordate, with a basal protuberance and tridentate apex. 2n=26. Four species distributed in SE Asia, the Philippines and western Malaysia (Fig. 37).

18. PAPHIOPEDILUM DIANTHUM T.Tang & F.T.Wang

Paphiopedilum dianthum T. *Tang & F.T. Wang* in Bull. Fan Mem. Inst. Biol. Ser. 10: 24 (1940); Cribb, Gen. Paphiopedilum, ed. 2: 189 (1998); Braem, C. Baker & M. Baker, Gen. Paphiopedilum 1: 126 (1998). **Type:** China, Yunnan, Mengtze, *Wang & Liu* 83446 (holotype PE!).

Paphiopedilum parishii (Rchb.f.) Stein var. *dianthum* (Tang & Wang) Karasawa & Saito in Bull. Hiroshima Bot. Gard. 5: 38 (1982); Cribb & C.Z. Tang in Plantsman 4: 170 (1982); Cribb, Gen. Paphiopedilum: 121 (1987); Braem, Paphiopedilum: 105 (1988).

DESCRIPTION. *Lithophytic* or rarely *epiphytic herb* normally forming clumps of up to 15 growths, each with 4–6 leaves. *Leaves* coriaceous, ligulate, obliquely obtuse at the apex, 20–50 cm long, 2–5 cm wide, dark green above, paler beneath. *Inflorescence* arching-pendent, horizontally spreading, or suberect, with (1)2–5 flowers; peduncle green, sparsely papillose below, glabrous above; bracts cucullate, broadly ovate, truncate, 1.2–3 cm long, green. *Flowers* very slightly papillose on outside of sepals; dorsal sepal white with a green base and venation; synsepal pale green or white with darker green veins; petals white or light yellowish-green, dark olive-green striped, with few black warts near the base and few clusters of black hairs along lower margin; lip light yellowish-brown to olive-green; staminode white with dark green venation in the center; pedicel and ovary glabrous, green. *Dorsal sepal* elliptic or obovate, shortly apiculate, 3.8–4.3 cm long, 2.2–2.5 cm wide, with reflexed basal margins, papillate at base on inner surface. *Synsepal* broadly elliptic to broadly ovate, obtuse, 3.5–3.8 cm long, 2.3–2.6 cm wide, with reflexed basal margins and inflexed apical ones. *Petals* falcate-pendent, strongly twisted, linear-tapering, rounded at apex, 8–10(11) cm long, 1–1.1 cm wide at the base, basal part incurved over base of column, with a few raised warts on inner side towards the base and few clusters of papillae-like hairs on the lower margin, not ciliate, papillose on inner surface at base and at apex. *Lip* deeply saccate, pointed at apex, 4–4.5 cm long, 2–2.5 cm wide, with inflexed triangular side lobes. *Column* 1–1.2 cm long, with a pubescent stalk and blunt pubescent filaments; staminode obcordate, with a blunt basal umbo, three-lobed at apex, 9–11 mm long, 6–7.5 mm wide, scarcely papillate on side margins. Figs. 198, 199.

DISTRIBUTION. Southern China (southeast Yunnan, southwest Guizhou and west Guangxi), northern Vietnam (Cao Bang, Ha Giang, Hoa Binh, Lai Chau, Lao Cai and Son La Provinces). Figs. 37, 200.

ECOLOGY. Primary, coniferous, mixed and broad-leaved, evergreen, mossy, wet forests on highly eroded, crystalline limestone from (600) 1000–1450 m. Figs. 196, 197.

FLOWERING IN THE WILD. September–November.

IUCN STATUS. EN.

AFFINITIES. *Paphiopedilum dianthum* is the only multi-flowered slipper orchid in the flora of Vietnam. This outstanding orchid has a long inflorescence which bears up to 5 large elegant flowers, 12–15 cm across. It is closely allied to *P. parishii* from Myanmar, Thailand and south-western Yunnan. *Paphiopedilum dianthum* differs, however, in its lithophytic habit, long glabrous inflorescences usually bearing two to five flowers, glabrous bracts, a glabrous ovary, a white dorsal sepal, white or light green petals striped with olive-green and its somewhat larger lip. In optimal conditions, large plants of *P. dianthum* often develop up to 15 growths but four to seven is more usual.

HISTORY. *Paphiopedilum dianthum* was originally described by the Chinese botanists T. Tang and F.T. Wang in 1940 based on the type collected by C.W.Wang at Mengtze in Yunnan and a second specimen from Tsingsi Hsien (Chinghsi), Kwangsi (Guangxi) Province of south-west China. The earliest collection of *Paphiopedilum dianthum* was made by W. Hancock (no. 582) at Mengtsz

Fig. 196. Primary coniferous, mixed and broad-leaved evergreen forests on highly eroded limestone, the usual habitat of *Paphiopedilum dianthum* in northern Vietnam.

N. Vietnam, Cao Bang Prov. (Photo.: L. Averyanov).

Fig. 197. Shady rocky outcrops on highly eroded karst limestone ridges in humid primary forest, the typical habitat of *Paphiopedilum dianthum* in northern Vietnam.

Chinese-Vietnamese border area. (Photo.: P.J. Cribb).

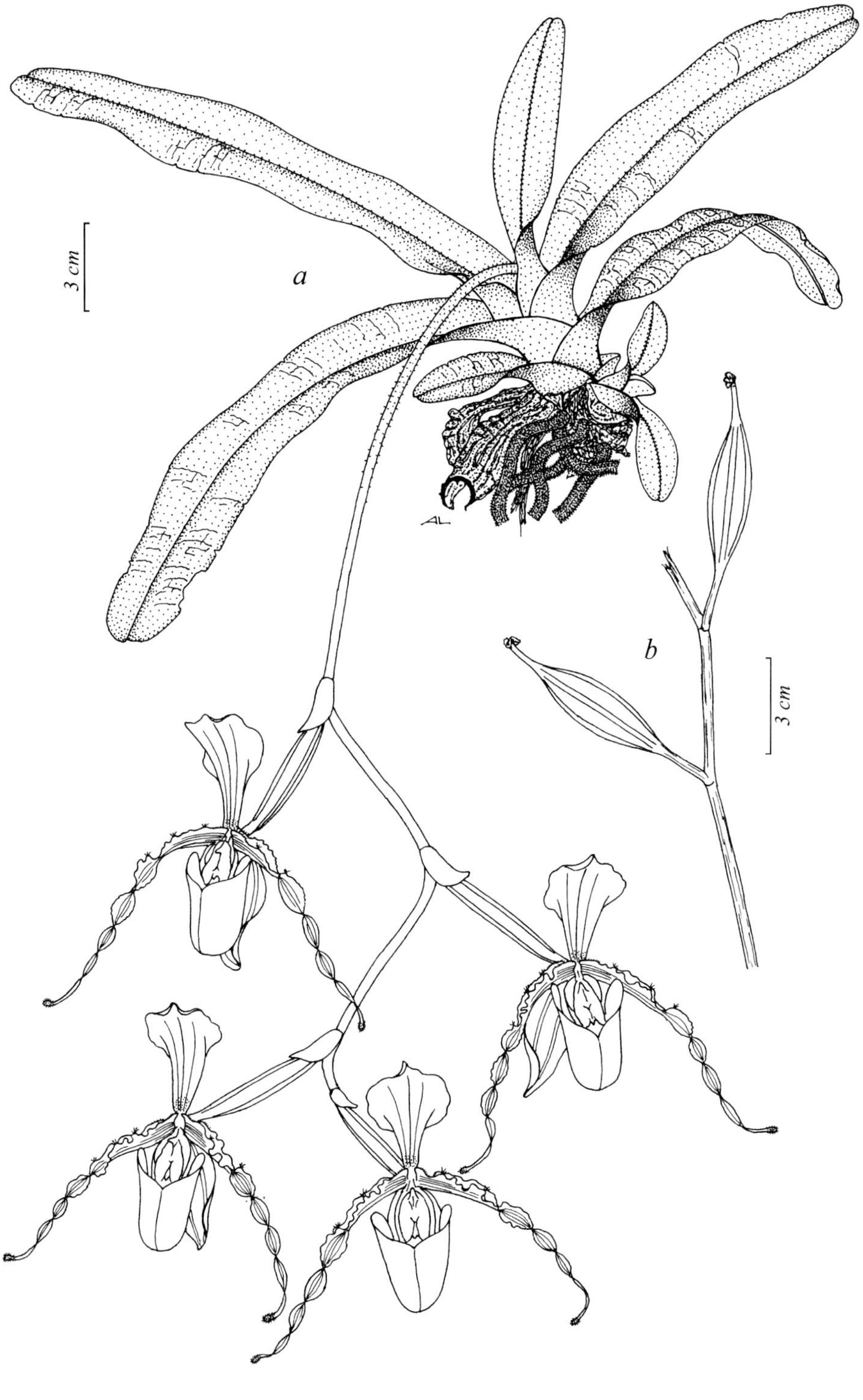

Fig. 198. *Paphiopedilum dianthum.* (Drawing: L. Averyanov).
a — flowering plant; **b** — portion of inflorescence axis with fruits. (Drawn from CBL 3364).

(Mengtze), Yunnan in November 1894 at 1800–2100 m, and is preserved in the Kew Herbarium. *Paphiopedilum dianthum*, previously considered as a Chinese endemic, has recently been discovered in limestone areas of northern Vietnam during botanical exploration by the Institute of Ecology and Biological Resources of the National Center for Natural Sciences and Technology of Vietnam (Averyanov & Averyanova, 2000; Averyanov *et al.*, 2000a,b). A little earlier, Tanaka (1998) mentioned that this species was a common plant in the north of Vietnam. His information came most probably from contacts in the orchid trade. For many years, numerous specimens of *Paphiopedilum dianthum*

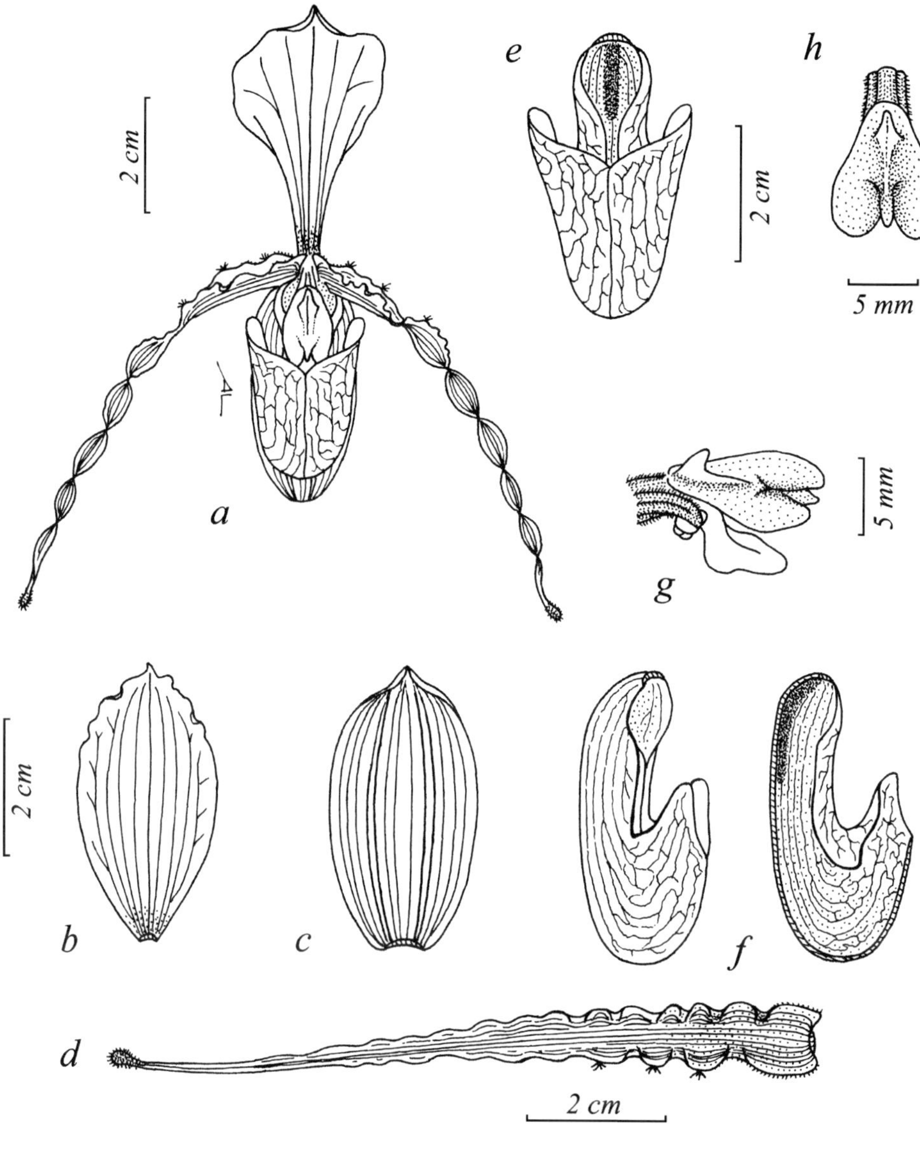

Fig. 199. *Paphiopedilum dianthum*. (Drawing: L. Averyanov).

a — flower; **b** — dorsal sepal; **c** — synsepal; **d** — petal; **e** — lip, frontal view; **f** — lip, side and section views; **g, h** — column, side and dorsal views. (Drawn from CBL 394).

have been cultivated in orchid nurseries in Sa Pa town. Some of them were certainly collected in limestone mountain regions of Hoang Lien Son ridge in Lai Chau and Lao Cai Provinces. However, the exact provenance of *P. dianthum* in this area remains a secret.

Formerly, *Paphiopedilum dianthum* was regarded as variety of *P. parishii* (Karasawa & Saito, 1982; Cribb, 1987; Braem, 1988), but it is now commonly accepted as a distinct species (Cribb, 1998; Braem, C.Baker & M.Baker, 1998).

HABITAT. *Paphiopedilum dianthum* usually grows on small shady cliffs and steep rocks of any exposition just below the summits of karst limestone mountains and hills usually between 1000–1450 m (Fig. 196, 197). It forms small or large colonies on very steep slopes. Mark (1987) reported similar habitats on limestone bluffs in China. Plants grow in the deep shade of trees such as laurels and oaks. Soil pH ranges from 7.5 to 7.86. In one Chinese locality it grew intermixed with *P. micranthum* and in another with *P. hirsutissimum* var. *esquirolei* (Cribb, 1998).

Paphiopedilum dianthum inhabitats areas of monsoon tropical climate associated with mountains. The dry cold winter lasts here from October to April. Median winter temperatures usually lie between 11–16°C (Fig. 201) with recorded absolute minima about -1.5°C. Cold wet NE winds coming from the sea that regularly form heavy fogs and mists are very common in February–March. At this time air humidity is very high. Summer is the rainy season with frequent torrential rain that reaches a maximum during June–September. Median temperatures at this time vary between 20 and 24°C (Fig. 201) with observed maxima about 31–35°C. Mean annual rainfall in the area ranges commonly between 1500–2000 mm.

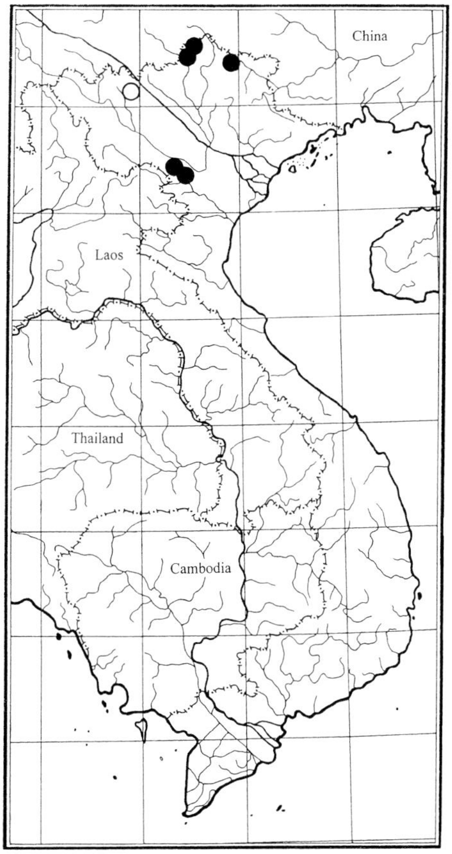

Fig. 200. Distribution of *Paphiopedilum dianthum* in Vietnam. (Drawing: L. Averyanov).

Studied populations are marked with dots, circle designates locality identified by local collectors.

In Vietnam, *Paphiopedilum dianthum* occurs in primary coniferous, mixed and broad-leaved, evergreen, wet, mossy forests on highly eroded, crystalline limestone. Nowadays it is usually found at 1000–1450 m (Fig. 196, 197) where habitats are much better preserved than at lower elevations. However, it may sometimes be seen as low as 600–700 m. In the past it certainly had a wider distribution at low elevations, but is extinct in many such sites due to anthropogenic degradation of primary limestone vegetation.

Common dominants in the forest here are conifers, such as *Pinus fenzeliana, Pseudotsuga brevifolia, Tsuga chinensis, Calocedrus macrolepis, Podocarpus brevifolius, P. neriifolius, P. pilgeri, Amentotaxus argotaenia, Cephalotaxus mannii, Fokienia hodginsii, Nageia fleuryi, N. wallichiana* and *Taxus chinensis*. A number of broad-leaved trees also often appear as co-dominants, among them species of *Acer* (Aceraceae), *Carpinus* (Betulaceae), *Garcinia* (Clusiaceae), *Lithocarpus* and *Quercus* (Fagaceae), *Madhuca* (Sapotaceae), *Ulmus* (Ulmaceae), *Myrsine kwangsiense* (Myrsinaceae), *Pistacia weinmannifolia* (Anacardiaceae) and *Platycarya strobilacea* (Juglandaceae). The tallest trees may reach 15–20 m in height.

Below these grow a variety of small trees, such as species of *Celastrus, Euonymus* (Celastraceae), *Cornus* (Cornaceae), *Elaeocarpus* (Elaeocarpaceae), *Eriobotrya, Photinia, Prunus, Sorbus* (Rosaceae),

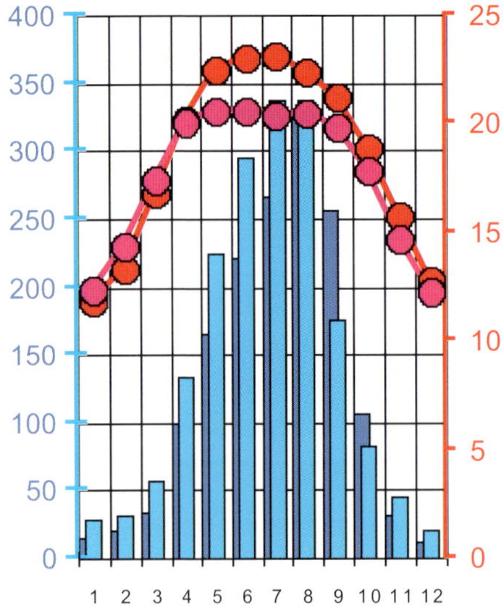

Fig. 201. Climatological data typical for the area of *Paphiopedilum dianthum* in Vietnam.

Black figures at the chart bottom are months. Blue axis represents atmosphere precipitation in mm (columns designate mean monthly rainfalls); red axis represents mean monthly temperature in °C. The deep blue and red are data from the Moc Chau weather station, 20°51'N 104°38'E, elev. 958 m; light blue and purple are data from the Pha Din weather station, 21°34'N 103°30'E, 1347 m, all according to Nguyen Khanh Van *et al.*, 2000.

Ficus (Moraceae), *Illicium jiadifengpi* (Illiciaceae), *Ixonanthes* (Ixonanthaceae), *Pittosporum* (Pittosporaceae), *Rhododendron*, *Vaccinium* (Ericaceae), *Schefflera pesavis* (Araliaceae), *Symplocos* (Symplocaceae), *Tirpitzia sinensis* (Linaceae) and *Wikstroemia* (Thymellaeaceae).

Lithophytic shrubs on rocky outcrops here include species of *Abelia, Lonicera* (Caprifoliaceae), *Agapethes, Rhododendron, Vaccinium* (Ericaceae), *Ardisia* (Myrsinaceae), *Berberis, Mahonia* (Berberidaceae), *Pyracantha, Rosa, Rubus, Spiraea, Stranvaesia* (Rosaceae), *Helwingia* (Helwingiaceae), *Ilex* (Aquifoliaceae), *Lespedeza*, *Sophora glauca* (Fabaceae), *Pittosporum* (Pittosporaceae) and *Polygala* (Polygalaceae). The dwarf palms, *Guihaia* and *Rhapis*, also occur here.

Fig. 202. *Paphiopedilum dianthum* in typical habitat.

N. Vietnam, Ha Giang Prov. (Photo.: P.J. Cribb).

Fig. 203 (above and next page). Typical flowers of *Paphiopedilum dianthum* in Vietnamese populations.
N. Vietnam, Ha Giang Prov. (Photos.: P.J. Cribb).

These forests support a very rich flora of lithophytic and epiphytic bryophytes, ferns and herbs. Amongst the herbs found here are species of Acanthaceae, Araceae (*Arisaema*), Asteraceae (*Ainsleya*), Balsaminaceae (*Impatiens*), Begoniaceae (*Begonia*), Convallariaceae (*Aspidistra, Disporum, Ophiopogon, Peliosanthes, Polygonatum, Tupistra*), Gesneriaceae, Liliaceae (*Lilium*), Primulaceae (*Lysimachia*), Rubiaceae (*Ophiorrhiza*), Urticaceae (*Elatostema, Pellea, Pellionia, Procris*), Violaceae (*Viola*), and numerous sedges (*Carex, Scleria*) and selaginellas (*Selaginella*). Species of *Adiantum, Asplenium, Cyrtomium, Colysis, Crypsinus, Davallia, Diplasium, Grammitis, Neocheiropteris, Polypodium, Polystichum, Pyrrosia, Tectaria* and *Trichomanes* are the commonest ferns in these forests.

A great number of orchids also occur here, including the lithophytic species *Bulbophyllum ambrosia, B. macraei, B. purpureifolium, Coelogyne fimbriata, C. lockii, C. malipoense, Eria coronaria, E. diversicolor, E. rhombodalis, Liparis distans, L. mannii, L. stricklandiana, Monomeria dichroma, Pholidota roseans, P. rubra* and *P. yunnanensis* which cover mossy rocks on the forest floor.

Common terrestrial orchids found here include *Anoectochilus calcareus, Bulbophyllum maculosum, Calanthe aleizettii, Cheirostylis bipunctata, C. yunnanensis, Cleisostoma paniculatum, Cryptochilus luteus, Cymbidium ensifolium, C. floribundum, C. lancifolium, Dendrobium chrysanthum, D. nobile, D. thyrsiflorum, Eria boniana, E. globulifera, E. pusilla, E. sutepensis, Goodyera hispida, G. schlechtendaliana, G. viridiflora, Liparis averyanoviana, Malaxis acuminata, Oberonia cavaleriei, Podochilus cultratus, P. microphyllus* and *Rhomboda petelotii.* The epiphytic orchids commonly seen here are *Bulbophyllum reptans, Ceratostylis himalaica, Cleisostoma striatum, Dendrobium acinaciforme, D. longicornu, Epigeneium amplum, E. chapaense, Eria foetida, E. pannea, E. siamensis, Eria thao, Holcoglossum wangii, Luisia appressifolia, Monomeria barbata, Ornithochilus difformis, Pholidota missionariorum, P. yunnanensis, Rhomboda petelotii, Schoenorchis gemmata, Sunipia scariosa, Thelasis khasiana* and *Vanda concolor.*

In Vietnam, *Paphiopedilum dianthum* usually grows as a lithophyte on mossy ledges or in cracks in the rock or cliff face (Fig. 202). It forms a stout root ledge that catches leaf litter, building up its own soil mound which is dark brown and crumbly in structure. In some populations this species grows intermixed with other slipper orchids such as *P. hirsutissimum, P. malipoense, P. micranthum* or *P. henryanum.* In wet conditions of cloud forests it may occasionally be seen as a humus epiphyte on mossy gnarled trees usually near the ground.

In highland primary forests, *Paphiopedilum dianthum* is still a common plant in many localities in Vietnam. In intact habitats it can form large colonies with numerous free-flowering and magnificent specimens. In the wild, it flowers from mid September to the end of November. Like other slipper species in Vietnam, *P. dianthum* is disappearing rapidly. After logging of primary forest it vanishes very quickly. Occasionally, a few plants may survive in shady gorges and karst holes on north-facing mountain slopes. However, they usually soon disappear due to progressive degradation after forest clearance. Few accessible localities boast large populations of *P. dianthum* nowadays. Commercial collecting is other factor that eliminates plants in nature all over the country, including officially protected areas. All this is leading to the rapid extinction of this fine slipper orchid in the wild.

VARIATION. According to our observations, *Paphiopedilum dianthum* exhibits little variation in form and colouration of the flowers in Vietnamese populations. The typical flowers are shown in Fig. 203a,b. Plants of less than five growths seldom bear fruits while larger plants have from one to five pods per plant.

Natural intraspecific *Paphiopedilum* hybrids reported from Vietnam

19. PAPHIOPEDILUM × AFFINE De Wild.

Paphiopedilum × affine *De Wild.* in Trib. Hort. 1: 57 (1906); Guillamin in Bull. Soc. Bot. France. 71: 550 (1924); id., Fl. Gen. Indo-Chine 6: 640 (1934). **Type:** Northern Vietnam (Tonkin), cult. Bruxelles (holotype not located).
Paphiopedilum villosum var. *affine* (De Wild.) Braem, Paphiopedilum: 119 (1988).

DESCRIPTION (fide Guillaumin, 1934). *Leaves* linear to oblong 15 cm long, 3 cm wide, marble-colored above, shortly bilobed at the apex. *Scape* about 15 cm long, pubescent, pale green with purple marks; bract ovate, conduplicate, half shorter than ovary, pale green, glabrous, ciliate at the margin. *Flower* 12 cm across. *Dorsal sepal* ovate, 5 cm long, 2.8 cm wide, deflexed backwards at the margin, green at the base and in the middle, white at the apex, with maroon spots on the lower third. *Synsepal* ovate, deflexed backwards along the margin near the base, glabrous inside except the base, white-haired outside. *Petals* spatulate, 6 cm long, 2.3 cm wide, light green to pale brown, glabrous, ciliate along the margin. *Lip* 4.3 cm long, light green to pale brown, glabrous, ciliate along front margin. *Staminode* indistinctly cordate at the base, broadly acute at the apex, light yellowish-green to light brown, with green spot and hardly visible umbo at the centre, shortly haired.

IUCN STATUS. DD.

AFFINITIES AND HISTORY. Very little is known about this plant. No information exists about its ecology, distribution and flowering period in nature. It was described on the basis of living plants imported supposedly from northern Vietnam or northern Laos. In the first edition of his *Paphiopedilum* monograph, Cribb (1987) included *P. affine* as a possible synonym of *P. gratrixianum* based on the description which mentions that the leaves are "d'un vert plus ou moins fonce ... , striees-tachetees de vert plus fonce". Having now seen what is probably an authentic specimen (hort. Bruxelles 21 Nov. 1912), he is inclined to think it a natural hybrid of *P. villosum* and *P. appletonianum*. We have not seen such plants among recent materials collected in Vietnam.

20. PAPHIOPEDILUM × ASPERSUM Aver.

Paphiopedilum × aspersum *Aver.* in Komarovia 2: 17 (2002). **Type:** N. Vietnam, Son La Prov., Moc Chau Distr., Dong Sang municipality, 1000–1100 m, south-east slope of limestone mountains, vertical bluffs near top, cult. Hanoi, 4 Sept. 2001,. *Phan Ke Loc* P-9895 (holotype HN!, photo. LE).

A natural hybrid of *P. barbigerum* var. *lockianum* and *P. henryanum*.

DESCRIPTION. *Lithophytic herb. Leaves* 4–6, lanceolate to ligulate, commonly shortly acute at unequally bilobulate apex, 7–112 cm long, 0.8–1.2 cm wide, glossy green above with a distinct paler margin, paler green on underside, sharply keeled beneath on mid-vein. *Inflorescence* suberect to horizontal, arcuate or pendent, 1-flowered, peduncle 7–12 cm long, dull purple-green, shortly deep purple-violet-pubescent; bract conduplicate, ovate, 2–2.5 cm long, up to 1.4 cm wide, green with brown-maroon tint, with short purple-violet hairs on the mid-vein and apical margin. Flower 5.5–6 cm across. *Pedicel and ovary* 4–5 cm long, green, densely covered with short purple-violet hairs. *Dorsal sepal* broadly ovate to semicircular, obtuse, 2.5–3.5 cm long, 2.5–3.5 cm wide, somewhat hooded above, with reflexed basal margins, white, yellowish green at the base and along mid-vein, with numerous purple-maroon spots. *Synsepal* ovate, obtuse, 2–2.5 cm long, 1.2–1.3 cm wide, white, dull light yellow at the base. *Lateral petals* spreading, spathulate, obtuse at apex, 4–4.5 cm long, 1–1.5 cm wide, glossy, shortly pubescent on outer surface and sparsely long purple-hairy at the base, undulate along margins, pink-brown with yellowish-green margins. *Lip* glossy, urceolate, light pink-violet to pink-brown, 3–4 cm long, about 1.6 cm wide, rounded at the apex, with a prominent claw-like base. *Staminode* obcordate-obovate, obtuse, 7–8 mm wide, 8–10 mm long, sides sometimes somewhat incurved, pustulate, with a glossy yellow umbo. *Column* white. Fig. 204.

DISTRIBUTION. Northern Vietnam (Son La Province).

ECOLOGY. Primary broad-leaved evergreen forests on very steep slopes and cliffs of mountains composed of crystalline highly eroded limestone at elevations of 1000–1100 m.

FLOWERING IN THE WILD. September– November.

IUCN STATUS. EN

AFFINITIES. *Paphiopedilum × aspersum* is morphologically intermediate between its putative parental species — *P. barbigerum* var. *lockianum* and *P. henryanum*. It has a white or light yellowish-pink dorsal sepal with numerous small purple-chestnut dots, light brownish-pink lateral petals and a light purple-pink or pink-brownish lip. Examples of this hybrid from wild sources show a clear and obvious resemblance to artificial hybrids of *P. barbigerum* and *P. henryanum* known in cultivation.

HISTORY AND HABITAT. The first example of *Paphiopedilum × aspersum* was seen by Prof. Phan Ke Loc in a private orchid nursery near Hanoi. According to the local collectors, it was discovered growing on a limestone mountain near Ba Phach village of Son La Province in the northern Vietnam. Two months later, Loc and his colleagues discovered a few further plants of this hybrid in nature during a botanical expedition of the Institute of Ecology and Biological Resources of the National Center for Natural Sciences and Technology of Vietnam. These rare plants were found growing in a large population of *Paphiopedilum barbigerum* var. *lockianum* on rocky limestone in Son La Province of Moc Chau District (Fig. 128). Specimens of *Paphiopedilum × aspersum* were found scattered or clustered in small groups growing lithophytically in cracks or on narrow shelves of vertical, south-east-facing, lightly shaded, limestone cliffs near the ridge top at elevations of 1000–1100 m in heavily logged, evergreen, tropical, seasonal, broad-leaved, submontane primary forest (Fig. 129). Climatic conditions and the character of the vegetation in this habitat are the same as were described above for *P. barbigerum*.

Fig. 204. *Paphiopedilum × aspersum*

Type example. N. Vietnam, Son La Prov. (Photo.: Phan Ke Loc).

18. PAPHIOPEDILUM × DALATENSE Aver.

Paphiopedilum × dalatense *Aver.* in Orchid Digest. 65, 3: 134 (2001). **Type:** Southern Vietnam, Lam Dong province, Don Duong, cult. Dalat City Institute of Biology, Nov. 1996, *N.V.Duy* P1 (holotype LE!).

Paphiopedilum purpuratum sensu Averyanov, Vasc. Pl. Synopsis Vietnam. Fl. 1: 133 (1990); id., Identif. Guide Vietnam. Orch.: 21 (1994); Seidenfaden, Orch. Indochina: 20 (1992), *non* (Lindl.) Stein.

DESCRIPTION. *Terrestrial herb. Leaves* 5–7, narrowly or oblong-elliptic acute at apex, 15–20 cm long, 2.5–3.5 cm wide, finely indistinctly green/pale-green tessellated above, fine purple-violet speckled toward the base on lower surface. *Inflorescence* 1-flowered; peduncle erect or suberect, spreading at acute angle, green, deep purple-violet pubescent, 15–20 cm long; bract ovate to elliptic, acute to subacute, 2–2.5 cm long, 1–1.5 cm wide, green, usually marked with purple, deep purple-violet ciliate. *Flower* large, 9–12 cm across; dorsal sepal white at the base and in lower half, flushed with pink, pink-brown, purple-brown, purple or olive-green, veined with pink-green, olive-green, brown or purple; synsepal white with pale greenish, brownish or purplish veins in basal half; petals white to dull pink, yellowish-green and pale brownish-green with pink, brownish-purple or purple stripes, sometimes slightly spotted with small black-purple flat warts along upper margin toward the base; lip pink, yellowish-pink or light yellowish-brown with more deep coloured veins; staminode white to light

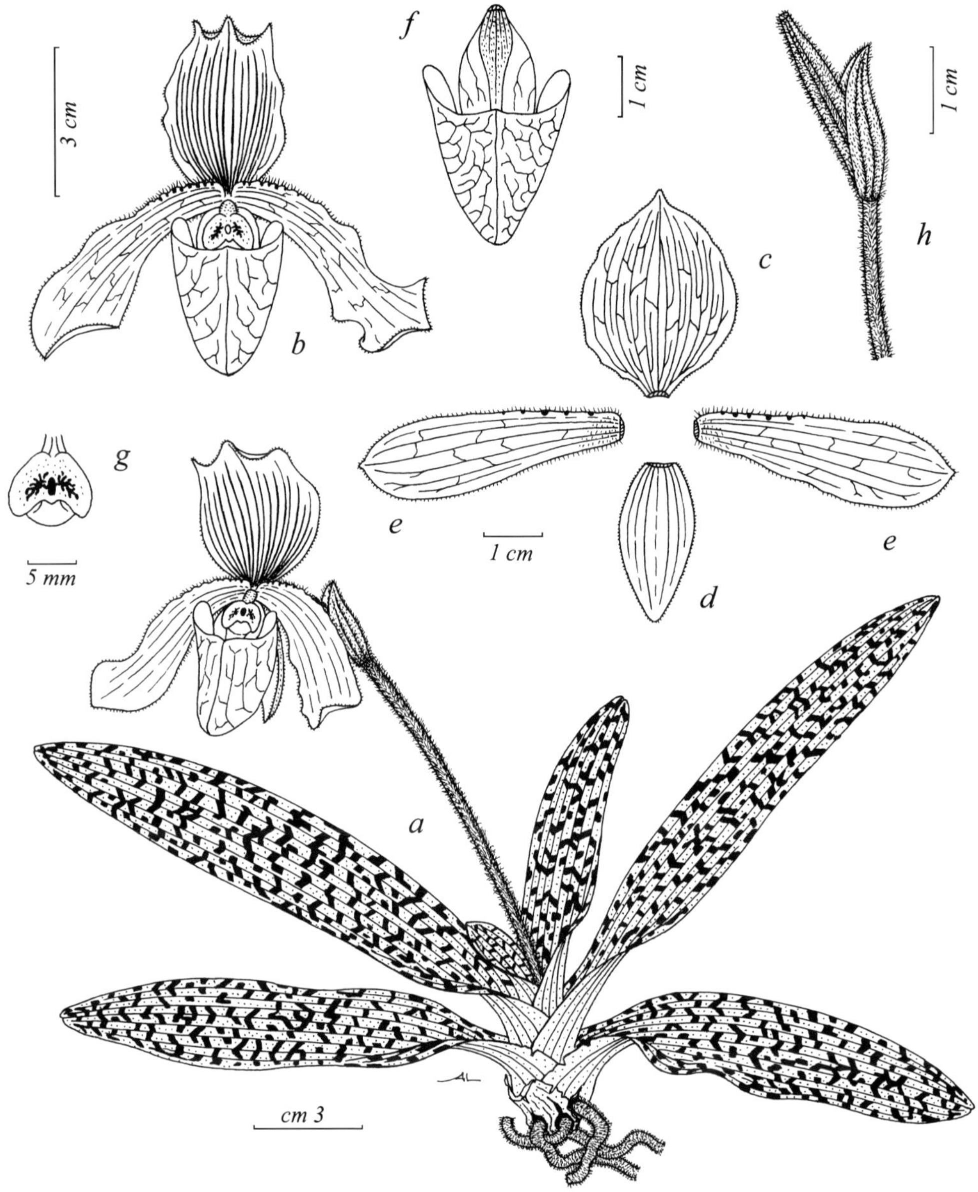

Fig. 205. *Paphiopedilum × dalatense.* (Drawing: L. Averyanov).

a — flowering plant; **b** — flower; **c** — dorsal sepal; **d** — synsepal; **e** — lateral petals; **f** — lip; **g** — staminode; **h** — ovary and flower bract. (Drawing from *N.V.Duy* P I 11).

yellowish-pink with an oblong-ovate green umbo and small irregular green spots in centre part; pedicel and ovary 3–5 cm long, green, deep purple-violet pubescent. *Dorsal sepal* ovate to broadly ovate, shortly apiculate, 3.5–5 cm long, 2.5–3.5 cm wide, with recurved ciliate margins. *Synsepal* concave, elliptic, obtuse at apex, 2.8–3 cm long, 1.4–1.8 cm wide. *Petals* reflexed, subsigmoid, ligulate to narrowly obovate, obtuse or shortly apiculate at apex, 4.5–6 cm long, 1.3–1.7 cm wide, deep purple-violet ciliate. *Lip* 3.5–4 cm long, 2–2.5 cm wide. *Staminode* broadly lunate, 7 mm long, 8 mm wide, with a small obtuse central tooth and oblong-ovate low umbo. Fig. 205.

DISTRIBUTION. Southern Vietnam (Lam Dong Province). Fig. 206.

ECOLOGY. Primary, broad-leaved, evergreen and mixed forests at elev. 900–1700 m.

FLOWERING IN THE WILD. June–July?

IUCN STATUS. DD.

AFFINITIES. *Paphiopedilum* × *dalatense* is a natural hybrid of *P. callosum* and *P. villosum* that is morphologically intermediate between its putative parental species (Fig. 205, 207). This remarkable plant has large flowers with variable colouration and indistinctly mottled leaves with fine tessellation. Some forms of this hybrid superficially resemble *P. purpuratum*.

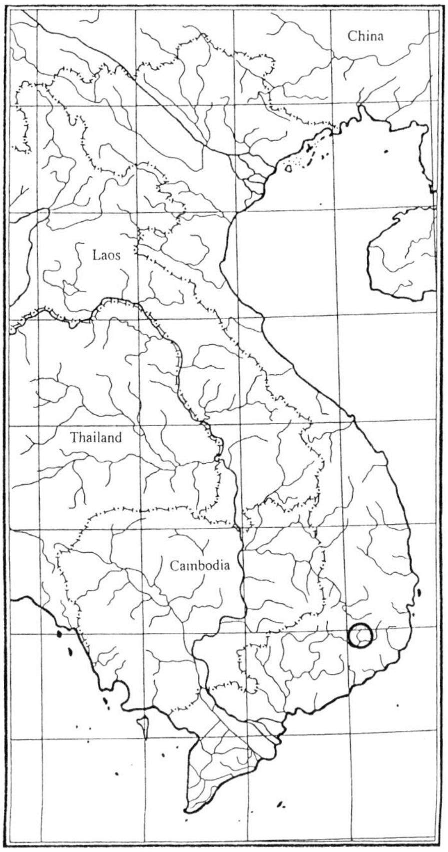

Fig. 206. Distribution of *Paphiopedilum* × *dalatense*. (Drawing: L. Averyanov).

HISTORY. Natural hybrids of *Paphiopedilum callosum* and *P. villosum* have been seen several times (Averyanov, 2001). They were found by collectors and dealers among wild collections of the parental species. The plants presumably come from Lam Dong Province in the southern Vietnam and were mostly grown on in private collections in Dalat city where the climate is quite favourable for their survival (Averyanov, 2001). *Paphiopedilum callosum* × *P. villosum* certainly has some superficial resemblance to *P. purpuratum* and plants were grown under the latter name in local orchid collections for many years. This resulted in the mistaken view that *P. purpuratum* occurred naturally in the southern Vietnam (Averyanov, 1990, 1994; Seidenfaden, 1992; Pham-Hoang Ho, 2000). Several wild-collected specimens were also successively cultivated in the orchid garden of the Institute of Biology of the National Center for Natural Sciences and Technology of Vietnam in Dalat city for some years after their discovery and collection in 1993–1994 at the Don Duong locality in Lam Dong province. Herbarium specimens of these plants were made in 1996 by N.V.Duy, a member of staff there. Examination of these and study of similar living plants in several private local collections leave no doubt that these interesting plants represent a natural hybrid between *P. callosum* and *P. villosum* (Averyanov, 2001).

HABITAT. The appearance of *Paphiopedilum* × *dalatense* in southern Vietnam is not too surprising as both putative parents occur in the area in very similar habitats, described under *P. callosum* and *P. villosum*. According to available information, *P.* × *dalatense* is a rare plant occurring in primary, broad-leaved, evergreen and mixed (with some coniferous trees) forests mainly between 900–1500 m. Besides Lam Dong Province its range may include other regions where both parental species occur together, e.g. in Dac Lac, Gia Lai, Khanh Hoa, Kon Tum and Ninh Thuan Provinces in the southern Vietnam and the mountains of Cambodia adjacent to Vietnamese border. However, to date it has not been reported from these areas.

VARIATION. Form, dimensions and colouration of the flowers of *Paphiopedilum* × *dalatense* are very variable, but very similar to those found in artificial hybrids of *P. callosum* × *P. villosum* registered under the name *Paphiopedilum* Indra in 1893 by R.I.Measures (Harold Koopowitz, pers. comm.). Observed patterns of plant variation are mentioned above in the description. Most common variants of the flower are shown in Fig. 207 a–c.

Fig. 207. **a–c** — Different variants in the shape and colouration of the flowers of cultivated wild-collected samples of *Paphiopedilum* × *dalatense* discovered near Dalat city.

(Photos: Mr.Son Ha; digital correction: L. Averyanov).

22. PAPHIOPEDILUM × HERRMANNII Fuchs & Reisinger

Paphiopedilum × herrmannii *Fuchs & Reisinger* in Linzer Biol. Beitr. 27, 2: 1213 (1995); Braem, C.Baker & M.Baker, Gen. Paphiopedilum 2: 208 (1999). **Type:** North Vietnam, border region to China, *Herrmann* 343/85 (holotype LI).

DESCRIPTION. *Lithophytic* and occasionally *terrestrial herb*, often forming clumps. *Leaves* 4–6, linear-ligulate, obtuse or shortly acute at unequally bilobulate apex, up to 10–14(18) cm long, 1–1.2(1.5) cm wide, glossy green above with a distinct paler margin, paler green on underside, sharply keeled beneath on mid-vein. *Inflorescence* suberect to arcuate, 1-flowered, 11–17 cm tall; peduncle 7–15 cm long, bright green, with short, dark purple-violet hairs; bract conduplicate, ovate, acute, 2–2.5 cm long, up to 1.6 cm wide, green, spotted with maroon at base, with short purple-violet hairs on the mid-vein and apical margin. *Flower* 5.5–6 cm across, 5.2–5.8 cm tall, shortly pubescent on outer surface of sepals; dorsal sepal glossy, dull yellowish-green with a dark brown base and veins; synsepal dull light yellow; petals glossy, dull yellowish more or less flushed and veined dark olive-brown; lip light pink-violet, yellowish along margin; staminode yellow, with a yellow-orange central umbo, column stalk purple-flushed; pedicel and ovary 4–5 cm long, green, densely covered with short purple-violet hairs. *Dorsal sepal* broadly ovate to semicircular, obtuse, 2.5–3.5 cm long, 2.5–3.5 cm wide, somewhat hooded above, with reflexed basal margins. *Synsepal* more or less appressed to ovary, narrowly ovate to ovate, subacute, 2.4–2.5 cm long, about 1.3 cm wide, both margins commonly incurved-undulate. *Petals* spreading, slightly deflexed, oblanceolate or spathulate, obtuse at apex, 4–4.5 cm long, 1–1.5 cm wide, glossy, brown-ciliate, slightly purple-hairy at base, with strongly undulate margins. *Lip* urceolate, rounded at the apex, with a prominent claw-like base, 3–4 cm long, about 1.6 cm wide; side lobes semi-elliptic, glossy. *Staminode* obcordate-obovate, obtuse, 6–8 mm wide, 8–10 mm long, sides sometimes somewhat incurved, pustulate, hirsute, with a central glossy umbo. Fig. 209.

DISTRIBUTION. Northern Vietnam (Cao Bang Province). Fig. 210.

ECOLOGY. Primary, coniferous, mixed and broad-leaved, evergreen, closed forests on very steep slopes of mountains composed of crystalline highly eroded limestone at elev. of 700–900 m. Fig. 208.

FLOWERING IN THE WILD. August–September.

IUCN STATUS. EN.

AFFINITIES. *Paphiopedilum × herrmannii* represents a more or less stabilized natural hybrid of *P. hirsutissimum* and another slipper species of *P. insigne* complex, most probably *Paphiopedilum helenae*. The morphology of *P. × herrmannii* is obviously intermediate between its putative parents, leaving little doubt as to its origin.

HISTORY. *Paphiopedilum herrmannii* was described by Fuchs and Reisinger in 1995 based on a cultivated plant that was found among a batch of *P. hirsutissimum* var. *esquirolei* at the Municipal Botanic Gardens of Linz, Austria. The plants were reportedly collected in northern Vietnam close to the Chinese border by Czech botanists in 1985 (Braem, C.Baker & M.Baker, 1999).

The authors compared *Paphiopedilum × herrmannii* with *P. henryanum*, noting that it differed in having an unspotted dorsal sepal and narrowly lanceolate, acute petals, a dorsal sepal with an undulate margin reflexed in the basal part, an acute synsepal and a column with a hairy base. Cox (in Cribb, 1998) compared its molecular fingerprint using leaf material extracted from a cultivated plant and concluded that it was probably of hybrid origin. He suggested that its parents were *P. hirsutissimum* var. *esquirolei*, a common species in the region, and the Chinese *P. barbigerum*. Averyanov and Hiep have found *P. × herrmannii* growing in the wild in between colonies of *P. hirsutissimum* var. *esquirolei* and *P. helenae* (Averyanov, 1999a; Averyanov *et al.*, 2000a,b), the latter a close ally of *P. barbigerum*. We are, therefore, confident that *P. × herrmannii* is a natural hybrid of these two species. In nature their flowering periods scarcely overlap in August, but obviously enough to produce the occasional cross pollination event.

HABITAT. *Paphiopedilum × herrmannii* sporadically occurs in intermixed populations of *P. hirsutissimum* and *P. helenae*, commonly at elevations of 700–900 m. Distribution of this plant probably coincides with range of *P. helenae*. Occasionally it may be seen along the border of populations of both parental species in primary, coniferous, mixed and broad-leaved, evergreen, closed

Fig. 208. Locality of intermixed populations of *Paphiopedilum hirsutissimum* and *P. helenae* in primary mixed forests on limestone mountains is home of *P.* × *herrmannii*.

N. Vietnam, Cao Bang Prov. (Photo.: L. Averyanov).

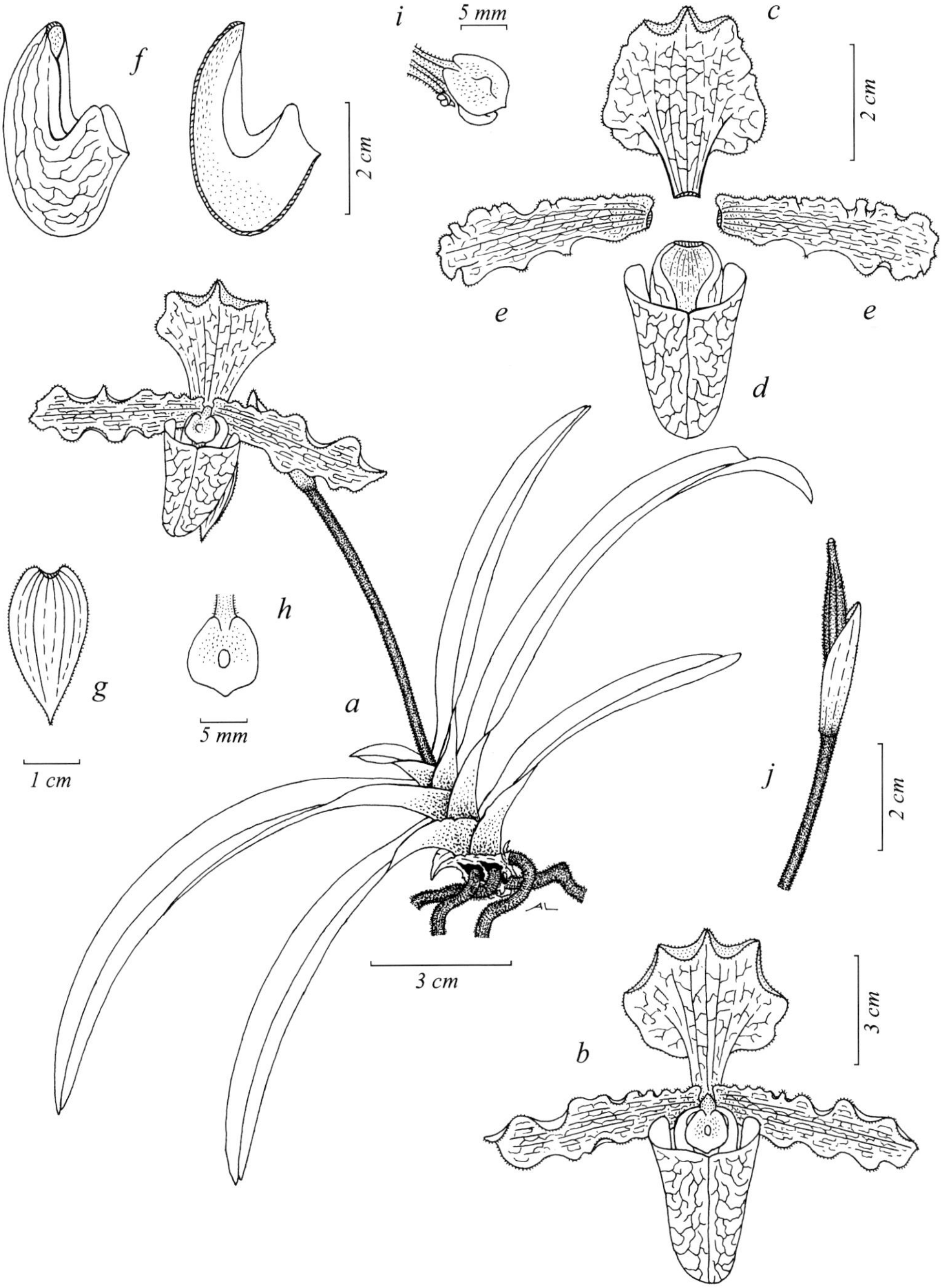

Fig. 209. *Paphiopedilum × herrmannii.* (Drawing: L. Averyanov).

a — flowering plant; **b** — flower; **c** — dorsal sepal; **d** — lip; **e** — petals; **f** — lip, side and section views; **g** — synsepal; **h, i** — column, dorsal and side views; **j** — ovary and flower bract. (Drawn from 25–27 May 1997, L. Averyanov *et al.* VH s.n.).

forests on very steep slopes of mountains composed of highly eroded crystalline limestone (Fig. 211). Typically plants grow as lithophytes, but occasionally they may be found as terrestrials among rocks on steep slopes. The habitat of *P.* × *herrmannii* is identical to those of *P. hirsutissimum* and *P. helenae* as described in earlier chapters.

VARIATION. The flowers of *Paphiopedilum* × *herrmannii* are quite uniform (Fig. 211, 212). However, plants from hybrid complexes that have appeared on the international market under this name show considerable variation. This wide variation undoubtedly reflects the different origin of such plants. At least four local species, *Paphiopedilum helenae, P. henryanum, P. barbigerum* and *P. tranlienianum*, in Vietnam occasionally hybridize with the common and widely distributed *Paphiopedilum hirsutissimum*, resulting in the formation of a spectrum of more or less rare natural reciprocal hybrids. Plants of this hybrid complex are usually fertile and in a number of cases introgress with one or other of their parental species. Often such plants show different combinations of parental features and may be difficult to identify. Meanwhile, the type of *P.* × *herrmannii*, the plant illustrated by Gruss and Perner (Gruss & Perner, 1998), undoubtedly belongs to *P. hirsutissimum* × *P. helenae*, but imported plants illustrated under the name *P. herrmannii* by Braem, C.O.Baker & M.L.Baker (1999) represent rather hybrids of *P. hirsutissimum* × *P. henryanum* and *P. hirsutissimum* × *P. barbigerum*.

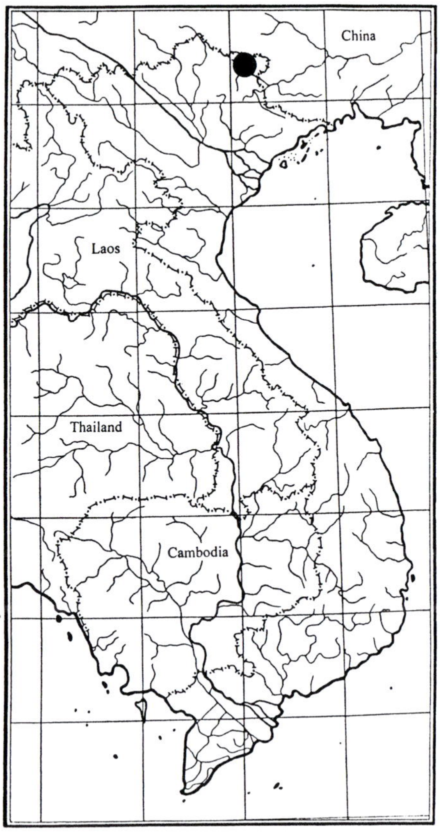

Fig. 210. Distribution of *Paphiopedilum* × *herrmannii*. (Drawing: L. Averyanov).

Fig. 211 (left). *Paphiopedilum* × *herrmannii* cult. Germany. (Photo.: E. Grell).

Fig. 212. Type specimen of *Paphiopedilum × herrmannii.* (Photo.: F. Fuchs).

Some indirect evidence exists for the occurrence of the following natural *Paphiopedilum* hybrids in Vietnam:

Paphiopedilum appletonianum × P. callosum,
P. barbigerum × P. hirsutissimum,
P. henryanum × P. hirsutissimum,
P. hirsutissimum × P. tranlienianum,
P. malipoense × P. micranthum (P. × fanaticum)

Unfortunately, no materials are available that would allow us to confirm the occurrence of these taxa. Further field studies of natural hybridization in the genus is urgently required.

PART IV. CONSERVATION

THE DECLINE AND EXTINCTION OF SLIPPER ORCHIDS IN THE WILD

It is difficult to find any other plants in the indigenous flora of Vietnam that are moving so rapidly to extinction in the wild as are the slipper orchids. All of the *Paphiopedilum* species in Vietnam are disappearing with catastrophic speed. On the basis of recent field studies, their present status in the wild, according to the IUCN categories of threat, is summarized below in Table 10.

TABLE 10

GROUPS OF VIETNAMESE SLIPPER ORCHIDS RANGED ACCORDING TO IUCN CONSERVATION CATEGORIES.

Extinct (EX):	*P. dianthum*	**Vulnerable (VU):**
P. malipoense var. *hiepii*	*P. emersonii*	*P. appletonianum*
	P. gratrixianum	*P. concolor*
Extinct in the wild (EW):	*P. hangianum*	*P. hirsutissimum* var. *chiwuanum*
P. vietnamense	*P. helenae*	
	P. henryanum	*P. hirsutissimum* var. *esquirolei*
Critical (CR):	*P.* × *herrmannii*	
P. delenatii	*P. malipoense* var. *malipoense*	*P. villosum* var. *annamense*
Endangered (EN):	*P. malipoense* var. *jackii*	**DD (data deficient):**
P. × *aspersum*	*P. micranthum*	*P.* × *affine*
P. barbigerum var. *lockianum*	*P. purpuratum*	*P.* × *dalatense*
P. callosum	*P. tranlienianum*	*P. villosum var. boxallii*

As can be seen from Table 10, taxa such as *Paphiopedilum malipoense* var. *hiepii* and *P. vietnamense* are probably already extinct in Vietnam. The situation with *Paphiopedilum delenatii* is estimated to be critical and it may well become extinct in the near future. Most Vietnamese slipper orchids fall into the category of endangered taxa. These include *Paphiopedilum* × *aspersum, P. barbigerum var. lockianum, P. callosum, P. dianthum, P. emersonii, P. gratrixianum, P. hangianum, P. helenae, P. henryanum, P.* × *herrmannii, P. malipoense* var. *malipoense, P. malipoense* var. *jackii, P. micranthum, P. purpuratum* and *P. tranlienianum*. It should be certainly emphasised that most of these taxa approach the critical criteria. Few slipper orchids are still relatively common in Vietnam but all of them are rapidly becoming rarer. This group of vulnerable taxa includes *Paphiopedilum appletonianum, P. concolor, P. hirsutissimum* var. *chiwuanum, P. hirsutissimum* var. *esquirolei* and *P. villosum* var. *annamense.* It is important to recognize that the status in the wild of some slipper orchids cannot yet be evaluated due to the current lack of reliable scientific data. *Paphiopedilum* × *affine, P.* ×

dalatense and *P. villosum* var. *boxallii* fall into this group in Vietnam. It has been noticeably that the picture of slipper orchid threat in Vietnam outlined here has changed dramatically in recent times with the rapid decline of all known populations. The complete extinction in Vietnam of most of the native slipper orchids is quite possible in the very near future.

The transformation of landscapes by man and widespread commercial collecting of slipper orchids are the main factors in their rapid present day decline.

The widespread destruction of natural habitats is, of course, connected intimately to the rapid economic development of the country and rapid rise of human population. Expansion of rural and agricultural areas, deforestation and logging activity, road building, and other kinds of agrarian and industrial land-utilization in the modernisation of the national economy has resulted in an acceleration in the collapse of natural habitats all over the country (Fig. 213). *Paphiopedilum* species, like many other orchids, are an integral element of the primary tropical forests that have been degraded throughout the country during recent years. Slipper orchids, as with most orchids, are sensitive to environmental change. They are a particularly vulnerable group of plants which are amongst the first to vanish during degradation of primary native habitats at the first. Only a few slipper orchid species, especially *Paphiopedilum concolor* and *P. hirsutissimum*, can survive for some time in secondary forest communities. However, the vast majority of species are exterminated completely along with successive destruction of primary forests. The survival of native slipper orchids in Vietnam is, therefore, directly connected with the survival of the remaining areas of primary forest, whose total area in 1982 did not exceed 17% of total territory (MacKinnon & Cox, 1991). With an annual deforestation rate of up to about 3110 sq. km, the remaining area is much smaller now. At present, pristine primary forests cover much less than 1% of the territory in many areas.

Official lumbering activity, widespread uncontrolled logging of the last forest patches for construction and fuel-wood, slash-and-burn shifting agriculture by national minorities, and frequent anthropogenic fires, are the main causes of the rapid loss of primary tropical forests in Vietnam (Fig. 213).

In the course of forest exploitation the largest trees are selectively felled first, mainly for construction timber. This opens up the forest so that greater levels of sunlight reach the ground and dry out the habitats. At this stage, the number of the flowering orchids, including species of *Paphiopedilum*, can often temporarily increase, giving the superficial impression that plants in such conditions reach their optimum growth. However, obvious senescence of orchid populations can be observed through this process. Fewer young plants can be found in these conditions, leading to a reduction in natural seed or vegetative propagation. Progressive drying out of the habitat steadily leads to the disappearance of the slipper orchids. This process is most noticeable on south-facing slopes and cliffs were the extinction of most delicate orchid species, including slipper orchids, is rapid. Complete or considerable logging of the forests results in a catastrophically rapid and irreversible degradation of the primary vegetation, successive rapid mineralization and washing away of fertile soils, and deep habitat drying (especially noticeable on karst limestone mountains). The rapid destruction of all components of primary forests and extinction of the indigenous humid-loving terrestrial and lithophytic herbs is commonly observed in such conditions.

Much faster and profound destruction of pristine forest happens during forest fires. In densely populated areas, extensive areas of montane woodland is occasionally burnt by man. In some cases, forest on isolated limestone mesas is burnt by the local people to flush out game. During the dry season, uncontrolled hardwood fires often develop on a wide front from burnt agricultural fields and logged areas (Fig. 214). Such fires usually completely destroy the plant communities (Fig. 215).

Fig. 213 (above and following pages). Widespread anthropogenic destruction of primary forests is a main factor of extinction of all indigenous species of Vietnamese flora including slipper orchids. (Photos: L. Averyanov).

a — Lumbering area in submontane primary forest of Gia Lai Province, very rich in orchids; **b** — pine forest clearance for mineral excavations in Bidup Mts. foothills, Lam Dong Prov.), locus classicus for many species of endemic Vietnamese orchids; **c** — areas of slash-and-burn shifting agriculture of Bana national minority in rich primary forests of Gia Lai Prov.

Partial regeneration of some elements of the primary plant communities does occasionally occur, when tree felling is stopped in time and habitats are not seriously damaged by fire. The success of this process depends on the degree of destruction of the forest structure and its species' composition. Some orchids survive longer than others as habitats dry out, for example, *Bulbophyllum ambrosia, B. retusiusculum, Coelogyne fimbriata, Coelogyne* spp., *Cymbidium aloifolium, C. lancifolium, Dendrobium lindleyi, Epigeneium amplum, Eria corneri, Flickingeria* spp., *Liparis mannii, L. viridiflora, Luisia* spp., *Oberonia ensiformis,*

Pelatantheria insectifera, Pholidota articulata, P. leveilleana, P. roseans and *P. yunnanensis.* They also appear to be the pioneer orchids during regeneration of the forest community. However, most delicate orchids, including *Paphiopedilum* species, very soon disappear completely and never regenerate (Fig. 215).

The regeneration of orchid populations after a forest fire is always problematic. Very often, a climax community comprising only a few weedy grasses and secondary xerophytic shrubs forms on completely burnt rocky limestone ridges and lasts for many decades. However,

Fig. 214. Uncontrolled forest fires during the dry season may completely destroy the primary vegetation on a limestone ridge in half an hour.

Cao Bang Prov., the richest area for *Paphiopedilum* species in Vietnam. (Photo.: L. Averyanov).

epiphytic orchids can survive in the canopies of occasional trees that survive where forest fires are less intense. Some lithophytic species, growing occasionally as epiphytes, may also survive and these start the regeneration of the lithophytic orchid flora as the forest recovers.

The regeneration of lithophytic orchid communities is usually extremely slow and may take many decades. The first orchids to appear after fire on open rock outcrops in limestone areas of northern Vietnam are *Bulbophyllum reptans, B. retusiusculum, Coelogyne fimbriata, Cymbidium aloifolium, Dendrobium fimbriatum, Epigeneium amplum, Eria coronaria, E. rhombodalis, Liparis mannii, L. stricklandiana, L. viridiflora, Pelatantheria insectifera, Pholidota leveilleana, P. roseans, P. yunnanensis* and *Podochilus microphyllus* (Averyanov *et al.*, 2000a). It is significant, that the rarest obligate lithophytic orchids, for example, most species of *Paphiopedilum*, die out completely during fires and never regenerate.

In recent years, botanical exploration has revealed catastrophic widespread and rapid destruction of the remaining areas of primary forest in Vietnam, mainly in rocky limestone areas (Nguyen Tien Hiep, Averyanov & Phan Ke Loc, 1998; Nguyen Tien Hiep, Phan Ke Loc & Averyanov, 1999; Phan Ke Loc, Nguyen Tien Hiep & Averyanov, 1999a; Phan Ke Loc *et*

Fig. 215. Hardwood forest fires completely destroy primary vegetation, regeneration of which does not occur for a many decades after burning.

Limestone cliffs of *Paphiopedilum* species habitats in Cao Bang Prov. after a forest fire. (Photo.: L. Averyanov).

Fig. 216. Local people collect slipper orchids to meet the demands of the marketplace.

Young hunters from Ede national minority in a forest camp with collected samples of *P. villosum*, Bidup mountains, Lam Dong Prov. (Photo: L. Averyanov).

Fig. 217. Small street markets for wild-collected ornamental plants are very common in many towns of Vietnam. (Photos: L. Averyanov).

a — wild orchid on sale in street market in suburbs of Hanoi; **b** — wild-collected cycads and slipper orchids (lot of *P. hirsutissimum* behind street dealer) are the most popular and best selling plants in such markets.

al., 1999; Averyanov *et al.*, 2000a). Without urgent measures for their effective protection, these extremely rich primary woods will certainly face extinction in the very near future. This will inevitably result in the extinction of large numbers of rare and endemic species unique to these highly specific environments. Like many other indigenous plants most of the native orchids, including *Paphiopedilum* species, are an integral component of the primary plant communities, which are presently on the verge of extinction.

The second very serious factor leading to the extinction of *Paphiopedilum* species in Vietnam in recent years is their widespread collection by local people engaged by orchid dealers. Wide-scale collecting and export of large quantities of wild-collected *Paphiopedilum* species has been observed over the past few years. This plundering stands in sharp contrast to the strong national legal protection of these plants in neighbouring countries. The absence of any real control of collecting of wild plants and the outstanding richness of the Vietnamese endemic *Paphiopedilum* species, is reflected in the large influx of international orchid dealers and companies in recent years. They encourage wide-scale collecting of the plants from the wild all over the country and have formed an extensive network of local dealers and collectors (Figs. 216, 217). This system has proved most effective. Commercial collecting of *Paphiopedilum* species is extensive throughout the country and any rarities or novelties are immediately illegally exported abroad.

At the present day, there is a very high level of demand internationally for wild-collected Vietnamese orchids. It is apparent that some orchid growers in developed countries are willing to pay high prices for wild collected plants, especially for novelties that occasionally appear. Most plants exported from Vietnam move via Hongkong and Taiwan or directly to the profitable plant markets in Japan, the U.S.A. and Europe. The trade in all these plants is illegal, lacking any official permission for export by the Vietnamese CITES agency (Gabel & Lyke, 2001).

Numerous recent publications describing novelties are based on illegally exported slipper orchids from Vietnam identify Germany as a major recipient of illegal plant importations. They present precise information and illustrations of cultivated plants illegally exported from Vietnam. Several popular orchid journals publish articles based on illegally held plants and this certainly encourages plant smuggling and supports the black market demands for wild-collected plants. Very large quantities of rare Vietnamese orchids have recently been imported by a number of countries. The recently described new species, such as *Paphiopedilum hangianum*, *P. hilmari*, *P. mirabile*, *P. tranlienianum*, *P. vietnamense*, *P. coccineum* and others, are all based on such plants. The purchasers of these plants in the developed world are the final links in the chain of individuals and companies who illegally collect and trade in rare wild orchids. The competition for new species between nurseries and growers involved in orchid breeding and the desire of many growers for wild-collected novelties support an illegal trade that still continues. This has led directly to a rapid rise in plant collecting and the extinction of many populations of Vietnamese slipper orchids, even from protected territories. Apart from *Paphiopedilum* species, the greatest demand in the domestic and international markets is for Vietnamese species of genera such as *Aerides, Arachnis, Ascocentrum, Calanthe, Christensonia, Coelogyne, Cymbidium, Dendrobium, Holcoglossum, Hygrochilus, Papilionanthe, Phalaenopsis, Renanthera, Rhynchostylis, Vanda* and *Vandopsis*. All these need protection in their natural habitats throughout the country.

THE FUTURE?

Survival of *Paphiopedilum* species in most areas of Vietnam is problematic. Conservation and protection of wild slipper orchid populations in Vietnam is complicated, and involves a basic understanding of the flora, organization of protected areas, effective monitoring of known populations, effective legal protection and prevention of their commercial collecting. These take a back seat in a country where economic exploration is extensive and rapid, urbanization is increasing, land use is expanding and the human population is expected to rise to 168 million by 2125 from 66.8 millions reported for 1989, with an annual population increase of about 2.4% (MacKinnon & Cox, 1991).

Nevertheless, nature protection is a current concern of the Vietnamese government and is receiving attention. A large network of protected areas has been organized in Vietnam in recent years as part of a broad governmental programme for nature protection (Tab. 11, Fig. 218). This positive process is directed at the conservation of the largest and most important areas of primary forest and other primary plant communities that are home to numerous endemic plant and animal species. By 1997, 55 officially protected areas, including ten national parks and 45 nature reserves, were recognized in the country. About 40 areas have been designated as regions of very high plant diversity which need urgent official protection but do not yet have such status (Tab. 11, Fig. 218). Unfortunately, the current excellent governmental programme for protection of primary forests cannot effectively prevent the extinction of slipper orchids and other plants that are in high demand in the market-place because of their widespread selective collecting for sale by local people. Another problem for plant protection is that the current state of knowledge about the floras of protected areas and areas of high diversity provides no information that allows the timely detection and monitoring of populations of rare species. Little is known about the flora of about 84 % of the protected regions and 41% of these areas have not been studied at all. More or less complete reliable floristic data are now available for only 5.3% of the protected areas (Tab. 11, Fig. 218). It should be emphasized that Vietnam is distinguished now as a country with the world's fastest rates of deforestation (MacKinnon & Cox, 1991; Rundel, 1999). As a result pristine habitats that still support populations of many strictly endemic plant species in unprotected areas may be destroyed in coming years. This is a particular concern for a number of locally endemic *Paphiopedilum* species that occur in the limestone areas of northern Vietnam.

TABLE 11.

REGIONS OF VIETNAM WITH HIGHEST PLANT DIVERSITY AND DEGREE OF THEIR BOTANICAL EXPLORATION.

NN	Region name and province where it situated	Level of botanical exploration and degree of specimen collecting in areas
	National Parks:	
1	Ba Be (Bac Can prov.)	2
2	Tam Dao (Vinh Phuc, Tuyen Quang and Thai Nguyen prov.)	3
3	Ba Vi (Ha Tay prov.)	4
4	Cat Ba (Hai Phong city)	2
5	Cuc Phuong (Ninh Binh, Hoa Binh and Thanh Hoa prov.)	4
6	Ben En (Thanh Hoa prov.)	2
7	Bach Ma (Thua Thien-Hue prov.)	3
8	Yok Don (Dac Lac prov.)	3
9	Cat Thien (Dong Nai prov.)	3
10	Con Dao (Ba Ria — Vung Tau prov.)	3
	Nature Reserves:	
11	Muong Nhe (Lai Chau prov.)	1
12	Hoang Lien Son (Lao Cai and Lai Chau prov.)	3
13	Sop Cop (Son La prov.)	1
14	Xuan Nha (Son La prov.)	1
15	Nam Don (Son La prov.)	1
16	Phong Quang (Ha Giang prov.)	1
17	Na Hang (Tuyen Quang prov.)	2
18	Trung Khanh (Cao Bang prov.)	2
19	Pia Oac (Cao Bang prov.)	2
20	Huu Lien (Lang Son prov.)	1
21	Ba Mun (Quang Ninh prov.)	2
22	Yen Tu (Quang Ninh prov.)	2
23	Xuan Son (Phu Tho prov.)	1
24	Thuong Tien (Hoa Binh prov.)	1

25	Pa Co — Hang Kia (Hoa Binh prov.)	2
26	Tam Quy (Thanh Hoa prov.)	2
27	Pu Huong (Nghe An prov.)	2
28	Pu Mat (Nghe An prov.)	2
29	Vu Quang (Ha Tinh prov.)	1
30	Phong Nha (Quang Binh prov.)	2
31	Ban Dao Son Tra (Da Nang city)	2
32	Ba Na — Nui Chua (Quang Nam prov.)	3
33	Deo Ca — Hon Ron (Phu Yen prov.)	2
34	Krong Trai (Phu Yen prov.)	2
35	Ngok Linh (Kon Tum prov.)	2
36	Mom Ray (Kon Tum prov.)	1
37	Kon Cha Rang & Kon Ka Kinh (Gia Lai prov.)	2
38	Chu Yang Sinh (Dac Lac prov.)	2
39	Nam Nung (Dac Lac prov.)	1
40	Ea Sup (Dac Lac prov.)	3
41	Ho Lac (Dac Lac prov.)	2
42	Nui Ba — Bi Dup (Lam Dong prov.)	2
43	Thuong Da Nhim (Lam Dong prov.)	2
44	Deo Ngoan Muc (Lam Dong prov.)	2
45	Nui Dai Binh (Lam Dong prov.)	1
46	Rung Kho Phan Rang (Ninh Thuan prov.)	3
47	Ta Kou (Ninh Thuan prov.)	2
48	Bien Lac — Nui Ong (Binh Thuan prov.)	2
49	Bu Gia Map (Binh Phuoc prov.)	2
50	Lo Go — Sa Mat (Tay Ninh prov.)	2
51	Binh Chau — Phuoc Buu (Ba Ria-Vung Tau prov.)	3
52	Tam Nong (Dong Thap prov.)	4
53	Rung Tram U Minh (Kien Giang prov.)	4
54	Rung Duoc Nam Can (Ca Mau prov.)	4
55	Phu Quoc (Kien Giang prov.)	3
	Unprotected areas:	
56	Phu Si Lung — 3076 m (Lai Chau prov.)	1
57	Phu Ta Tong — 2109 m (Lai Chau prov.)	1
58	Phu Tra — 2504 m (Lai Chau prov.)	1
59	Phu Huoi Long — 2178 m (Lai Chau prov.)	1
60	Bach Moc-Luong Tu — 2098 m (Lao Cai prov.)	1
61	Pu Sam Sao — 1897 m (Son La prov.)	1
62	Phu Sung Mon — 2445 m (Son La prov.)	1
63	Pha Luong — 1880 m (Son La prov.)	1
64	Lang Cung — 2913 m (Yen Bai prov.)	1
65	Phu Luong — 2985 m (Yen Bai prov.)	1
66	Tay Con Linh — 2419 m (Ha Giang prov.)	1
67	Phu Tha Ca — 2274 m (Ha Giang prov.)	1
68	Pho Bang — 1500 m (Ha Giang prov.)	2
69	Tra Linh — 1000 m (Cao Bang prov.)	2
70	Pia La — 1980 m (Cao Bang prov.)	1
71	Mau Son — 1541 m (Lang Son prov.)	2
72	Sam Cuoi Tien — 1508 m (Quang Ninh prov.)	2
73	Phu Pha Phong — 1587 m (Hoa Binh prov.)	1
74	Bu Rinh — 1183 m (Thanh Hoa prov.)	1
75	Bu Cho — 1563 m (Nghe An prov.)	1

76	Pu Hoat — 2452 m (Nghe An prov.)	1
77	Phu Xai La Leng — 2711 m (Nghe An prov.)	1
78	Tay Bac Deo Ngang (Ha Tinh prov.)	1
79	Ba Ren (Quang Binh prov.)	1
80	Dong Voi Mep — 1701 m (Quang Tri prov.)	2
81	Dong Ngai — 1774 m (Thua Thien-Hue prov.)	2
82	Nui Mang — 1708 m (Thua Thien-Hue prov.)	1
83	Song Thanh — Dac Pring (Quang Nam prov.)	1
84	Ngok Go Le Rang — 1855 m (Quang Nam prov.)	1
85	Kon Plong (Kon Tum prov.)	2
86	Chu To Ryan — 1331 m (Gia Lai prov.)	1
87	Bu Co Rac — 867 m (Dac Lac prov.)	1
88	Dac Nong (Dac Lac prov.)	2
89	Bac Cong Troi — 2004 m (Lam Dong prov.)	2
90	So Ro Lung — 1545 m (Lam Dong prov.)	2
91	Chu Ho Mu — 2051 m (Khanh Hoa prov.)	3
92	Hon Ba — 1348 m (Khanh Hoa prov.)	3
93	Khanh Vinh (Khanh Hoa prov.)	2
94	Nui Chua — 1040 m (Ninh Thuan prov.)	1
95	Nui Be (Binh Thuan prov.)	1

Explanation for the Table. Degrees of regional collecting, samples of which are presented in scientific herbaria and available for reliable floristic inventory work:

1. No specimens collected.
2. Less than a third of the species occurring in the area have been collected.
3. Between a third and two-thirds of the species occurring in the area have been collected.
4. More than two-thirds of the species of the local flora have been collected.

TABLE 12.

OCCURRENCE OF *PAPHIOPEDILUM* SPECIES IN VIETNAM IN PROTECTED AREAS AND AREAS PROPOSED FOR PROTECTION.

***Paphiopedilum* species name**	**Occurrence in designated areas**
P. × *affine*	None
P. appletonianum	**Nature reserves**: 35. Ngok Linh (Kon Tum prov.), 38. Chu Yang Sinh (Dac Lac prov.), 42. Nui Ba — Bi Dup (Lam Dong prov.). **Not yet protected areas:** 91. Chu Ho Mu — 2051 m (Khanh Hoa prov.)?
P. × *aspersum*	None
P. barbigerum var. *lockianum*	None
P. callosum	**Nature reserves:** 36. Mom Ray (Kon Tum prov.), 43. Thuong Da Nhim (Lam Dong prov.)?

P. concolor	**National parks:** 1. Ba Be (Bac Can prov.), 4. Cat Ba (Hai Phong city). **Nature reserves:** 20. Huu Lien (Lang Son prov.), 25. Pa Co — Hang Kia (Hoa Binh prov.)?
P. × dalatensis	**Nature reserves:** 43. Thuong Da Nhim (Lam Dong prov.)?
P. delenatii	**Not yet protected areas:** 91. Chu Ho Mu — 2051 m (Khanh Hoa prov.)?, 92. Hon Ba — 1348 m (Khanh Hoa prov.)?
P. dianthum	**Nature reserves:** 12. Hoang Lien Son (Lao Cai and Lai Chau prov.)?, 16. Phong Quang (Ha Giang prov.)?, 25. Pa Co — Hang Kia (Hoa Binh prov.). **Not yet protected areas:** 63. Pha Luong — 1880 m (Son La prov.), 68. Pho Bang — 1500 m (Ha Giang prov.)?
P. emersonii	**Nature reserves:** 17. Na Hang (Tuyen Quang prov.)?
P. gratrixianum	**National parks:** 2. Tam Dao (Vinh Phuc, Tuyen Quang and Thai Nguyen prov.). **Nature reserves:** 12. Hoang Lien Son (Lao Cai and Lai Chau prov.)?
P. hangianum	**Nature reserves:** 17. Na Hang (Tuyen Quang prov.).
P. helenae	**Nature reserves:** 18. Trung Khanh (Cao Bang prov.)? **Not yet protected areas:** 69. Tra Linh — 1000 m (Cao Bang prov.).
P. henryanum	**Nature reserves:** 16. Phong Quang (Ha Giang prov.)?. **Not yet protected areas:** 68. Pho Bang — 1500 m (Ha Giang prov.)?
P. × herrmannii	**Not yet protected areas:** 69. Tra Linh — 1000 m (Cao Bang prov.).
P. hirsutissimum var. *chiwuanum*	**Nature reserves:** 14. Xuan Nha (Son La prov.)?, 25. Pa Co — Hang Kia (Hoa Binh prov.). **Not yet protected areas:** 63. Pha Luong — 1880 m (Son La prov.)?
Paphiopedilum hirsutissimum var. *esquirolei*	**National parks:** 5. Cuc Phuong (Ninh Binh, Hoa Binh and Thanh Hoa prov.). **Nature reserves:** 14. Xuan Nha (Son La prov.)?,

	18. Trung Khanh (Cao Bang prov.)?, 25. Pa Co — Hang Kia (Hoa Binh prov.). **Not yet protected areas:** 63. Pha Luong — 1880 m (Son La prov.)?, 68. Pho Bang — 1500 m (Ha Giang prov.), 69. Tra Linh — 1000 m (Cao Bang prov.), 73. Phu Pha Phong — 1587 m (Hoa Binh prov.)?
P. malipoense var. *jackii*	**Nature reserves:** 17. Na Hang (Tuyen Quang prov.).
P. malipoense var. *hiepii*	**Nature reserves:** 17. Na Hang (Tuyen Quang prov.)?
P. malipoense var. *malipoense*	**National parks:** 1. Ba Be (Bac Can prov.)? **Nature reserves:** 17. Na Hang (Tuyen Quang prov.), 20. Huu Lien (Lang Son prov.), 25. Pa Co — Hang Kia (Hoa Binh prov.), 30. Phong Nha (Quang Binh prov.)? **Not yet protected areas:** 63. Pha Luong — 1880 m (Son La prov.), 67. Phu Tha Ca — 2274 m (Ha Giang prov.)?, 68. Pho Bang — 1500 m (Ha Giang prov.)?, 70. Pia La — 1980 m (Cao Bang prov.)?, 73. Phu Pha Phong — 1587 m (Hoa Binh prov.).
P. micranthum	**Nature reserves:** 16. Phong Quang (Ha Giang prov.), 17. Na Hang (Tuyen Quang prov.). **Not yet protected areas:** 67. Phu Tha Ca — 2274 m (Ha Giang prov.)?, 68. Pho Bang — 1500 m (Ha Giang prov.), 69. Tra Linh — 1000 m (Cao Bang prov.), 70. Pia La — 1980 m (Cao Bang prov.)?
P. purpuratum	**Nature reserves:** 12. Hoang Lien Son (Lao Cai and Lai Chau prov.)?
P. tranlienianum	**Nature reserves:** 17. Na Hang (Tuyen Quang prov.)?
P. vietnamense	No one
P. villosum var. *annamense*	**Nature reserves:** 36. Mom Ray (Kon Tum prov.), 38. Chu Yang Sinh (Dac Lac prov.), 42. Nui Ba — Bi Dup (Lam Dong prov.). **Not yet protected areas:** 91. Chu Ho Mu — 2051 m (Khanh Hoa prov.)?
P. villosum var. *boxallii*	**Nature reserves:** 12. Hoang Lien Son (Lao Cai and Lai Chau prov.)?

? = Area where species may be found with high probability but not yet reported

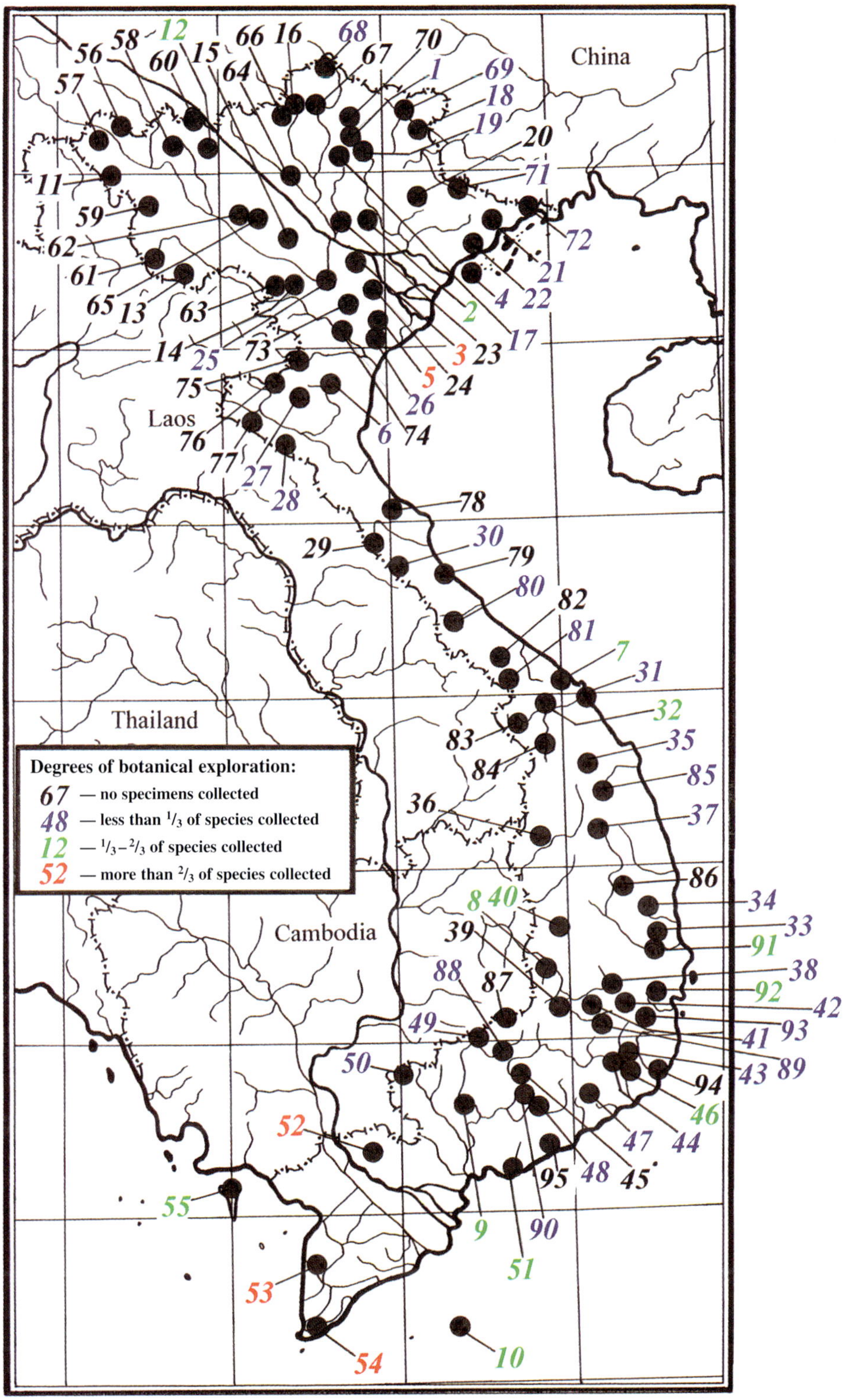
China
Laos
Thailand
Cambodia
Degrees of botanical exploration:
67 — no specimens collected
48 — less than 1/3 of species collected
12 — 1/3–2/3 of species collected
52 — more than 2/3 of species collected

Few *Paphiopedilum* species in Vietnam have been reported from officially protected areas or areas proposed for protection (Tab. 12). Most of their known populations have been found in places that have no official conservation status. All these places need urgent effective protection to save slipper orchid habitats. Without such action, the survival of the endemic Vietnamese *Paphiopedilum* species is unlikely.

The following measures may help them survive in nature:

- An urgent accurate taxonomic inventory of all protected areas to discover if *Paphiopedilum* populations exist there for subsequent monitoring and effective protection.
- Discovery of the remaining *Paphiopedilum* populations in unprotected regions of the country, giving those areas effective protection and preventing slipper orchid species collection.
- Ratification by the government of laws to protect all native Vietnamese *Paphiopedilum* species. All *Paphiopedilum* species should be included in the forthcoming revised edition of the Red Data Book of Vietnam as officially protected plants.
- Widespread public education on the goals of conservation and the need to protect rare and endemic plant species as part of the nation's heritage. Broadcasting information about the official protection and prohibition of the free sale and uncontrolled export of indigenous Vietnamese slipper orchids will assist this. Such information should also seriously diminish the scale of plant collecting by local collectors.
- Effective prohibition of uncontrolled commercial exploitation of wild slipper orchids by private companies and individuals that leads directly to the rapid selective loss of the natural resources.
- Development in the country of a clear understanding of the significance of effective nature protection and conservation for the development of ecotourism, education, horticulture, breeding, selection, biodiversity and other aspects of the national interest.
- Comprehensive and fundamental investigations of the distribution, biology, ecology and natural requirements of native slipper orchids as a necessary scientific base for their effective conservation.
- Legal collection of limited numbers of plants of native slipper orchids from the wild to act as a germ-plasm source for the legal propagation of seedlings for trade, reintroduction and restoration of natural populations.
- Inspection and prevention of illegal importations of wild-collected Vietnamese *Paphiopedilums* by CITES agencies in importing countries. Repatriation of confiscated plants.
- Official explanation by IUCN and the CITES secretariat to editorial boards of popular orchid journals about the negative effects for nature protection of any publication or advertisement of illegally imported plants.

Fig. 218 (opposite). Protected areas of Vietnam and areas of high conservation priority which have not yet been given this status. Names of areas are presented in Tab. 17.

Degrees of botanical exploration are marked in different colours (black — no specimens collected; blue — less than a third of the species collected; green — a third to two-thirds of the species collected; red — more than two-thirds of the species collected and represented in scientific collections).

ACKNOWLEDGEMENTS

Much of the information presented in this book was obtained during a number of expeditions supported by the following organisations:

- **U.S.A. National Geographic Society** (grants — "Flora of Highlands of South Vietnam", # 5094-93, 5803-96, 6383-98; "Limestone Flora of Cao Bang Province of northern Vietnam", # 6300-98; "Botanical Inventory of Unexplored Areas in northern Vietnam", # 6733-00);
- **U.S.A. National Science Foundation** (grant "Collaborative Research: A Multi-Taxa Inventory of Threatened Conservation areas in Vietnam", # DEB-9870231);
- **American Orchid Society** (grants "Exploration of endangered Vietnamese *Paphiopedilums*"; "Exploration of endangered calcium dependent orchid flora in inaccessible rocky limestone areas of northern Vietnam");
- **Fauna & Flora International** (grant "The distribution of *Paphiopedilum vietnamense* and its current status in the wild");
- **San Diego County Orchid Society Conservation Committee** (grant "Exploration of endangered calcium dependent *Paphiopedilum* species in northern Vietnam").
- **TOBU Department Store, Tokyo** (grant "Exploration of endangered *Paphiopedilum* species in northern Vietnam" 1999 expedition to Ha Giang Province).

We would like to cordially thank the Director and authorities of the Institute of Ecology and Biological Resources of the National Center for Natural Sciences and Technology of Vietnam for their kind help in organizing and in gaining permits for our botanical explorations and for allowing us to study the excellent herbarium collections deposited at the Institute. L.Averyanov also wishes to thank particularly the authorities of the Institute of Biology in Dalat and all Institute staff members for their kind hospitality, endless friendship and assistance in studies of the Institute's living orchid collection during his many visits. We are especially grateful to Mr. N.V.Duy, staff member of the Institute, and manager of its orchid collection for collecting alcohol material on interesting orchid species.

We would particularly like to thank Carol Woodin for the superb paintings for the cover and frontispiece and we would like to thank the US CITES authority for allowing her to paint confiscated plants for this project. We express our great thanks to E.Grell, F.Fuchs, Mr. E.Mokhov, H.Okuyama and Son Ha for kind permission to use their excellent photographs for illustration of this book; and Mark Chase for the kind use of Figure 29.

The completion of this book was made possible with the assistance of a great number of colleagues and friends including local authorities, officers of Forest Protection Department of Ministry of Agriculture and Rural Development of Vietnam, local forest explorers, hunters, woodmen, guides, porters and many other people that helped us during our work in remote forest areas and to whom we would like to express our sincere thanks.

We are grateful to our close friends and colleagues D.Harder, N.X.Tam, N.Q.Hieu, D.T.Doan, P.H.Hoang, N.Q.Binh, N.T.Vinh, N.T.Ban, and D.D.Huyen, who have participated in various of our numerous and difficult botanical expeditions all over the country.

We are also would like to thank Eric Christenson, Helen Congleton, Lois Duffin, Olaf Gruss, Norito Hasegawa, Harold Koopowitz, Holger Perner, Paul Phillips, Norris Powell, Chuck Spain and Pat Worthington for discussing many aspects of *Paphiopedilum* and for help in *Paphiopedilum* investigations; A. Farjon and P. Boyce who have read and commented on the manuscript; and Suzy Dickerson, John Harris and Chris Beard for preparing the book for publication.

We also thank the authorities of Komarov Botanical Institute of the Russian Academy of Science and the Royal Botanic Gardens, Kew, for allowing us to use their facilities and letting us complete this monograph during official hours.

LITERATURE CITED

Atwood, J.Y. 1984. The relationships of the slipper orchids (Subfamily Cypripedioideae). *Selbyana* 7: 129–247.

Aubreville, A., Leroy, J.F & Morat, Ph. (eds.). 1960–1997. *Flore du Cambodge, du Laos et du Viet-Nam.* Fasc. 1–29.

Averyanov, L.V. 1988a. New species and nomenclatural changes in the family Orchidaceae of the Vietnamese flora. *Bot. Journ.* (Leningrad) 73, 1: 100–107 (in Russian).

Averyanov, L.V. 1988b. New taxa and nomenclatural changes in the family Orchidaceae of the Vietnamese flora. *Bot. Journ.* (Leningrad) 73, 3: 423–432 (in Russian).

Averyanov, L.V. 1988c. New and rare species of orchids (Orchidaceae) in the Vietnamese flora. *Bot. Journ.* (Leningrad) 73, 5: 720–729 (in Russian).

Averyanov, L.V. 1988d. New and rare species of the family Orchidaceae in the Vietnamese flora. *Bot. Journ.* (Leningrad) 73, 6: 892–898 (in Russian).

Averyanov, L.V. 1988e. Three new species of the orchids (Orchidaceae) from Vietnam. *Bot. Journ.* (Leningrad) 73, 7: 1022–1025 (in Russian).

Averyanov, L.V. 1988f. Preliminary list of the Vietnamese orchids. *Leningrad, Bot. Inst. Ac. Sci. USSR* 1: 1–247; 2: 1–152 (in Russian).

Averyanov, L.V. 1989. New and rare species of the family Orchidaceae in the Vietnamese flora. *Bot. Journ.* (Leningrad) 74, 9: 1347–1355 (in Russian).

Averyanov, L.V. 1990a. Systematics of the Vietnamese orchids (Orchidaceae). Subfamilies Apostasioideae, Cypripedioideae, Neottioideae and Orchidoideae. *Bot. Journ.* (Leningrad) 75, 7: 1013–1028 (in Russian).

Averyanov, L.V. 1990b. Systematics of the Vietnamese orchids (Orchidaceae). Subfamily Epidendroideae, tribes Vanilleae, Gastrodieae, Epipogieae, Arethuseae and Coelogyneae. *Bot. Journ.* (Leningrad) 75, 12: 1757–1767 (in Russian).

Averyanov, L.V. 1990c. *Introduction. Vascular plants synopsis of Vietnamese flora* 1. Leningrad. Nauka, Leningrad branch. Pp. 5–19.

Averyanov, L.V. 1990d. Fam. Orchidaceae Juss. *Vascular plants synopsis of Vietnamese flora* 1. Leningrad. Nauka, Leningrad branch. Pp. 20–167.

Averyanov, L.V. 1991a. Systematics of the Vietnamese orchids (Orchidaceae). Subfamily Epidendroideae, tribes Malaxideae and Epidendroideae. *Bot. Journ.* (Leningrad) 76, 1: 120–128 (in Russian).

Averyanov, L.V. 1991b. Systematics of the Vietnamese orchids (Orchidaceae). Subfamily Vandoideae. *Bot. Journ.* (Leningrad). 76, 6: 880–895 (in Russian).

Averyanov, L.V. 1991c. The general lines of morphological evolution in the orchid family (Orchidaceae). *Bot. Journ.* (Leningrad) 76, 7: 921–935 (in Russian).

Averyanov, L.V. 1991d. The origin and features of evolution, biology and ecology of the orchids (Orchidaceae). *Bot. Journ.* (Leningrad) 76, 10: 1345–1359 (in Russian).

Averyanov L.V. 1991e. Orchids of Vietnam. Diss. Thes. for Dr. Sci. Degree in Biol. (03.00.05). *Leningrad. Komarov Bot. Inst. USSR Academy of Sciences.* 61 pp. (in Russian).

Averyanov, L.V. 1994. *Identification guide to Vietnamese orchids* (Orchidaceae Juss.). St. Petersburg. World & Family. 432 pp. (in Russian).

Averyanov, L.V. 1996a. *Paphiopedilum helenae* (Orchidaceae) — new slipper orchid from North Vietnam. *Bot. Journ.* (St.-Petersburg) 81, 9: 105–110.

Averyanov L.V. 1996b. New species of orchids (Orchidaceae) from Vietnam. *Bot. Journ.* (St.-Petersburg) 81, 10: 73–83.

Averyanov, L.V. 1996c. *Acanthephippium simplex* Aver., *Acanthephippium odoratum* Aver. - zwei neue Arten der Gattung *Acanthephippium* Blume aus Vietnam. *Die Orchidee* 47, 4: 192–193.

Averyanov, L.V. 1997a. New species of orchids (Orchidaceae) from Vietnam. *Bot. Journ.* (St.-Petersburg) 82, 3: 131–148.
Averyanov L.V. 1997b. *Paphiopedilum hiepii* Aver. *New Orchids* 86: 58–60 (in Japanese).
Averyanov, L.V. 1997c. *Renanthera citrina* Aver., new yellow flowering species of the genus from limestone areas of North Vietnam. *Orchids. Mag. Amer. Orch. Soc.* 66, 12: 1284–1290.
Averyanov, L.V. 1998a *Paphiopedilum hiepii:* a new species from Vietnam. *Orchids. Mag. Amer. Orch. Soc.* 67, 3: 260–263.
Averyanov, L.V. 1998b. Familiae Orchidaceae species novae et rarae in flora Vietnamica. *Syst. Pl. Vasc.* (St.-Petersburg) 31: 39–67 (in Russian).
Averyanov, L.V. 1999a. Nomenclatural changes and new orchids (Orchidaceae) in the flora of Vietnam. *Bot. Journ.* (St.-Petersburg) 84, 10: 126–130 (in Russian).
Averyanov, L.V. 1999b. Florogenetic analysis of Orchidaceae in the flora of Vietnam. XVI *International Botanical Congress, Abstracts*, St.Louis: 418.
Averyanov, L.V. 1999c. New species of orchids from Vietnam. *Lindleyana* 14, 4: 214–224.
Averyanov, L.V. 2000a. Rare species of Orchidaceae in the flora of Vietnam. 1. The genera *Acanthephippium-Didymoplexiopsis. Bot. Journ.* (St.-Petersburg) 85, 3: 128–138 (in Russian).
Averyanov, L.V. 2000b. Rare species of Orchidaceae in the flora of Vietnam. 2. The genera *Eria-Zeuxine. Bot. Journ.* (St.-Petersburg) 85 4: 137–147 (in Russian).
Averyanov, L.V. 2000c. New species of Orchidaceae from northern Vietnam. *Lindleyana.* 15, 2: 73–80.
Averyanov, L.V. 2001. New natural interspecific hybrid — *Paphiopedilum dalatense* from Vietnam. *Orchid Digest* 65, 3: 133–134.
Averyanov, L.V. & Averyanova, A. 2000. Rare species of orchids (Orchidaceae) in the flora of Vietnam. *Turczaninowia* 3, 2: 5–86 (in Russian).
Averyanov, L.V. & Christenson, E.A. 1998a. Orchids of Vietnam. *Papilionanthe pedunculata.* A rare *Vanda* relative under cultivation at an orchid garden in Dalat. *Orchids. Mag. Amer. Orch. Soc.* 67, 2: 148–149.
Averyanov, L.V. & Christenson, E.A. 1998b. A new *Vanda* species from Indochina. *Orch. Dig.* Jan.-Feb.-Mar.: 27–29.
Averyanov, L.V. & Christenson, E.A. 1998c. Orchids of Vietnam. *Phaius tankervilleae. Orchids. Mag. Amer. Orch. Soc.* 67, 4: 368–370.
Averyanov, L.V. & Christenson, E.A. 1998d. Orchids of Vietnam. *Cymbidium schroederi.* Expanding the diversity of a classic genus of orchids. *Orchids. Mag. Amer. Orch. Soc.* 67, 7: 712–713.
Averyanov, L.V. & Christenson, E.A. 1998e. Orchids of Vietnam. *Dendrobium christyanum.* The requirements of a miniature cool-growing *Dendrobium. Orchids. Mag. Amer. Orch. Soc.* 67, 11: 1148–1149.
Averyanov, L.V. & Duong Duc Huyen. 1989. Rare species of the genus *Dendrobium* (Orchidaceae) in the Vietnamese flora. *Bot. Journ.* (Leningrad) 74, 11: 1667–1668 (in Russian).
Averyanov, L.V. & Duong Duc Huyen. 1990. New and rare species of the Orchidaceae family in the Vietnamese flora. *Bot. Journ.* (Leningrad) 75, 5: 721–724 (in Russian).
Averyanov, L.V. & Dzuong Duc Huyen. 1993. New and rare species of the orchids (Orchidaceae) in the Vietnamese flora. *Bot. Journ.* (St.-Petersburg) 78, 7: 81–85.
Averyanov, L.V., Nguyen Tien Ban & Kudriavtzeva, E.P. 1988. Contributions to the Vietnamese islands flora and vegetation. *Leningrad. Bot. Inst. Ac. Sci. USSR.* 190 pp. (in Russian).
Averyanov, L.V., Nguyen Tien Hiep, Phan Ke Loc & Averyanova, A.L. 2000a. Preliminary orchid checklist of Cao Bang Province (northern Vietnam). *Lindleyana* 15, 3: 130–164.
Averyanov, L.V., Nguyen Tien Hiep, Phan Ke Loc & Averyanova, A.L. 2000b. Preliminary orchid checklist of Cao Bang Province. Problems of basic research in biology. *Proc. National Conf. Biol.* Aug. 2000 (Hanoi): 177–180 (in Vietnamese).
Averyanov, L.V., Nguyen Tien Hiep, Phan Ke Loc & Dzuong Duc Huyen. 1996a. Endangered Vietnamese *Paphiopedilums*. Part 1. *Paphiopedilum helenae. Orchids. Mag. Amer. Orch. Soc* 65, 10: 1062–1068.

Averyanov, L.V., Nguyen Tien Hiep, Phan Ke Loc & Dzuong Duc Huyen. 1996b. Endangered Vietnamese *Paphiopedilums*. Part 2. *Paphiopedilum delenatii. Orchids. Mag. Amer. Orch. Soc* 65, 12: 1302–1308.

Averyanov, L.V., Nguyen Tien Hiep, Phan Ke Loc & Dzuong Duc Huyen. 1997a. Endangered Vietnamese *Paphiopedilums*. Part 3. *Paphiopedilum malipoense. Orchids. Mag. Amer. Orch. Soc.* 66, 2: 150–155.

Averyanov, L.V., Nguyen Tien Hiep, Phan Ke Loc & Dzuong Duc Huyen. 1997b. Endangered Vietnamese Paphiopedilums. Part 4. *Paphiopedilum hirsutissimum* var. *esquirolei* and var. *chiwuanum. Orchids. Mag. Amer. Orch. Soc.* 66, 4: 378–383.

Averyanov, L.V. & Vu Ngok Long. 1989. Rare species of the genus *Eria* (Orchidaceae) in the Vietnamese flora. *Bot. Journ.* (Leningrad) 74, 10: 1518 (in Russian).

Averyanov, L.V. & Vu Ngok Long. 1990. The genus *Eria* (Orchidaceae) in Vietnam. Sections *Bambusifoliae, Mycaranthes, Trichosma, Polyura, Dendrolirium, Strongyleria, Pinalia* and *Urostachya. Bot. Journ.* (Leningrad) 75, 11: 1583–1593 (in Russian).

Averyanov, L.V. & Vu Ngoc Long. 1991. The genus *Eria* (Orchidaceae) in Vietnam. Sections *Hymenaria, Secundae, Cylindrolobus* and *Conchidium. Bot. Journ.* (Leningrad) 76, 3: 442–449 (in Russian).

Averyanov, L.V. et al. (eds.). 1990–1996. *Vascular plant synopsis of Vietnamese flora* 1–2.

Baenziger, H. 1996. The mesmerizing wart: the pollination strategy of epiphytic lady slipper orchid *Paphiopedilum villosum* (Lindl.) Stein (Orchidaceae). *Journ. Linn. Soc. Bot.* (London) 121: 59–90.

Birk, L.A. 1983. *The Paphiopedilum Grower's Manual.* Pisang Press, Santa Barbara (California, U.S.A.).

Braem, G.J. 1987. *Paphiopedilum henryanum*, a new species from Indo-China. *Schlechteriana* 1, 1: 3–6.

Braem, G.J. 1988a. *Paphiopedilum*. Hildesheim. Brucke-Verlag, Kurt Schmersow. 249 pp.

Braem, G.J. 1988b. *Paphiopedilum henryanum*. A new species from China. *Orchid Digest* 52, 1: 11–13.

Braem, G.J. 1991a. Some comments regarding the Rules of Botanical Nomenclature discussed in respect to two case studies. *Schlechteriana* 2: 9–15.

Braem, G.J. 1991b. *Paphiopedilum henryanum* var. *christae. Schlechteriana* 4. S. 157–162.

Braem, G.J. 1999. A new taxon for the genus *Paphiopedilum* Pfitzer, *Paphiopedilum cerveranum* Braem. *Orch. Cult. Protect.* 38: 28–30.

Braem, G.J., Baker, C.O. & Baker, M.L. 1998. *The genus Paphiopedilum. Natural History and Cultivation* 1. Florida. Bot. Publ. Inc. 182 pp.

Braem, G.J., Baker, C.O. & Baker, M.L. 1999. *The genus Paphiopedilum. Natural History and Cultivation* 2. Florida. Bot. Publ. Inc: 185–363.

Brieger, F.G. 1973. 3. *Paphiopedilum*. In Schlechter, R., *Die Orchideen*. Ed. 3. Berlin. Paul Parey: 171–185.

Bruzon, E. & Carton, P. 1930. *Le climat de l'Indochine et les typhoons de la mer de Chine.* Hanoi. Imprimerie D'extreme-Orient. 310 pp.

Cavestro, W. & Chiron, G. 1999. Un nouveau *Paphiopedilum* du Vietnam, *P. mirabile* W.Cavestro & G.Chiron. *Orchidées. Culture et protection* 38, 2: 31–34.

Chen, S.C., Liu F.Y. 1982. Notes on some species of *Paphiopedilum* from Yunnan. *Acta Bot. Yunnanica* 4: 163–167.

Chen, S.C., Liu, Z.J. & Zhang, J.Y. 2001. A new species and a new variety of *Paphiopedilum* (Orchidaceae). *Acta Phytotax. Sinica* 39 (5): 455–458.

Chevey, P. & Carton, P. 1935. *Les courants de la Mer de Chine méridionale et leurs rapports avec le climat de l'Indochine.* Institut oceanographique de l'Indochine 26. Hanoi.

Cox, A.V., Pridgeon, A.M., Albert, V.A. & Chase, M.V. 1997. Phylogeny of the slipper orchids (Cypripedioideae: Orchidaceae): nuclear r DNA ITS sequences. *Pl. Syst. Evol.* 208: 197–223.

Cribb, P.J. 1987. *The Genus Paphiopedilum*. London. Collingridge. 222 pp.

Cribb, P.J. 1998. *The Genus Paphiopedilum.* R.B.G. Kew & Natural History Publ. (Borneo), Kota Kinabalu. 427 pp.

Cribb, P.J. 1999. Wild Paphiopedilums. *Orchids. Mag. Amer. Orchid. Soc.* 68, 4: 340–349.

Cribb, P., Averyanov, L. & Nguyen Tien Hiep. 2000. *Paphiopedilum henryanum* in North Vietnam. *Orchid Review* 108, 1232: 124–127.

Crocker, D. 1962. *Soils of Cambodia, exploratory survey.* Direction d'Agriculture. Phnom Penh. 83 pp.

Dalrymple, W. 1990. Raiders of the Lost Orchids. *Orchid Review* 98, 1156: 43–48.

De Wildeman, E. 1906. Plantes nouvelles ou interessantes. *La Tribune Hortic.* (1906) 8: 57.

Dovzikov, A.E., et al. 1965a. In Dovzikov, A.E. (ed.). *Geological map of Vietnam* 1:500 000. Hanoi. Main Geological Department of DRV.

Dovzikov, A.E., et al. 1965b. In Dovzikov, A.E. (ed.). *Geology of the North Vietnam. Description for geological map of the North Vietnam* 1:500 000. Hanoi. Main Geological Department of DRV. 668 pp. (in Russian).

Dressler, R.L. 1981. *The Orchids: Natural History and Classification.* Cambridge, Massachusets. Harvard University Press. 332 pp.

Dressler, R.L. 1993. *Phylogeny and Classification of the Orchid Family.* Portland, Oregon. Dioscorides Press. 314 pp.

Duong Duc Huyen & Averyanov, L.V. 1989. New variety of *Dendrobium nobile* var. *alboluteum* (Orchidaceae) from the territory of South Vietnam. *Bot. Journ.* (Leningrad) 74, 7: 1039–1040 (in Russian).

Duong Duc Huyen & Averyanov, L.V. 1991. The genus *Dendrobium* Sw. (Orchidaceae) in flora of Vietnam. *Syst. Pl. Vasc.* (Leningrad) 28: 42–51 (in Russian).

Duong Duc Huyen & Averyanov, L.V. 1992. The genus *Dendrobium* Sw. (Orchidaceae) in flora of Vietnam. *Syst. Pl. Vasc.* (St.-Petersburg) 29: 26–54 (in Russian).

Dy Phon, P. 1970. La vegetation du sud-ouest du Cambodge. *Ann. Fac. Sci. Phnom Penh. T.* 3: 1–136.

Dy Phon, P. 1982. *Vegetation du Cambodge: endemisme et affinites de sa flore avec regions voisines.* C.R.Société Biogeographie 58: 135–144.

Farjon, A., Nguyen Tien Hiep, Harder, D.K., Phan Ke Loc & Averyanov, L. (2002). A new genus and species of Cupressaceae (Coniferales) from northern Vietnam, *Xanthocyparis vietnamensis. Novon* 12: 179–189.

Fontaine, H. & Workman, D.R. 1978. Review of the geology and mineral resources of Kampuchea, Laos and Vietnam. In Nutalaya, P. (ed.) *Geology and Mineral Resources of Southeast Asia.* Bangkok. Asian Institute of Technology: 538–603.

Fowlie, J.A. 1966. An annotated checklist of the species of *Paphiopedilum*, 1966. *Orchid Digest* 30: 307–313.

Fowlie, J.A. 1969. Some notes on *Paphiopedilum appletonianum* and its natural hybrids, *P. siamense. Orchid Digest* 33: 307–313.

Fowlie, J.A. 1974. Malaya revisited: part 4. *Orchid Digest* 38: 90–94.

Fowlie, J.A. 1976. Malaya revisited: part 9. *Orchid Digest* 40: 164–169.

Fowlie, J.A. 1977. Malaya revisited: part 12. *Orchid Digest* 41: 190–197.

Fowlie, J.A. 1978. Malaya revisited: part 13. *Orchid Digest* 42: 72–76.

Fowlie, J.A. 1989. China: awash in the Bitter Sea. Part III. The habitat of *Paphiopedilum micranthum* amongst limestone on the southeast border of Guizhou Plateau. *Orchid Digest* 53: 132–139.

Fowlie, J.A. 1990a. China: awash in the Bitter Sea. Part IV. The habitat of *Paphiopedilum emersonii* on limestone concretions of southest of the Guizhou Plateau. *Orchid Digest* 54: 41–47.

Fowlie, J.A. 1990b. The *Paphiopedilum* Species associated with the Southeastern Guizhou Plateau. *Orchid Digest* 54: 72–73.

Fowlie, J.A. 1990c. China: awash in the Bitter Sea. Part V. *Paphiopedilum esquirolei* on cliffs over limestone caves in Guangxi. *Orchid Digest* 54: 137–142.

Fowlie, J.A. 1991a. China: awash in the Bitter Sea. Part VI. *Paphiopedilum malipoense* in a distinct limestone bluff habitat. *Orchid Digest* 55: 5–15.

Fowlie, J.A. 1991b. China: awash in the Bitter Sea. Part VII. A habitat of *Paphiopedilum barbigerum* on limestone boulders in Guizhou. *Orchid Digest* 55: 137–139.

Fowlie, J.A. 1993. China: awash in the Bitter Sea. Part X. A trip to the Yunnan-Guangxi Frontier to Study a new subspecies of *Paphiopedilum micranthum* and the discovery of a new species of *Vanda. Orchid Digest* 57: 183–188.

Fowlie, J.A. & Tang, C.Z. 1987. The rediscovery of *Paphiopedilum barbigerum* Tang et Wang in an importation by Richard Jack Topper. *Orchid Digest* 51, 1: 45–46.

Fridland, B.M. 1961. *Nature of the North Vietnam.* Moscow. USSR Academy of Science Editorial House. 175 pp (in Russian).

Fromaget, J. 1952. Aperçu de nos connaissances sur la geologie de l'Indochine en 1948. In International Geological Congress. Report of the 18th session. London. Pt. 13.

Fuchs, F. & Reisinger, H. 1995. *Paphiopedilum herrmannii* Fuchs et Reisinger. Eine neue Frauenshuhart aus Nord-Vietnam. *Linzer Biol. Beitr.* 27, 2: 1213–1215.

Gabel, R. & Lyke, J. 2001. Orchids and CITES. Understanding the basics of this international trade agreement. *Orchids. Mag. Amer. Orch. Soc.* 70, 7: 660–666.

Gagnepain, F. 1951. Deux collections précieuses d'Orchidées d'Indochine. *Bot. Syst.* (Paris). 14, 2: 114–132.

Garay, L.A. 1972. On the origin of the Orchidaceae, II. *J. Arnold Arboret.* 53: 202–215.

Gaussen, H., Legris, P. & Blasco, F. 1967. *Bioclimats du Sud-Est Asiatique.* Travaux de la Section Scientifique et Technique. Institute Français de Pondichery 3, 4: 1–114.

Ghatak, J. & Devi, R.K.J. 1986. Orchids of Nagaland. In Vij, S.P. (ed.). *Biology, Conservation and Culture of Orchids.*

Good, R. 1964. *The geography of the flowering plants.* 3 ed. London & Colchester. Longmans. 518 pp.

Godefroy-Lebeuf, A. 1886. *Cypripedium concolor* var. *tonkinense. L'Orchidophile* 1886: 226.

Gruss, O. 1999. Nature is full of wonders: new species of the genus *Paphiopedilum. Caesiana* 13: 39–44.

Gruss, O. 2000. New species for the genus *Paphiopedilum.* Part 1: *Paph. tranlienianum - Paph. vietnamense. Oasis* 1, 2: 3–6, 9–10.

Gruss, O. & Perner, H. 1998. *Paphiopedilum tranlienianum*, a new species of *Paphiopedilum* from Vietnam. *Caesiana* 11: 63–73.

Gruss, O. & Perner, H. 1999a. *Paphiopedilum vietnamense.* A new species of Vietnam. *Die Orchidee Suppl.* 5: 1–8.

Gruss, O., Perner, H. 1999b. *Paphiopedilum vietnamense* Gruss et Perner. *Die Orchidee* 51, 4, *Orchideenkart*: 927–930.

Gruss, O. & Röllke, L. 1996. Little known species of the genus *Paphiopedilum* (part 2). *Paphiopedilum barbigerum - Paphiopedilum henryanum - Paphiopedilu jackii - Paphiopedilum sangii - Paphiopedilum schoseri. Caesiana* 6: 27–40.

Gruss, O. & Röth, J. 1999a. *Paphiopedilum helenae* f. *aureum*, eine neue Farbvariante. *Die Orchidee* 50, 1: 2–3.

Gruss, O. & Röth, J. 1999b. *Paphiopedilum hirsutissimum* var. *esquirolei* f. *viride*, a new variation in colour and recombinations of the albino varieties within the genera *Paphiopedilum* and *Phragmipedium. Caesiana* 12: 57–65.

Guillamin, A. 1924. Les Cypripediées d'Indo-Chine. *Bull. Soc. Bot. France.* 71: 548–558.

Guillamin, A. 1934. 100. *Paphiopedium* Kerch. (*Paphiopedilum* Pfitz.). In Lecomte, H. & Humbert, H. eds., *Flore generale de l'Indo-Chine* 6. Fasc. 5: 636–646.

Hozumi, K., Yoda, K., Kokawa S., & Kira, T. 1969. Production ecology of tropical rainforests in southeastern Cambodia. I. Plant Biomass. *Nature and Life in Southeast Asia* 6: 1–54.

Hu Song Hua. 1994. Eine Geschichte uber *Paphiopedilum barbigerum. Die Orchidee* 45, 4: 165–166.

Hu Song Hua. 1999. *Paphiopedilum barbigerum* var. *aureum.* An exciting new color phase of the miniature Chinese species. *Orchids. Mag. Amer. Orch. Soc* 68, 3: 242–243.

Hydro-Meteorological Data. 1989. *So lieu khi tuong - thuy van Viet Nam.* Tap. I - So lieu khi tuong. Tong cuc Khi tuong Thuy Van. Ha Noi (in Vietnamese).

Hydro-Meteorological Data Used for Construction — Vietnam Standard. 1997. *Tieu chuan Viet Nam. So lieu khi hau dung trong xay dung.* Vien Khi tuong Thuy Van. Ha Noi (in Vietnamese).

Karasawa, K. & Saito, K. 1982. A revision of the genus *Paphiopedilum* (Orchidaceae). *Bull. Hiroshima Bot. Gard.* 5: 1–69.

Khoang Tjyung. 1988. The classification of mountain meadows of the plant cover in north Vietnam. *Bull. Moscow Obschz. Ispit. Prirod. Biol. Dept.* 93, 4. 107–119. (in Russian).

Khuong Bun Luk. 1971. *Contribution a l'étude economique de la foret dense des terres basses du versant occidental des Monts Cardomomes.* Université de Sciences Agronomiques. Phnom Penh. 72 pp.

Koopowitz, H. & Cribb, P. 1986. *Paphiopedilum emersonii* — a remarkable new slipper orchid from China. *Orchid Advocate* 12, 3: 84–86.

Koopowitz, H. & Hasegawa, N. 1992. *Paph.* × *fanaticum*: the natural hybrid of *Paphiopedilum malipoense* and *Paphiopedilum micranthum. Orchid Advocate* 18: 48–51.

Kores, P.J., Cameron, K.M., Molvray, M. & Chase, M.W. 1997. The phylogenetic relationships of *Orchidoideae* and *Spiranthoideae* (Orchidaceae) as inferred from rbcL plastic sequences. *Lindleyana* 12(1): 1–11.

Kubitzki, K. & Krutzsch, W. 1998. Origin of east and south east Asian plant diversity. In "Floristic characteristics and diversity of east asian plants". *Proceedings of the First Intern. Symp. on Flor. Charact. and Diversity of E.Asian Pl. Kunming*, 1996: 56–70. China Higher Ed. Press & Springer-Verlag.

Lay Khim & Taylor-Hunter, D. 1995. *Kirirom general survey.* Ministry of Environment. Phnom Penh. 48 pp.

Lecomte, M.H. & Humbert, H. (eds.). 1907–1951. *Flore Generale de l'Indochine* 1–7, Supplements.

Lee, T.C. 1975. *Paphiopedilum purpuratum* (Lindl.) Pfitz. *Orchid Digest* 39: 192–194.

Linden, E. 1994. Ancient Creatures in a Lost World. *Time.* June 20 1994: 52–54.

Linden, L. & Rodigas, E. 1886. *Cypripedium concolor* var. *tonkinense. Lindenia* 2: Tab. 77.

Liu Zhong-Jian & Zhang Jian-Yong. 2001. *Paphiopedilum singchii* sp.nov., an addition to the subgenus *Brachypetalum* of *Paphiopedilum* (Orchidaceae). *Acta Phytotax. Sinica* 38, 5: 467–470.

Loureiro, J. 1790. *Flora Cochinchinensis* 1–2. Lisboa. 744 pp.

Loureiro, J. 1794. *Flora Cochinchinensis*, ed. 2, 1–2. Berolini. 882 pp.

Lückel, E. 1987. Ein neuer Frauenschuh, *Paphiopedilum dollii. Die Orchidee* 38, 5: 266–268.

MacKinnon, J. & Cox, R. 1991. 28 Vietnam. In N.M., Sayer, J.A., Whitmore, T.C. eds., *The conservation atlas of tropical forests. Asia and the Pacific*: 232–239. Collins, IUCN & Simon & Schuster.

Mark, F. 1987. A preliminary introduction to and cultivation of the Chinese slipper orchids, genus *Paphiopedilum. Orchid Digest* 51, 2: 63–82.

Martinetto, E. 1998. East Asian elements in the Plio-Pleistocene floras of Italy. In *Proc. Intern. Sympos. Flor. Charact. Divers. East As. Pl. Kunming*, 1996: 71–87. China High. Ed. Press & Springer-Verlag.

Merrill, E.D. 1946. On Loureiro's "*Flora cochinchinensis*". *Chronica Bot.* 10. N. 3/4 (*Merilliana*): 243–274.

Nguyen Khanh Van, Nguyen Thi Hien, Phan Ke Loc & Nguyen Tien Hiep. 2000. *Bioclimatic diagrams of Vietnam.* Hanoi. Vietnam Nat. Univ. 126 pp.

Nguyen Nghia Thin. 1994. Diversity of the Cuc Phuong flora a primary forest of Vietnam. *Proc. Nat. Cent. Nat. Sci. Techn. Vietnam* 6, 2: 77–82.

Nguyen Nghia Thin. 1997a. The vegetation of Cuc Phuong National park. *SIDA* 17: 713–751.

Nguyen Nghia Thin. 1997b. *Manual on research of biodiversity.* Hanoi. Agr. Publ. House. 223 pp. (in Vietnamese).

Nguyen Nghia Thin. 1998. The Fansipan flora in relationship to the Sino-Japanese floristic region. In D. E.Boufford, H.Obha, eds., *Sino-Japanese Flora: Its Characteristics and Diversification*: 111–122. Tokyo. Univ. of Tokyo.

Nguyen Nghia Thin. 1999. *Key to taxa and classification of Euphorbiaceae of Vietnam.* Hanoi. Nha xuat Ban Nong Nghiep. 138 pp.

Nguyen Nghia Thin & Harder, D. 1996. The diversity of the flora of Fansipan - the highest mountain in Vietnam. *Ann. Missouri Bot. Gard.* 83: 404–408.

Nguyen Nghia Thin & Nguyen Thi Thoi. 1998. *Diversity of vascular plants of high mountain area: Sa Pa - Phan Si Pan.* Hanoi. Nha Xuat Ban Dai Hoc Quoc Gia Ha Noi. 115 pp. (in Vietnamese).

Nguyen Thien Tich. 1998a. "Phat hien moi. Cay lan hai huong lan hoa trang huong thom! [New discovery: The white fragrant Huong lan *Paphiopedilum*]. *Hoa Canh* 1998, 3: 10–11 (in Vietnamese).

Nguyen Thien Tich. 1998b. [More details of the Huong lan *Paphiopedium*]. *Hoa Canh* 1998, 4: 12 (in Vietnamese).

Nguyen Thien Tich. 1999. Cay lan hai *Paphiopedilum caobangense* moi phat hien o Viet Nam. [*Paphiopedilum caobangense* a newly discovered species in Vietnam] *Hoa Canh* 1999, 1: 14 (in Vietnamese).

Nguyen Thien Tich. 2001. *The Orchids of Vietnam* 1. Ho Chi Minh. Nha Xuat Ban Nong Nghiep. 424 pp. (in Vietnamese).

Nguyen Tien Ban. 1997. *Handbook of reference and identification of the families of Angiospermae plants in Vietnam*. Hanoi. Agricult. Publ. House. 532 pp. (in Vietnamese).

Nguyen Tien Hiep. 1998. The gymnosperms of Vietnam. In *Proceedings of the First Intern. Symp. on Flor. Charact. and Diversity of E.Asian Pl. Kunming*, 1996: 91–103. China Higher Ed. Press & Springer-Verlag.

Nguyen Tien Hiep, Averyanov, L.V. & Phan Ke Loc. 1998. Conservation of the population of *Keteleeria fortunei* on limestone at Thang Heng (Tra Linh, Cao Bang) is needed. *Vietnam Forestry Review* 1998, 6: 38–40 (in Vietnamese).

Nguyen Tien Hiep, Phan Ke Loc & Averyanov, L.V. 1999. Some new data on the Pinopsida in the north Truong Son range. Selected papers of seminar on north Truongson biodiversity. *40 year celebr. Vinh Univ. Hanoi Nat. Univ.*: 104–108 (in Vietnamese).

Nguyen Tien Hiep, Phan Ke Loc & L.V.Averyanov. 2000. Some new species for the flora of Vietnam collected from the limestone mountains of Cao Bang Province. *Journ. Biol.* (Hanoi) 22, 4: 1–11 (in Vietnamese).

Nguyen Tien Hiep & Vidal, J.E. 1996. Gymnospermae. In Morat, Ph. ed., *Flore du Cambodge du Laos et du Vietnam* 28. Paris. Mus. Nat. Hist. Nat. 166 pp.

Nilsson, L.A. 1979. Anthecological studies on the lady's slipper, *Cypripedium calceolus*. *Bot. Tiddskr.* 132: 329–349.

Pain, S. 1989. The case of the stolen slippers. *New Scientist* 1670: 48–49.

Perner, H. & Gruss, O. 1999. *Paphiopedilum hangianum*, a new species of the genus from Vietnam. *Die Orchidee Suppl*. 6: 3–7.

Perner, H. & Herrmann, R. 2000. *Paphiopedilum coccineum* Perner et Herrmann, ein neuer Frauenschuh aus Vietnam. *Die Orchidee* 51, 5: 622–624.

Petelot, A. 1955. *Bibliographie botanique de l'Indochine*. Arch. Réch. Agron. Past. Vietnam (Saigon) 24: 99 pp.

Pfitzer, E.H. 1903. Orchidaceae - Pleonandrae. In A. Engler, *Das Pflanzenreich* IV, 50: 1–132.

Pham-Hoang Ho. 1970–1972. *An illustrated flora of South Viet-Nam*. Vols. 1–2. Saigon. Bo Giang-Duc.

Pham-Hoang Ho. 1972. Orchidaceae: ho Lan. In *An illustrated flora of South Viet-Nam*. Saigon. Bo Giang-Duc. Trung-Tam Hoc-Lieu 2: 998–1101.

Pham-Hoang Ho. 1991–1993. *An illustrated flora of Vietnam*. Vols. 1–3. Santa Ana. Mecong Printing.

Pham-Hoang Ho. 1993. Orchidaceae: ho Lan. In *An illustrated flora of Vietnam*. Santa Ana. Mecong Printing 3: 943–1162.

Pham-Hoang Ho. 1999–2000. *An illustrated flora of Vietnam*. Vols. 1–3. Ho Chi Minh. Nha Xuat Ban Tre.

Pham-Hoang Ho. 2000. Orchidaceae: ho Lan. In *An Illustrated Flora of Vietnam*. Ho Chi Minh. Nha Xuat Ban Tre 3: 760–968.

Pham Ngoc Toan & Phan Tat Dac. 1993. *The climate of Vietnam* (ed. 2). Hanoi. Nha Xuat Ban Khoa Hoc Va Ky Thuat ["Science and Technical" Publ. House]. 312 pp.

Phan Ke Loc. 1984. Pinopsida in the flora of Vietnam. *Journ. Biol.* (Hanoi) 4, 4: 5–10 (in Vietnamese).

Phan Ke Loc. 1998. On the systematic structure of the Vietnamese flora. In *Proceedings of the First Intern. Symp. on Flor. Charact. and Diversity of E.Asian Pl. Kunming*, 1996: 120–129. China Higher Ed. Press & Springer-Verlag.

Phan Ke Loc, Averyanov L.V., Nguyen Tien Hiep & Nguyen Quoc Binh. 1999. New data on the flora of Ngoc Linh area and its neighbourhood in the district of Dac Glei, Con Tum province. *Vietnam Forestry Rev.* (Hanoi) 8: 35–37 (in Vietnamese).

Phan Ke Loc & Nguyen Tien Hiep. 1999. Is there *Cunninghamia konishii* Hayata growing wild in Vietnam, and what is the scientific name of the Sa Moc Dau. Selected papers of seminar on north Truong Son biodiversity. *40 year celebr. Vinh Univ. Hanoi Nat. Univ.*: 61–64 (in Vietnamese).

Phan Ke Loc, Nguyen Tien Hiep & Averyanov, L.V. 1999a. Some threatened plant communities and species of the limestone mountains of Cao Bang province needed to be protected in new proposed protected areas. *Vietnam Forestry Rev.* (Hanoi) 12: 35–36 (in Vietnamese).

Phan Ke Loc, Nguyen Tien Hiep, Averyanov, L.V. 1999b. Are there new plant records from limestone mountains of Cao Bang? In *Conservation and sustainable development of forest and biodiversity on limestone areas of Vietnam*: 32–41. Hanoi. Forest Inventory and Planning Inst. 132 pp. (in Vietnamese).

Phan Ke Loc, Nguyen Tien Hiep, Averyanov, L.V. 1999c. Du sam da voi — *Keteleeria davidiana* (Bertrand) Beissn. var. *davidiana* mot loai thuc vat hat tran moi duoc ghan o bac Viet Nam. In *Conservation and sustainable development of forest and biodiversity on limestone areas of Vietnam*: 25–28. Hanoi. Forest Inventory and Planning Inst. 132 pp. (in Vietnamese).

Pradhan, G.M. 1972. *Paphiopedilum hirsutissimum. Paph. World* 2: 84–85.

Pradhan, U.C. 1976. *Indian Orchids: guide to identification and culture* 1. Kalimpong, India.

Quang Dau. 1994. Hunting for "Lan Hai Do" in Vietnam. *Vietnam Economic News* 4, 13: 40.

Rasmussen, F.N. 1985. Orchids. In Dahlgren, R.M., Clifford, H.T. & Yeo, P.F. *The families of the Monocotyledons. Structure, Evolution and Taxonomy*: 249–274. Springer-Verlag. Berlin, Heidelberg, NY, Tokyo.

Red Data Book of Vietnam 2. *Plants*. 1996. Sci. Tchn. Publ. House. Hanoi. 482 pp.

Rolfe, R.A. 1893. *Cypripedium bullenianum* var. *appletonianum* Rolfe. *Orchid Rev.* 1: 135.

Rolfe, R.A. 1895. *Cypripedium poyntzianum. Orchid Rev.* 3: 55.

Rolfe, R.A. 1896a. The *Cypripedium* group. *Orchid Rev.* 4: 327–334, 363–367.

Rolfe, R.A. 1896b. *Cypripedium appletonianum. Orchid Rev.* 4: 18.

Rundel, P.W. 1999. Conservation Priorities In Indochina - WWF Desk Study. *Forest habitats and flora in Lao PDR, Cambodia, and Vietnam.* Hanoi. World Wide Fund for Nature, Indochina Programme Office. 194 pp.

Schlechter, R. 1919. *Orchidaceae Sino-Japonicae Prodromus. Repert. Sp. Veg. Regni Veg. Beih.*: 4: 1–319.

Schmid, M. 1974. *Vegetation du Vietnam: le massif sud-annamitique et les regions limitrophes.* ORSTOM. 74. Paris: 243.

Schmid, M. 1989. Vietnam, Kampuchea and Laos. In: D.G. Campbell, H.D. Hammond (eds.). *Floristic Inventory of Tropical Countries*: 83–90. New York Bot. Gard.

Schuiteman, A. & de Vogel, E.F. 2000. *Orchid Genera of Thailand, Laos, Cambodia, and Vietnam.* Leiden. Nation. Herb. Nederl. 118 pp.

Schzeglova, T.N. 1957. *Vietnam.* Moscow. USSR Academy of Sciences. Governmental editorial house of Geographical literature. 183 pp. (in Russian).

Seidenfaden, G. 1972. Contributions to the orchid flora of Thailand IV. *Bot. Tiddskr.* 67, 1–2: 76–127.

Seidenfaden, G. 1975. *Contributions to the orchid flora of Cambodia, Laos and Vietnam.* Fredensborg. 117 pp.

Seidenfaden, G. 1992. The orchids of Indochina. *Opera Botanica* 114: 502 pp.

Senghas, K. & Schettler, R. 1999a. *Paphiopedilum hilmari*, eine ungewohnliche Entdeckung aus Vietnam. *J. Orchideenfr.* 6, 1: 4.

Senghas, K. & Schettler, R. 1999b. *Paphiopedilum hilmari - Paphiopedilum vietnamense - Paphiopedilum mirabile. J. Orchideenfr.* 6, 3: 156–161.

Skvortsova, N.T. & Averyanov, L.V. 1994. New genus and species — *Grushvitzkya stellata* (Araliaceae) from the North Vietnam. *Bot. Journ.* (St.-Petersburg) 79, 7: 108–112 (in Russian).

Szlachetko, D.L. 1995. *Systema Orchidalium. Fragm. Flor. Geobot. Suppl.* 3: 152 pp.

Takhtajan, A. 1978. *The floristic regions of the world.* Leningrad. Nauka. 247 pp. (in Russian).

Takhtajan, A. 1986. *Floristic Regions of the World.* Berkeley, Los Angeles, London. Univ. California Press. 522 pp.

Tanaka, Toshinori. 1998–1999. Homepages: http://www.orchid.or.jp/orchid/people/tanaka/ indexe.html).

Thai Van Trung. 1978. Tham thuc vat rung Viet Nam. Tren quan diem he sinh thai. In *lan thu hai, co sua chua. Nxb Khoa hoc va Ky thuat.* Hanoi (in Vietnamese).

Tixier, P. 1979. *Bryogeographie du Mont Bokor (Cambodge).* J. Cramer. Vaduz. 121 p.

Tolmachev, A.I. 1974. *Introduction to the plant geography.* Leningrad. Leningrad Univ. Publ. House. 244 pp. (in Russian).

Tran Hop. 1998. *Phong lan Viey Nam (The Orchids of Vietnam).* Ho Chi Minh. Saigon Orchidex. Nha Xuat Ban Nong Nghiep. 703 pp. (in Vietnamese).

Tsi, Z.H., Luo, Y.B., Cribb, P.J., McGough, N.H., Siu & G., Chau, L. 1999. A preliminary report on the population size, ecology, and conservation status of some *Paphiopedilum* species in southwest China. *Lindleyana* 14, 1: 12–23.

Vidal, J. E. 1972. Bibliographie botanique indochinoise (1955–1969). *Bull. Soc. Etud. Indoch.,* n.s. 47, 4: 657–748.

Vidal, J. E., Vidal, Y. & Pham Hoang Ho. 1988. *Bibliographie botanique indochinoise* (1970–1985). Paris. Mus. National Hist. Natur. Lab. Phanerogamie. 132 pp.

Vidal, J. E., Falaise, H, Phan Ke Loc & Nguyen Thi Ky. 1994. *Bibliographie botanique indochinoise* (1986–1993). Paris. Mus. National Hist. Natur. Lab. Phanerogamie. 105 pp.

Vo Quy. 1995. *Conservation of flora, fauna and endangered species in Vietnam.* Tropical Forest Ecosystems/ BIOTROP Special Publication 55: 139–146.

Vu Van Can, Vu Van Dung & Le Van Cham. 1999a. Phat hien mot loai moi thuoc ho hoang dan (Cupressaceae): cav che - *Thuja quanbaensis* sp.nov. o vung nui da voi tinh Ha Giang. In *Conservation and sustainable development of forest and biodiversity on limestone areas of Vietnam*: 12–13. Hanoi. Forest Inventory and Planning Inst. 132 pp. (in Vietnamese).

Vu Van Can, Vu Van Dung & Le Van Cham. 1999b. Mot so loai cay nganh hat tran thuoc khu bttn Bat Dai Son, Ha Gieng. In *Conservation and sustainable development of forest and biodiversity on limestone areas of Vietnam*: 21–24. Hanoi. Forest Inventory and Planning Inst. 132 pp. (in Vietnamese).

Vu Van Dung. 1996. *Vietnam Forest Trees.* Hanoi. Agricult. Publ. House. 788 pp.

Walter, H. & Leith, H. 1960, 1964, 1967. *Klimatogram - Weltatlas.* Vols. 1–3. Gustav Fischer Verlag. Jena.

Wikramanayake, E., Dinerstein, H., Hedao, P. & Olson, O. 1997. *A conservation assessment of ecoregions of the Indo-Pacific region.* WWF-US. Washington.

Wu Zhengyi & Wu Sugong. 1998. A proposal for a new floristic kingdom (realm) — the E.Asiatic Kingdom, its delineation and characteristics. In *Proc. Intern. Sympos. Flor. Charact. Divers. East As. Pl.* 1996. *Kunming*: 3–42. China High. Ed. Press & Springer-Verlag.

INDEX OF SCIENTIFIC NAMES